Deep-Seated Inclusions in Kimberlites and the Problem of the Composition of the Upper Mantle

Deep-Seated Inclusions in Kimberlites and the Problem of the Composition of the Upper Mantle

by N. V. Sobolev

Translated from the Russian
by David A. Brown

English Translation
edited by F. R. Boyd

American Geophysical Union
Washington, D.C.

Deep-Seated Inclusions in Kimberlites and the Problem of the Composition of the Upper Mantle

Russian Edition Copyright © 1974
Izdatel'stvo Nauka

Standard Book Number: 0-87590-202-2
Library of Congress Catalog Card Number: 76-062627

Printed by
Edwards Brothers, Inc.
Ann Arbor, Michigan

Table of Contents

PREFACE

Almost immediately after the discovery of kimberlites, the attention of investigators was attracted to their unusual mineralogy. It has been established that, besides the fragments caught up from various crustal rocks, xenoliths of unusual types occur in the kimberlites, believed by most investigators to have been brought up from great depths. The interest in these xenoliths increased even further following the first discovery of a diamond-bearing eclogite in 1897, the study of which led to the hypothesis of an eclogite layer in the Earth.

An early survey of the mineralogy and petrography of the deep-seated xenoliths has been brought together in the authoritative monograph of Williams (1932).

The next stage in the investigation of the deep-seated inclusions began after the discovery of the Yakutian kimberlites. The geologists of the Amakinsk Expedition, the Scientific-Research Institute on the Geology of the Arctic (NIIGA), the All-Union Geological Institute (VSEGEI), and other organizations obtained numerous data on their petrography and mineralogy.

Work on the synthesis of the diamond, which had demonstrated the role of high pressures during the formation of the pyrope-bearing rocks and diamonds, and successes in the investigation of the Yakutian kimberlites established the study of deep-seated inclusions in kimberlites amongst the most pertinent problems in modern petrology. Most investigators see in these inclusions, along with xenoliths in basalts, the principal source of information on the composition of the upper mantle. However, independently of this hypothesis, which the author supports, the study of minerals formed along with diamond or under similar conditions, is of great interest in connection with the unusual nature of their formation. And whereas, during mineral formation in the Earth's crust, pressure only in rare cases reaches 15-20 kb, during deep-seated processes it may be greater, and for the diamond-bearing parageneses it exceeds 40-50 kb. In other words, we are here dealing with a study of minerals formed under super-high pressures.

Our goal was a detailed study of the minerals and parageneses of deep-seated xenoliths, especially the minerals of the diamond-bearing xenoliths, including the diamonds themselves. Thus, the patterns of change in the composition of the minerals and in the parageneses depending on the conditions of their formation have been clarified, in order to obtain additional information on the composition of the upper mantle. As a starting point, the proposition, recognized by the majority of investigators, was accepted that diamonds form under natural conditions in the thermodynamic region of their stability. This hypothesis has been emphasized by an investigation of the mineral-satellites of diamonds and their associations.

In accordance with the established goal (the most complete definition of all the types of deep-seated xenoliths), we selected a number of typical kimberlite pipes: 'Obnazhennaya' from the Lower Olenëk region, 'Udachnaya' and 'Zagadochnaya' from the Daldyn-Alakit region, and 'Mir' from the Malo-Botuoba region, where over the course of a number of years, field investigations have been carried out and a large amount of information has been amassed, which characterizes the entire complex of xenoliths for each pipe. Data were also obtained from certain other pipes in Yakutia and South Africa, and for comparison, we employed data from eclogites and pyrope peridotites, occurring in outcrops.

vii

During the sampling of the xenoliths, we excluded not only rocks of the sedimentary cover and the crystalline basement, but also partially eclogitized rocks, already studied in detail earlier by other investigators, and amongst the ultramafic rocks, we considered only the garnet-bearing varieties. In other words, we studied only those rocks formed at pressures in excess of 15-20 kb.

Diamonds with inclusions were collected by the author from the special collections, and these were also studied (like the diamond-bearing xenoliths) in company with colleagues from the 'Yakutalmaz' Combine, the Amakinsk Expedition, VSEGEI, and other organizations: Z. V. Bartoshinsky, A. I. Botkunov, S. I. Futergendler, M. A. Gnevushev, A. D. Khar'kiv, L. N. Mikhailovskaya, B. S. Nai, V. S. Pustyntsev, and E. S. Yefimova.

The investigations of mineral compositions involved the participation of I. K. Kuznetsova, who carried out the chemical analyses, Yu. G. Lavrent'ev, under whose direction L. N. Pospelova and the author carried out quantitative analyses in the electron-microprobe laboratory of the Institute of Geology and Geophysics of the Siberian Division of the Academy of Sciences of the USSR (IGG SO AN SSSR). (The author also assisted in the development of methods of silicate analysis on the microprobe.) Mass determinations of unit-cell parameters for garnets were undertaken by N. I. Zyuzin.

Great assistance in the author's work was rendered by N. A. Chernova, L. S. Gudina, V. K. Kirillov, S. G. Motorin, L. A. Panina, N. P. Pokhilenko, I. V. Prugova, and I. T. Sapozhnikov.

The author expresses his profound gratitude to his teacher Academician V. S. Sobolev for constant aid during the work, and to his colleagues in IGG SO AN SSSR, I. T. Bakumenko, N. L. Dobretsov, V. V. Khlestov, V. P. Kostyuk, G. V. Pinus, V. V. Reverdatto, and V. V. Vasilenko for their part in discussions of individual aspects of the present work, and also all those persons listed above for cooperation in completing the work.

The author is grateful to foreign colleagues F. R. Boyd, G. F. Claringbull, R. G. Coleman, D. S. Coombs, R. Danchin, J. B. Dawson, J. Fiala, E. D. Jackson, H. O. A. Meyer, G. W. Milledge, M. Prinz, G. Rösler, and G. Switzer, who provided certain samples for investigation, standards for X-ray microanalysis on the electron probe, and other materials.

CHAPTER I

METHODS OF INVESTIGATION

During the study of the deep-seated xenoliths, attention has primarily been directed at the rock composition and the relationships between the coexisting minerals. With this object in mind, we studied a large number of thin-sections and in many cases, polished samples (in order to clarify the structural-textural features of the coarse-grained rocks).

In examining the small xenoliths, and also intergrowths of minerals of deep-seated origin in the kimberlitic breccias, a large amount of material was observed under the binocular microscope. Special attention was paid to the garnets with inclusions of kyanite, garnets displaying the alexandrite effect, and also intergrowths of chrome-rich garnets and pyroxenes in concentrates of the heavy fraction of the kimberlites, of which large amounts were examined under the binocular microscope. In a number of cases, when there was a predominance of garnets in the concentrates, covered by kelyphitic rims (e.g. in the 'Zagadochnaya' pipe), they were given preliminary treatment in acids with the object of removing the rims.

The Study of the Physical Properties of the Minerals. During the mineralogical investigations of the deep-seated xenoliths, primary attention was paid to diagnosis of the garnets, because of the abundance of garnets in all the samples studied, and also because of the possibility of a relatively precise estimate of the principal features of the garnet composition based on their physical properties (refractive index (n) and unit-cell dimensions (a_0)).

The results of the study of the garnet parageneses (ultramafic and eclogitic associations) permitted an unequivocal assessment of the effect of increased amounts of Cr_2O_3 (in peridotites) and the iron index of the garnets (in the eclogites) on the value of the refractive index. The role of calcium in the garnets was estimated on the basis of the unit-cell parameter (a_0).

The refractive indices of the garnets were determined with the aid of a standardized set of immersion liquids, and also a standard set of highly-refractive mx immersion liquids, prepared on the basis of a solution of sulfur and arsenic sulfide in arsenic tribromide, using methylene iodide as a solvent. The values of the refractive indices of the immersion liquids employed (in the range of 1.732-1.805 to 0.005-0.007) were verified with the aid of a IRF-23 refractometer.

Use of the method of focal screening (Cherkasov, 1957) permitted us to determine the value of n with a precision of 0.001-0.002. These investigations were used as a basis for the primary diagnosis of the garnets and their number significantly exceeded 1000.

In order to estimate the amount of CaO in the garnets, X-ray analysis was carried out. The parameters of the unit cell were determined by N. I. Zyuzin by the powder method on a URS-50I diffractometer in iron radiation using reflections (10.4.0), (10.4.2), and (880). The error in measuring the reflection angles was $\pm 0.05°$, that is, the parameter was determined with a precision down to 0.004 Å. On the basis of the method indicated, the value of a_0 for about 300 samples was determined.

Chemical Analysis. During the process of investigating the minerals of the deep-seated xenoliths, about 100 wet chemical analyses were

1

carried out, mainly of garnets and pyroxenes (analyst I. K. Kuznetsova, IGG SO AN SSSR). The principal object was a study of the composition of the coexisting minerals. A preliminary study of the physical properties of the garnets, enabled us to select for more detailed studies, the samples of different composition, in most cases differing from those known in the literature. These came from eclogites, grospydites, and peridotites. Substantial difficulties arose in separating pure monomineralic fractions both of the garnets and the pyroxenes, as a result of alteration in many of the samples or an exceptionally small amount of garnet and pyroxene in many of the peridotites.

Most samples of diamond-bearing eclogites presented special difficulties. In them, the amount of all the material available for investigation did not exceed 1 g. This material was used for the separation of the monomineralic fractions of garnet and pyroxene. The residue after separation was also analyzed, and then the results of the analyses were recalculated, allowing for the known weights of the minerals and their composition, and thus, the complete composition of the xenolith fragment was determined.

The limited amount of material, prepared for analysis, required the most careful preliminary preparation in order to avoid contamination by secondary impurities. The total weight of samples analyzed did not exceed 200-300 mg, and in the completion of partial analyses of individual grains of garnets and chrome-bearing kyanites, 10-30 mg, which required the use of a number of semi-micro- and microanalytical methods.

The classical methods of silicate analyses were employed. The results of analysis were controlled on the basis of data from duplicate determinations, and determinations carried out by different methods. The latter applied particularly to the control of the determination of Al_2O_3. The amount of the alkali metals (Na and K) was determined by flame photometry in IGG SO AN SSSR by analysts G. M. Gusev and A. S. Surzhko, and also in the Central Chemical Laboratory of IGEM AN SSSR.

X-ray-Spectral Quantitative Microanalysis. The development of methods of X-ray-spectral quantitative microanalysis using the electron probe has proved to have exceptional scope in mineralogical investigations. The local nature of the analysis enables us to study variations in the content and distribution of elements in minerals over intervals of a few microns. The use of these methods excludes contamination by mechanical additions of other phases in the volumes analyzed, which often fall into the material being prepared for chemical analysis, particularly concerning trace-elements, the amount of which in minerals is measured in a few tenths of a percent.

X-ray-spectral quantitative microanalysis is a uniquely favorable method for determining the composition of mineral inclusions in diamonds, that are of exceptionally small dimensions, from 0.05 up to 0.3 mm.

Because of the insignificant dimensions of the samples, special care is necessary during the process of preparation for their analysis. Such preparation in general includes the combined sectioning of samples and standards and their polishing in epoxy resin mounts. The author personally carried out all operations for the preparation of the samples for analysis.

Inclusions were extracted from diamonds by crushing or combustion of the latter in a platinum crucible. Dry epoxy-resin mounts

were prepared first with several inbuilt glass tubes (rings). The
grains, selected for analysis, were mounted individually, with constant control under a binocular microscope. For this, small holes
were made in the mount, which were impregnated with specially prepared epoxy resin. Suitable grains were mounted in the holes so
that they could be ground at the same time (along with the standards).
After complete hardening of the resin, the mount was ground in the
glass with abrasive powders having particle dimensions of 7 and 5
microns. Constant observation under the binocular microscope enabled
us to determine the appropriate moment for the end of grinding.
Before polishing, the sample was again subjected to fine grinding in
the glass. In some cases, diamond pastes were employed with particle
dimensions of from 3 to 1 micron. The samples were polished mainly
with a Cr_2O_3 powder. Only in special cases, for example, when determining the amount of chromium in the olivines, were all operations
carried out with a previously finely-ground powder of Al_2O_3. The
quality of the polishing was controlled with the aid of an ore microscope.

The mineral compositions from the deep-seated inclusions of the
kimberlite pipes were determined on a MS-46 X-ray microprobe manu-
factured by 'CAMECA' in the electron microprobe unit of IGG SO AN
SSSR under the direction of Yu. G. Lavrent'ev. The results of over
250 complete mineral analyses are given in the tables in this book.
A voltage acceleration of 15 kV was used for the analyses.

As standards, the homogeneity of which had been verified with the
microprobe, we used minerals with a composition quite similar on
average to that of the unknowns analyzed. Such selection of standards is convenient for choosing the conditions of analysis, which
guarantee the possibility of defining the closest values for the
intensity of the $K\alpha$-lines of radiation of the various elements for
the sample and standard being analyzed. This is also recommended
by the staff of the Geophysical Laboratory of the Carnegie Institution, who have successfully studied the inclusions in diamonds and
minerals from kimberlites (Meyer and Boyd, 1972) with this difference, that instead of minerals, they used uniform silicate glasses,
with the composition of pyrope, enstatite, and diopside.

As a standard for Mg, Si, and Ca, we employed diopside with a
composition close to the stoichiometric value (Gurulev et al., 1965;
Analysis 3), and for Al and Fe, a pyrope (Sample 0-145, Table 1) from
a xenolith of a pyrope websterite from the 'Obnazhĕnnaya' pipe. For
the determination of chromium, different standards were used in different cases. For the determination of Cr in garnets and pyroxenes,
we used a chrome-rich pyrope (Sample Ud-92, Table 2), containing 11%
Cr_2O_3. For its determination in chromites, containing 50-70% Cr_2O_3,
we used both metallic chromium, and also chromite sample 531-N8,
employed as a standard in the investigations of the U. S. Geological
Survey, kindly made available to us by Dr. E. D. Jackson.

The standard for sodium was an albite of stoichiometric composition,
and for potassium, orthoclase (Sample Or-1, from the standards of the
Geophysical Laboratory of the Carnegie Institution). The standard
used for Mn was metallic manganese, and for titanium, mainly rutile,
sometimes metallic titanium, and in a number of cases also, a glass
of diopside composition, containing 2.0% TiO_2 (Sample GL-6), kindly
made available to us amongst other standards used at the Geophysical
Laboratory, by J. F. Schairer and F. R. Boyd. It should be emphasized
that the repeated analysis of these synthetic glasses, which have, in

TABLE 1. Chemical Composition of Garnets

Oxides	'Obnazhënnaya' Pipe		'Mir' Pipe			BM 1050	'Obnazhënnaya' Pipe			
	O-145	O-309	M-160	M-99	M-114		235/586	O-172*	O-467	O-379
SiO_2	42,69	42,69	41,35	42,76	44,38	42,50	42,40	43,8	43,72	41,31
TiO_2	0,11	0,03	0,19	0,26	0,10	0,20	0,16	0,10	0,18	0,30
Al_2O_3	23,42	23,68	22,74	23,42	23,63	21,90	22,56	23,5	22,48	22,05
Cr_2O_3	0,08	0,08	0,26	0,27	0,24	0,17	0,29	0,55	0,58	0,71
Fe_2O_3	1,26	1,77	2,76	1,56	0,76	4,33	1,46	—	1,70	2,96
FeO	8,62	7,22	10,26	5,21	5,59	9,53	14,43	9,88	5,71	11,32
MnO	0,16	0,12	0,21	0,20	0,20	0,13	0,43	0,18	0,22	0,10
MgO	19,97	19,52	18,81	21,78	20,98	18,60	13,64	20,5	20,86	15,93
CaO	4,17	4,85	3,34	4,48	4,01	3,20	4,18	3,67	4,24	4,94
Total	100,48	99,96	99,92	99,94	99,89	100,56	100,19	102,18	99,69	99,62
Si	3,016	3,018	2,967	2,994	3,000	3,028	3,000	3,041	3,000	3,030
Ti	0,004	0,001	0,013	0,017	0,006	0,011	0,009	0,004	0,013	0,018
Al	1,943	1,970	1,923	1,926	2,002	1,842	2,014	1,919	1,909	1,908
Cr	0,005	0,004	0,017	0,025	0,014	0,009	0,017	0,033	0,035	0,044
Fe^{3+}	0,048	0,025	0,047	0,032	—	0,138	—	—	0,043	0,030
Fe^{3+}	0,020	0,068	0,108	0,052	0,042	0,094	0,083	0,044	0,052	0,137
Fe^{2+}	0,509	0,424	0,617	0,303	0,337	0,570	0,912	0,532	0,341	0,695
Mn	0,008	0,004	0,013	0,013	0,012	0,008	0,028	0,013	0,013	0,004
Mg	2,100	2,059	2,009	2,275	2,244	1,975	1,533	2,120	2,233	1,736
Ca	0,318	0,369	0,259	0,336	0,310	0,245	0,338	0,271	0,328	0,387
Pyrope	71,0	70,5	66,9	76,4	76,2	68,2	53,0	72,3	75,3	58,7
Almandine	17,9	16,8	24,1	11,9	12,9	23,0	34,3	18,1	13,2	28,1
Spessartine	0,3	0,1	0,4	0,4	0,4	0,3	1,0	0,4	0,4	0,1
Grossular	8,0	11,1	4,6	7,5	9,5	0,5	10,5	5,1	6,5	8,5
Andradite	2,4	1,3	2,4	1,6	—	6,9	—	2,2	2,1	1,5
Ti-Andradite	0,2	—	0,7	0,9	0,3	0,6	0,4	0,2	0,7	0,9
Uvarovite	0,2	0,2	0,9	1,3	0,7	0,5	0,8	1,7	1,8	2,2
f	21,6	20,1	27,8	14,5	14,4	28,9	39,4	21,3	16,3	33,2
Ca-component	10,8	12,6	8,6	11,3	10,5	8,5	11,7	9,2	11,1	13,1
Mg-component	71,0	70,5	66,9	76,4	76,2	68,2	53,0	72,3	75,3	58,7
Cr-component	0,2	0,2	0,9	1,3	0,7	0,5	0,8	1,7	1,8	2,2
Ca/Ca+Mg	13,2	15,2	11,4	12,9	12,1	11,0	18,1	11,3	12,8	18,2
Excess Si	—	—	—	—	0,202	—	0,203	--	0,145	—

Note: A-3: NiO=0.02; CoO=0.004; P_2O_5=0.018; Na_2O=0.04; 235/586: Na_2O=0.10; K_2O=0.036; Calcination loss=0.18; BLT=1; K_2O=0.01; P_2O_5=0.05; Na_2O=0.03; for samples 0-802, BLT-1, E-10, and Knorringite 1, 7, 4, 0, 9, and 7 respectively. Analyser of Samples M-114, According to Lutts and Marshintsev, 1963; BM-1050-Frantsesson, 1968; 253/586-Milashev et al., 1963; E-3, E-11, E-10-Nixon et al., 1963; A-3-O'Hara and Mercy, 1963.

(Table 1)

From Websterite and Lherzolite Xenoliths

'Mir' Pipe		'Obnazhënnaya' Pipe				South Africa				
M-63	M-201	O-800	O-802	O-466c	O-466т	E-3	E-11	A-3	BLT-1	E-10
42,48	41,24	41,72	41,58	44,12	44,07	42,44	42,77	41,78	42,45	41,90
0,18	0,19	0,12	0,20	0,35	0,39	0,52	0,30	0,17	0,13	0,11
21,31	19,64	20,54	20,51	22,72	21,98	22,24	21,91	22,26	19,65	16,92
2,68	3,31	3,59	4,03	0,26	0,63	1,98	1,90	2,32	5,41	7,52
2,15	1,27	0,86	1,08	2,85	2,30	1,56	1,25	1.91	0,61	1,24
7,03	10,06	7,90	6,73	5,34	5,20	4,75	6,79	7,08	5,80	6,17
0,45	0,44	0,24	0,52	0,21	0,30	0,16	0,26	0,45	0,28	0,59
19,36	17,5	18,69	20,58	20,68	20,72	21,53	20,70	20,13	20,74	19,64
4,82	6,03	6,16	4,67	4,04	4,28	4,90	4,65	4,60	5,21	6,27
100,46	99,68	99,82	99,90	100,58	99,87	100,08	100,53	100,78	100,37	100,36
3,019	3,015	3,028	2,980	3,000	3,000	2,987	3,015	2,965	3,022	3,031
0,010	0,013	0,004	0,013	0,021	0,022	0,025	0,017	0,009	0,009	0,004
1,785	1,685	1,754	1,731	1,910	1,881	1,845	1,821	1,858	1,650	1,444
0,154	0,193	0,206	0,224	0,017	0,035	0,110	0,106	0,130	0,304	0,431
0,051	0,070	0,036	0,032	0,052	0,062	0,030	0,056	0,003	0,034	0,070
0,064	—	0,016	0,028	0,102	0,060	0,055	0,012	0,098	—	—
0,418	0,614	0,480	0,405	0,317	0,313	0,279	0,402	0,418	0,346	0,374
0,027	0,026	0,013	0,030	0,013	0,017	0,008	0,017	0,027	0,017	0,035
2,050	1,905	2,024	2,196	2,197	2,234	2,260	2,173	2,122	2,197	2,118
0,367	0,474	0,401	0,357	0,308	0,335	0,368	0,352	0,350	0,397	0,487
70,1	63,1	69,0	71,1	74,8	75,5	76,1	73,5	70,3	70,3	60,5
16,5	20,3	16,9	14,4	14,3	12,6	11,2	14,0	17,1	11,7	12,4
0,9	0,9	0,4	1,0	0,4	0,6	0,3	0,6	0,9	0,6	1,2
1,7	1,6	1,4	—	5,9	5,3	4,1	2,9	4,5	—	—
2,6	3,6	1,8	1,6	2,6	3,1	1,5	2,8	0,2	1,7	3,6
0,5	0,7	0,2	0,7	1,1	1,1	1,3	0,9	0,5	0,5	0,2
7,7	9,8	10,3	9,5	0,9	1,8	5,5	5,3	6,5	11,2	12,4
20,6	26,4	20,8	18,4	17,7	16,3	13,9	17,8	19,7	14,7	17,3
12,5	15,7	13,7	11,8	10,5	11,3	12,4	11,9	11,7	13,4	16,2
70,1	63,1	69,0	71,1	74,8	75,5	76,1	73,5	70,5	74,3	70,3
7,7	9,8	10,3	11,2	0,9	1,8	5,5	5,3	6,5	15,2	22,1
15,2	19,9	16,5	14,0	12,8	13,0	14,0	13,9	14,1	15,3	18,7
—	—	—	—	0,192	0,143	—	—	—	—	—

All analyses carried out in the microprobe, have been recorded in this and subsequent tables with an asterisk beside the sample number or beside the table number.

TABLE 2. Chemical Composition of Pyropes From

Oxides	Uv-196*	Uv-126*	Ud-93	Ud-40	Ud-3	Ud-111	Ud-98	Ud-94	Ud-5	Ud-101
SiO_2	40,7	40,9	42,06	41,34	41,75	41,08	41,46	41,34	41,68	41,52
TiO_2	0,04	0,82	0,80	0,57	0,16	0,00	0,38	0,70	0,80	0,22
Al_2O_3	13,5	16,3	18,94	19,49	19,66	18,6	18,80	18,20	16,58	17,84
Cr_2O_3	12,7	7,35	5,03	5,06	5,07	5,78	6,16	6,46	7,15	7,64
Fe_2O_3	—	—	1,69	1,60	1,99	—	1,63	1,61	2,30	1,64
FeO	6,40	7,18	5,45	6,87	5,82	7,01	5,93	5,71	5,39	5,83
MnO	0,28	0,25	0,49	0,47	0,38	0,41	0,49	0,42	0,27	0,38
MgO	21,6	19,4	20,32	20,41	19,24	19,7	20,06	20,26	20,40	20,01
CaO	2,86	6,53	4,52	5,09	5,68	5,77	5,50	5,09	5,90	5,58
Total	98,08	98,73	99,30	99,90	99,75	99,07	100,41	99,79	100,47	100,66
Si	3,027	3,008	3,024	2,966	3,004	3,040	2,979	2,983	3,005	2,987
Ti	0,002	0,044	0,043	0,034	0,009	0,000	0,021	0,038	0,043	0,012
Al	1,180	1,424	1,603	1,647	1,669	1,590	1,588	1,552	1,412	1,513
Cr	0,751	0,427	0,286	0,293	0,291	0,330	0,354	0,369	0,407	0,432
Fe^{3+}	0,067	0,104	0,068	0,026	0,031	0,078	0,37	0,041	0,125	0,043
Fe^{3+}	—	—	0,024	0,060	0,077	—	0,049	0,047	—	0,043
Fe^{2+}	0,331	0,341	0,329	0,354	0,350	0,345	0,354	0,345	0,325	0,350
Mn	0,018	0,016	0,030	0,030	0,023	0,026	0,030	0,026	0,016	0,026
Mg	2,396	2,148	2,184	2,182	2,062	2,136	2,150	2,177	2,191	2,144
Ca	0,228	0,516	0,351	0,392	0,437	0,450	0,423	0,395	0,455	0,432
Pyrope	47,2	59,3	67,0	67,9	68,0	66,9	64,9	63,5	59,6	61,6
Almandine	11,1	11,3	12,1	13,4	14,5	11,3	13,4	13,1	10,9	13,1
Spessartine	0,6	0,5	1,0	1,0	0,8	0,9	1,0	0,9	0,5	0,9
Andradite	3,4	5,2	3,4	1,3	1,6	3,9	1,9	2,1	6,3	2,2
Ti-Andradite	0,1	2,3	2,2	1,7	0,5	0,0	1,1	1,9	2,2	0,6
Uvarovite	4,2	9,6	6,4	10,1	12,7	11,3	11,2	9,2	6,7	11,7
Knorringite	33,4	11,8	7,9	4,6	1,9	5,3	6,5	9.3	13,8	9,9
f	14,2	17,2	16,2	16,8	18,2	16,5	17,0	16,6	17,0	16,9
Mg-component	80,6	71,1	74,9	72,5	69,9	78,2	71,4	72,8	73,4	71,5
Ca-component	7,7	17,1	12,0	13,1	14,8	15,2	14,2	13,2	15,2	14,5
Cr-component	37,6	21,4	14,3	14,7	14,6	16,6	17,7	18,5	20,5	21,6
n measured	1,781	1,765	1,755	1,757	1,755	—	1,758	1,759	1,759	1,763
n calculated	1,778	1,769	1,757	1,756	1,756	—	1,759	1,761	1,768	1,764

the main, the stoichiometric composition of pyrope, enstatite, and a
solid solution of diopside-jadeite, carried out by Yu. G. Lavrent'ev,
as compared with our standards (0-145, Ud-92, diopside, albite) showed
good reproducibility for the amounts of all the elements (within 2%
relatively).

The measured ratios of intensities were corrected for absorption
of X-ray radiation in the sample by Philibert's method (1963), for
the effect of the atomic number of the emitter, by the method of

Periodite Xenoliths of the 'Mir' Pipe   (table 2)

Ud-20*	Ud-91	Ud-17*	Uv-155*	Ud-1	Ud-18	209/576*	Uv-161*	Ud-14*	Ud-92	Uv-175*	Ud-4
41,0	41,35	41,6	41,6	40,81	40,40	40,5	40,2	40,1	40,72	40,3	41,46
1,20	0,22	0,90	0,85	0,75	0,54	1,31	1,56	0,39	0,42	1,41	0,80
14,8	16,40	14,1	14,7	14,85	15,08	13,1	13,2	13,5	14,02	11,6	16,86
8,44	9,06	9,29	9,79	9,90	9,94	10,0	10,1	10,8	11,02	12,6	8,46
—	1,15	—	—	2,09	1,67	—	—	—	2,49	—	1,37
7,32	5,84	6,95	6,44	5,44	5,82	7,48	7,68	6,84	5,07	7,40	5,16
0,33	0,51	0,32	0,26	0,28	0,10	0,35	0,35	0,38	0,34	0,30	0,21
20,0	19,72	19,9	18,8	19,60	19,37	18,3	18,9	18,9	18,50	18,0	19,31
6,75	6,35	6,85	6,74	6,58	6,78	7,25	7,22	7,52	7,72	7,62	6,43
99,84	100,60	99,91	99,18	100,30	99,70	98,29	100,21	98,43	100,30	99,27	100,06
3,012	2,995	3,053	3,064	2,983	2,972	3,047	3,042	3,012	2,996	3,028	3,001
0,066	0,012	0,049	0,047	0,041	0,030	0,072	0,089	0,023	0,023	0,081	0,043
1,281	1,402	1,218	1,275	1,283	1,309	1,154	1,144	1,192	1,212	1,028	1,444
0,486	0,522	0,538	0,567	0,571	0,580	0,597	0,585	0,641	0,642	0,748	0,484
0,167	0,063	0,195	0,111	0,105	0,081	0,177	0,182	0,144	0,123	0,143	0,029
—	—	—	—	0,009	0,012	—	—	—	0,015	—	0,046
0,283	0,353	0,233	0,283	0,333	0,358	0,293	0,293	0,285	0,312	0,321	0,312
0,022	0,031	0,018	0,018	0,018	0,006	0,023	0,022	0,027	0,021	0,018	0,013
2,191	2,129	2,179	2,063	2,135	2,123	2,053	2,080	2,118	2,031	2,018	2,083
0,530	0,492	0,538	0,531	0,514	0,535	0,583	0,568	0,605	0,609	0,614	0,499
54,0	57,4	52,5	53,2	52,4	53,0	47,0	46,7	49,1	48,9	40,2	59,5
9,4	11,7	7,9	9,8	11,1	12,2	10,1	9,7	9,4	10,9	10,6	11,5
0,7	1,0	0,6	0,6	0,6	0,2	0,8	0,7	0,9	0,7	0,6	0,4
8,4	3,2	9,6	5,6	5,3	4,1	8,6	9,3	7,2	6,2	7,3	1,5
3,3	0,6	2,5	2,4	2,1	1,5	3,7	4,3	1,2	1,2	3,9	2,2
5,8	12,6	6,0	10,3	9,7	12,0	7,4	5,7	11,5	13,0	9,6	13,2
18,4	13,5	20,9	18,1	18,8	17,0	22,4	23,6	20,7	19,1	27,8	11,0
17,0	16,3	16,4	16,0	17,3	17,5	18,6	18,6	16,8	18,1	18,7	15,7
72,4	70,9	73,4	71,3	71,2	70,0	69,4	70,3	69,8	68,0	68,0	70,5
17,5	16,4	18,1	18,3	17,1	17,6	19,7	19,2	19,9	20,4	20,8	16,9
24,3	26,1	26,9	28,4	28,5	29,0	29,8	29,3	32,2	32,1	37,4	24,2
1,778	1,764	1,781	1,780	1,778	1,779	1,784	1,788	1,779	1,782	1,796	1,764
1,776	1,769	1,778	1,776	1,776	—	1,785	1,786	—	—	1,794	—

Duncumb and Reed (1968), and for the fluorescence from the character-istic spectrum, by Reed's method (1965). Calculation of corrections was carried out on a computer by the method of successive approxima-tions.

The reproducibility of the measurements is defined by a variation coefficient, which does not exceed 2% for most of the elements.

The small amounts of K (in clinopyroxenes), Na (in garnets and enstatites), and Cr (in olivines) were determined with special care.

The determination of Cr was carried out, in contrast to all the other elements, with a voltage of 25 kV, which significantly increased the sensitivity of determination of Cr (in recalculations to Cr_2O_3--0.007%). For K (recalculated to K_2O), the limiting sensitivity of detection was 0.008%. For the determination of Na, as for K, the voltage was 15 kV, and the current for the sample (per standard), 15×10^{-8} amps. During the analysis of garnets for traces of Na (enstatites were investigated in a similar way), the $K\alpha$-line for sodium was recorded at ten different points for the samples examined during a continuous 20-second reading. In a similar way, the background at a distance of 0.16 Å (3 mm on the scale) was measured from the long-wave side of the line.

The determination of small amounts of sodium is quite a complicated matter. One of the factors affecting the accuracy of the analysis, is the possibility of vaporization of the sodium from the microvolume under investigation under the influence of electron radiation. With the relatively large current, necessary for obtaining a sufficient number of counts, and a probe diameter of 5-10 microns, an insignificant fall in the intensity of sodium radiation during the time of exposure has been noted for the albite, used as a standard. Slow movement of the sample under the probe completely removed this phenomenon. The small size of the garnet grains in most cases restricted the possibility of movement. However, a repeated analysis of one of the largest garnet grains with maximum amount of Na_2O (0.20%), carried out with movement of the sample under the probe, demonstrated perfect agreement with an analysis carried out on a steady sample, and consequently, the absence of vaporization of the sodium in the garnets. This supports the view (Smith, 1965; Siivola, 1969) that vaporization does not exert a significant influence on the results of analyses of silicates, characterized by large amounts of Mg, Ca, and Fe.

Special precautions were instituted in order to avoid the effect of possible inclusions in the garnet (and enstatite) of sodium-bearing minerals. The recorded readings were tested for homogeneity on the basis of 3σ-criteria. Then the dispersion of measurements for each sample were compared on the basis of Fisher's criteria with dispersion, relative to Poisson's statistic (Nalimov, 1960). In practice, in all cases, the dispersions compared could be regarded as equal to a probability of more than 5%. Thus, the results of the measurements may be regarded as free from the influence of possible inclusions of extraneous minerals.

Verification on the basis of F-criteria enables us to calculate the sensitivity of detection for Na in garnets and enstatites allowing only for some of the statistical errors of recording, and to regard the instrumental error (Lavrent'ev and Vainshtein, 1965) as small enough to be ignored. Under these conditions of analysis, the background value is 1.4 cps, and for 1% sodium corresponded to 60 cps, which agrees with a limiting detection concentration of Na_2O of about 0.01%.

Corrections for absorption were introduced into the directly measured values of sodium concentration (K) on the basis of Philibert's method (1963) and for atomic number, on the basis of the method of Duncumb and Reed (1968). Where complete analyses of the garnets have been carried out with the aid of electron microprobe, then corrections have been performed by the method of successive approximations calculated on a computer. In a number of cases, obtaining a

complete analysis was inappropriate, and corrections were made on
the basis of empirically determined dependence of the amount of iron
in the garnet (Fig. 1). For the chrome-bearing garnets, additional
allowance was made for the concentration of chromium, corrected to
the level of absorption of iron.

Order of Recalculation of Analyses. When recalculating mineral
analyses, we employed the widely-used oxygen method. Starting with
the atomic amounts of oxygen, we calculated the amount of cations,
which was controlled by the sum of the cations. Calculation of components was employed only for minerals of the garnet group as a
result of the uncertain recognition of the end members in such a complex group as the clinopyroxenes.

For the garnets of the deep-seated inclusions, containing Mg, Ca,
Fe, and a little Mn in eightfold coordination in the cation group, no
doubts arose as to the reality of the existence of such end members,
as pyrope, grossular, and almandine, and also spessartine. Complica-
tion arises from the substitution for aluminum by chromium and ferric
iron, and also the uncertain coordination of titanium. We have demonstrated unequivocally that titanium in the pyrope-almandine garnets of
the kimberlites is involved in the group of R^{VI} cations (see §12) and
it must not be treated as the schorlomite component (Sobolev, 1964a;
Rickwood, 1968), but as the titanium-andradite component.

Recently, the appearance of a general scheme for recalculating the
composition of garnets to components with a single order of recalculation for all types of garnets, assumed to have universal application (Rickwood, 1968), compels us to dwell on this question in more
detail. It seems to us that the separation of uvarovite before andradite, and recalculation of titanium in the schorlomite component
are unfounded. The former leads, in many cases, to the appearance
of an hypothetical koharite component ($MgFe^{3+}$ garnet), and the
second is disproved by our new data (see §12). In avoiding the
choice of a universal scheme for recalculating all garnets, we shall
limit ourselves to those belonging to the pyrope-almandine-grossular
solid - solution series, which contain traces of Cr, Fe^{3+}, and Ti.
In commencing the recalculation with the recognition of components,
which do not contain Al, we shall in that way eliminate the necessity to introduce artificial end members, not known in the pure form
as natural minerals. We have accepted the following order in the
recalculations (see §12): (1) titanium-andradite; (2) andradite;
(3) uvarovite; (4) grossular (if Ca > Cr); (5) pyrope; (6) almandine;
(7) spessartine; and (8) knorringite (if Cr > Ca) with the deduction
of an equivalent amount of Mg from the pyrope.

We shall regard the excess of Si in certain wet chemical analyses
of garnets as accidental introduction during preparation of the
material for analysis by contamination from the agate crucible
(Lebedev, 1959; Sobolev, 1964) and we shall recalculate it as free
SiO_2. Excess Fe^{3+} (over 2.00) in the R^{VI} group of cations will be
regarded as almandine component (Sobolev, 1964).

For the analyses, carried out with the aid of an electron microprobe, the amount of Fe^{3+} has been established by the summation of
ΣR^{VI} to 2.000. The role of Fe^{3+} has been estimated in a similar way
for the chrome-spinels, making allowances, however, for traces of Ti
(the ulvöspinel component). Since it is necessary in this case to
allow for the substitution Ti ⇄ $2Fe^{2+}$, the amount of ΣR^{VI} is taken
as equal to 2 — Ti. Such a scheme of compensation isomorphism has
also been accepted for recalculating analyses of ilmenites (Mikheev

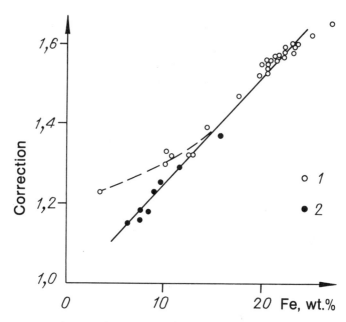

Fig. 1. Dependence of the correction for the amount of Na in the pyropealmandine garnets on the amount of Fe (1) and on the total amount of Cr + Fe in pyropes (2).

and Kalinin, 1961) allowing for the addition of $\Sigma R^{3+}(Fe^{3+}$, Cr, and Al) in the form of the component $R^{3+}R^{3+}O_3$ and the substitution FeTi $\rightleftarrows 2R^{3+}$.

It is considerably more difficult to assess the probable role of Fe^{3+} in the clinopyroxenes, which have been analyzed with the aid of the microprobe. For the pyroxenes of eclogites, this problem has been eased by the specific features of the especially deep-seated pyroxenes which we have studied, almost devoid of Al^{IV} in most cases. For such pyroxenes, the role of Fe^{3+} has been estimated by comparing the amounts of cations of Al (along with Cr) and Na.

When treating the analytical results we used certain methods of mathematical statistics (Urbakh, 1964), in particular the computation of the average arithmetic and standard deviations, the comparison of the averages, the calculation of correlation coefficients and equations of rectlinear regression.

CHAPTER II

BRIEF DESCRIPTION OF KIMBERLITE PIPES INVESTIGATED

The kimberlite pipes of the Siberian Platform are located mainly within its northeastern part, where the principal regions of manifestation of kimberlite volcanism are known. The problems of geologic definition and the structural position of the kimberlites have been discussed in numerous works (Bobrievich et al., 1959; Moor and Sobolev, 1957; Sobolev, 1962; Bobrievich and Sobolev, 1962; Milashev et al., 1963; Milashev, 1965; Krasnov et al. [in Russian], 1966; Frantsesson, 1968; Frantsesson and Prokopchuk, 1968; Vasil'ev et al., 1968; Odintsov and Strakhov, 1968; etc.), and we shall dwell only briefly on the features of the individual pipes that we have studied.

The data on the distribution of the kimberlite pipes within the Siberian Platform demonstrate a clear separation in the individual regions of their development. The position of the regions of kimberlite volcanism and their relationships to the structural elements of the Siberian Platform have been shown on the diagram (Fig. 2), taken from the work of Frantsesson and Prokopchuk (1968). These authors and other investigators of the Siberian kimberlites have emphasized the presence of two groups of regions of manifestation of kimberlite volcanism: the regions of the marginal areas of the platform and those confined to areas of ancient Archean-Proterozoic basement. The kimberlites of the second group, which are diamond-bearing in contrast to those of the first group, were formed, according to available data, two to three geologic periods earlier than the kimberlites of the marginal areas (Milashev, 1965). Geophysical data indicate a significant depth of occurrence of the surface of the crystalline basement in the areas of the second group (about 4 km), with decrease in the overall thickness of the sedimentary formations northwards (Frantsesson, 1968).

Detailed investigations of the xenoliths of deep-seated rocks have been carried out by us on the material from four kimberlite pipes, belonging to different regions of development of kimberlites: the Lower Olenëk region (the 'Obnazhënnaya' pipe), the Daldyn-Alakit region (the 'Udachnaya' and 'Zagadochnaya' pipes), and the Malo-Botuoba region (the 'Mir' pipe). Such a choice of pipes, rich in xenoliths, was aimed as much as possible at a more complete definition of the whole variation of these unique deep-seated formations, and also the features of the assemblage of xenoliths of each of the pipes studied.

In addition, certain xenoliths from the 'Aikhal' pipe (pyrope peridotites, including diamond-bearing types, and a kyanite eclogite), collected by B. S. Nai, have been studied.

There was special interest in the clarification of the features of mineral composition of the deep-seated inclusions of such typically diamond-bearing pipes as 'Mir', 'Udachnaya', and 'Aikhal', and pipes, not containing diamonds ('Obnazhënnaya' and 'Zagadochnaya'; see Chapter VIII). For comparison, we studied some samples of xenoliths from the kimberlite pipes of South Africa, especially the diamond-bearing eclogites from the 'Newland' pipe and kyanite eclogites from the 'Roberts Victor' pipe.

THE 'OBNAZHËNNAYA' PIPE is located in the lower course of the River Olenëk, near the mouth of the River Kuoika (see Fig. 2). This is an exceptionally rare example for the Siberian Platform of a beautifully exposed kimberlite pipe with a riverbank cliff outcrop, and it attracted attention immediately after its discovery in 1957. The

12

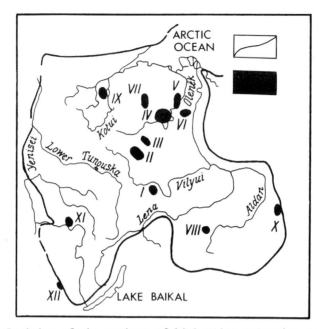

Fig. 2. Position of the regions of kimberlite volcanism on a map of
the relationships between the post-Early proterozoic Siberian Plat-
form and the structures of its surroundings (after Frantsesson &
Prokopchuk, 1968, simplified by the author). 1) boundary of region
described, including the unchanged ancient platform (craton) and
sectors of younger structural zones, regarded as part of the Siberian
Platform; 2) regions of kimberlite volcanism: I) Malo-Botuoba,
II) Daldyn-Alakit, III) Muna; IV) Middle Olenek; V) Lower
Olenek; VI) Near-Lena; VII) Kuonamka; VIII) Upper Aldan; IX)
Maimecha-Kotui, X) Ingily, XI) Chadobets, X11) Belozima.

kimberlitic breccia contains quite a lot of fresh olivine and a large
quantity of fragments of various rocks, mainly belonging to the sedimentary cover. The most detailed work on the pipe, carried out by
Milashev (1960) and Milashev et al. (1963), has revealed a number of
interesting features. The pipe is small, and is elliptical in shape
with axes about 30 and 45 m long. Bulk measurements of the orientation of the xenoliths have identified a primary 'flow texture' (Milashev et al., 1963). The first discoveries and descriptions of deep-
seated xenoliths, distinguished by their exceptionally fresh
minerals (Milashev, 1960), served as a preamble for special exploration, which we carried out in 1963. The representative collection
consisted of several hundreds of samples of the most varied sizes,
up to 15 cm in diameter and up to 6 kg in weight, allowed to identify
a whole series of new features, and also revealed corundum eclogites
and xenoliths of complex composition.
    It is interesting to note that there is a sector within the small
pipe, which has been locally enriched in deep-seated xenoliths. The
presence of such a zone indicates the unique removal of a 'jet' of
deep-seated material analogous to the features of local enrichment
in xenoliths, observed sometimes in the alkaline basaltoids.
    Interest in the 'Obnazhĕnnaya' pipe was still greater strengthened

after the establishment of its age on the basis of the discovery of
a belemnite in the breccia at the level of the erosional section into
the Sinian dolomites surrounding the pipe (Milashev and Shul'gina,
1959). This discovery not only narrowed down the interval of probable time of injection of the kimberlites, but also served as proof
of substantial remixing of the material in the body of the pipe with
significant remixing below the level of the original occurrence.
The late Mesozoic age of the pipe is also emphasized by the later
dating of a phlogopite from the groundmass of kimberlite (Sarsadskikh et al., 1966).

THE 'UDACHNAYA' PIPE, which is situated in the Daldyn-Alakit region of kimberlitic magmatism, was discovered in 1955. In consists
of two adjacent pipes ('Udachnaya East' and 'Udachnaya West')
(Bobrievich et al., 1959) with clearly distinguishable types of kimberlite, consisting of massive kimberlite, containing a lot of fresh
olivine for the first, and typical breccia, for the second. The
country rocks are Lower Ordovician limestones. The number of xenoliths in the 'Udachnaya East' pipe is half that in 'Udachnaya West'.
Data are presented below on the content of xenoliths in the kimberlite breccias of the Yakutian pipes (after Koval'sky and Yegorov,
1964) in volume percentages.

Depth interval, m	'Mir'	'Udachnaya East'	'Udachnaya West'
0-50	27.59	18.07	41.36
200-300	17.02	16.49	39.53
300-400	--	14.71	37.69
500-600	15.03	--	--
1000-1200	9.00	--	--

A characteristic feature of both pipes, like all those of the
Daldyn-Alakit region, is the presence of a large number of xenoliths
of crystalline schists, including eclogite-like rocks with plagioclase (Bobrievich and Sobolev, 1957).

Both portions of the 'Udachnaya' pipe are characterized by the
presence of xenoliths of unusual pseudo-porphyritic, pyrope peridotites (see §1). These xenoliths in the 'Udachnaya West' pipe are,
as a rule, intensely serpentinized even at a depth of up to 500 m
and in drill-holes. In the 'Udachnaya East' pipe, even at a depth
of about 100 m, there are xenoliths of fresh, almost unserpentinized
peridotites.

We have found a significant number of xenoliths of pyrope peridotites during the detailed investigations of both pipes. The sizes
of the xenoliths are extremely varied: from 1-2 up to 10-12 cm in
diameter. The eclogites are represented by several samples including kyanite eclogites, some even with diamonds (Ponomarenko et al.,
1976). The age of the pipe is assumed to be late Paleozoic (Sarsadskikh et al., 1966).

THE 'ZAGADOCHNAYA' PIPE is located 30 km from the 'Udachnaya' pipe,
and it consists of two small outcrops (two pipes) at a distance of
about 150 m from each other amongst the surrounding Lower Ordovician
limestones. The kimberlite of this pipe contains a large amount
of chrome-diopside, which distinguishes it from many of the other
kimberlite pipes of the region. A substantial number of xenoliths
of crystalline schists has been found in the pipe, amongst which
there are individual fragments of eclogitic rocks, so that the assem-

blage of xenoliths belonging to the crystalline basement, is similar
to that of the 'Udachnaya' pipe and other pipes of the Daldyn-Alakit
region.

The 'Zagadochnaya' pipe differs significantly from all the other
pipes not only of this region, but also of Yakutia as a whole in the
assemblage of deep-seated xenoliths. A unique feature of this pipe
is the significant number of xenoliths of grospydites and kyanite
eclogites. In this very pipe, the first grospydite xenolith was discovered, characterized as a new type of metamorphic rock (Bobrievich
et al., 1960). Along with the xenoliths of these rocks in the pipe,
there are also eclogites without kyanite.

The possible presence of peridotite xenoliths is suggested only
by the discoveries of individual grains of pyrope and chrome-diopside and their intergrowths in the kimberlite concentrate of this
pipe. Peridotite xenoliths themselves have not so far been found,
in spite of careful search. As noted above, the 'Zagadochnaya' pipe
is markedly separated by the unique nature of its assemblage of deep-
seated xenoliths. Amongst the South African kimberlites, only a few
pipes including the 'Roberts Victor' Mine contain more than 90% of
eclogites, including kyanite types also, in the deep-seated material
(MacGregor and Carter, 1970), but pyrope peridotites are also known.
In the 'Zagadochnaya' pipe, olivine has not been found either in the
very altered kimberlite, or in the deep-seated xenoliths.

THE 'MIR' PIPE (the Malo-Botuoba region, valley of the River
Iirélyakh) consists of a funnel-shaped body, oval in plan, with dips
of 90° to 75° (Bobrievich et al., 1959). The pipe occurs amongst
Lower Ordovician carbonate deposits, and consists of kimberlite of
varied aspect and degree of alteration, separated into several types
(Botkunov, 1964). The various types of kimberlite are characterized
by different amounts of xenoliths of the country rocks, which decrease
with depth. Crystalline schists are very rare. Deep-seated xenoliths
are also rare. Amongst them, pyrope-bearing ultramafic rocks, both
lherzolites, and also websterites and harzburgites, predominate. The
eclogites occupy a markedly subordinate position (about 6%), and
amongst them, as a result of careful search, about 20 samples of diamond-bearing varieties have been found, including a corundum diamond-
bearing eclogite (Sobolev, Botkunov, and Lavrent'ev, 1976).

Features of the Xenogenic Material of the Kimberlite Pipes. Data,
based on a study of the xenoliths in the kimberlite pipes both from
South Africa and also Yakutia demonstrate that amongst the fragments
which form the kimberlite breccias, material of the country rocks predominates. For the Yakutian pipes, this material consists mainly of
fragments of sedimentary rocks, principally carbonate types, comprising the main part of the sedimentary formations of the Siberian Platform. To this group of xenoliths we may assign the fragments of dol-
erite often present in the pipes. In significantly lesser number are
the xenoliths of crystalline schists, similar to the rocks of the
Anabar and Aldan shields, and extracted, in all probability, from the
crystalline basement. The xenoliths of pyrope-bearing rocks and eclogites comprise an extremely insignificant percentage and belong to the
most rare formations. Such a pattern of occurrence of fragments of
different rocks in the pipes is associated with the depth of their
extraction.

An undoubted indicator of the depth of entrapment of the xenoliths is the nature of their abrasion during the process of eruption
of the kimberlites towards the surface. The shallowest xenoliths of

rocks of the sedimentary cover are characterized by an angular, irregular shape, sometimes very large dimensions, and termed in such cases, by analogy with the African pipes, 'floating reefs' (Williams, 1932). The xenoliths of crystalline schists, analogous in composition to the rocks of the basement, bear clearer traces of abrasion, and finally, the pyrope-bearing rocks are characterized by similar morphologic features (rounded or discoid shape), which enables us clearly to distinguish them amongst all the xenoliths (Photo 1)[*]. This, in equal degree, applies to the pyrope peridotites, eclogites, and grospydites of the various kimberlite pipes, and indicates their great depth in comparison with all the other types of xenoliths.

Special investigations based on the calculation of the total amount of various xenoliths in the Yakutian pipes, including the micro-xenoliths (Koval'sky and Yegorov, 1964), have shown that their amount in the kimberlites of the breccias varies widely: from 6 up to 41% in the near-surface portions of the pipes. Simultaneously, a clear trend towards a decrease in the overall amount of xenoliths with depth has appeared, especially clearly defined in the 'Mir' pipe, as a result of the large number of data based on depth.

It should be emphasized that the data presented reflect only variations in the amount of xenoliths of the sedimentary cover, which comprise the great majority of all the xenoliths. The fragments of basement rocks and pyrope-bearing xenoliths are evenly distributed in depth, which is supported by discoveries of deep-seated xenoliths in deep drill-cores from the 'Mir' pipe at different depths (up to 1150 m).

Data indicating a decrease in the number of xenoliths (of country rocks) with depth are of significant interest in explaining the degree of remixing of the material in the pipes, which may probably be different in different pipes. Thus, although there is completely verified information on the vertical displacement of the xenoliths in a number of pipes below the level of their original occurrence (Milashev and Shul'gina, 1959; Williams, 1932), the tendency towards decrease in their amount with depth, established quite definitely so far only for the 'Mir' pipe alone, indicates the limited manifestation of remixing.

---

*All of the photographs appear after page 120 of the book.

CHAPTER III

THE FACIES OF PYROPE PERIDOTITES AND ECLOGITES WITH GRAPHITE

(graphite-pyrope)

The present chapter discusses the compositional features and parageneses of the xenoliths of garnetiferous rocks of ultramafic and
basic composition, containing no plagioclase or aluminous spinel. In
these xenoliths, graphite is occasionally recorded. The principal reason for assigning these xenoliths to the facies under consideration is
the presence of pyrope (or a more ferriferous garnet in the eclogites)
and the absence of diamond. There is no doubt that the rocks of this
facies, which are stable under an extraordinarily wide range of temperatures and pressures in the mantle, are much more widely distributed amongst the deep-seated xenoliths of the kimberlite pipes of
Yakutia and Africa as compared with the diamond-bearing types. For
some of the series of deep-seated xenoliths considered by us from
the 'Obnazhĕnnaya' and 'Zagadochnaya' pipes, the impossibility of
appearance of diamond-bearing varieties has been proved quite unequivocally. This enables us to recognize features of differences in xenoliths of similar composition and of lesser pressures (the facies under
consideration) and of higher pressures (parageneses with diamond).

§1. PYROPE PERIDOTITES AND PYROXENITES

Xenoliths, containing an association of magnesian garnet with
diopside (chrome-diopside), enstatite, and forsterite, belong to the
most widely distributed types of rocks in the kimberlite pipes of
Yakutia and South Africa. In practically all the pipes, with rare
exceptions, these xenoliths comprise the overwhelming bulk of the
deep-seated material, which has been emphasized by a series of
detailed investigations into the kimberlite pipes of Yakutia
('Obnazhĕnnaya', 'Udachnaya', and 'Mir').
The great interest, directed towards the study of the peridotites,
included by the kimberlites, is explained, first, by the similarity
between their composition and model systems of the upper mantle, and
second, by the possibility of determining the temperatures of crystallization of these rocks mainly on the basis of compositional relations of the associated pyroxenes.
Below, along with the description of the chemical composition of
the peridotites and eclogites from the kimberlite pipes, quite wide
variations in the composition of the peridotites are shown, clearly
correlated with the features of their quantitative-mineral composition. These features, in particular, are most completely expressed
in the assemblage of xenoliths from the 'Obnazhĕnnaya' pipe amongst
which there are numberous samples, distinguished by complex composition, with individual zones, consisting of associations with a
marked predominance of olivine, on the one hand, and pyroxenes, on
the other. In the samples known to us, such zones bear not a closed
character, as assumed by Milashev (1960), but are arranged in the form
of bands, which emphasizes the non-uniform nature of the substrate,
from which the xenoliths came. This is also indicated by observations of the first samples of banded xenoliths of complex composition
(Bobrievich et al., 1959).

16

In defining the various xenoliths of pyrope-bearing ultramafic rocks, textural-structural features are of great importance, enabling us to recognize, following other authors who have studied the kimberlites (Bobrievich et al., 1959, 1964), two types of such rocks, almost indistinguishable on the basis of the rock-forming minerals. The first type may include the coarse-grained rocks with granoblastic or poikiloblastic texture, and the second, the porphyritic garnet peridotites with clearly defined porphyritic segregations and a fine-grained groundmass.

Rocks with a complex history of crystallization are recognized in the first type on the basis of features of paragenesis and relationships of minerals. In them, without doubt, a late reaction formation of garnet at the expense of spinel and pyroxenes, has been identified, which leads to the important conclusion about the possibility of garnetization of the spinel peridotites under natural conditions (Sobolev and Sobolev, 1964), actually at the same time as the experimental results of a study of the reaction of formation of pyrope at the expense of spinel and orthopyroxene (MacGregor, 1964).

In connection with the establishment of the reaction nature of the formation of pyrope, there is particular interest in the fine intergrowths of garnet in the pyroxenes in the ultramafic xenoliths of the 'Obnazhënnaya' pipe. Such intergrowths of garnet in clinopyroxene were recorded for the first time in xenoliths from South Africa (Williams, 1932), and the pattern of orientation of such intergrowths with respect to the pyroxene has been established for eclogites from the serpentinites of Saxony (Hentschel, 1937). It has been shown (Bobrievich et al., 1960; Milashev, 1960) that the garnet replaces pyroxene along the cleavage cracks and partings. However, a detailed study of the intergrowths in both clino- and orthopyroxene has shown that the intergrowths are not a replacement of grains along the cleavage and are not simple exsolution textures caused by lowering of temperature, as might be expected from the equilibrium diagram, but peculiar phenomena of replacement of exsolution textures of the pyroxenes (Sobolev and Sobolev, 1964). In the clinopyroxenes in the garnet bands, relicts of enstatite are always identified. On the other hand, in the enstatite, the garnet replaces the intergrowths of clinopyroxene. In replacing the pyroxenes by reaction, the garnet in the intergrowths is developed in the form of spindle-shaped bands along the parting cracks.

In the peridotites with a complex crystallization history, that is, secondary development of garnet as the result of reaction between spinel and the pyroxenes, significant fluctuations in the content of Cr_2O_3 are possible in these same rocks, apparently as a result of the primary precipitation of spinels during the magmatic phase. A fine example of this is the earlier-unknown xenoliths of complex composition with varying amount of chromium in the various sectors of a sample, discovered in Yakutia (Sobolev and Kuznetsova, 1965). Macroscopically, both portions of such a xenolith are distinguished by the color of the garnet and pyroxene.

In one of the samples investigated (O-466), the difference in the amount of Cr_2O_3 in the rock, garnet, and pyroxene from various portions of the sample is almost fourfold, with similarity in mineral compositions based on the amount of other components (see Tables 1 and 3). This indicates the low mobility of chromium during metamorphism and the enrichment in chromium of segregations in deep-seated rocks, like the segregations enriched in chromite in the ultramafic rocks.

The second type of pyrope peridotites (porphyritic) is characterized by a completely different aspect of the rocks. These xenoliths
in the first-described deep-seated rocks from the Yakutian kimberlites (Bobrievich et al., 1959a, 1964), in contrast to the holocrys-
talline types, have been termed porphyritic peridotites. Their principal feature is the porphyritic aspect with the fine-grained groundmass, consisting of olivine, and with porphyritic segregations of
garnet and pyroxenes. The garnet in these peridotites is surrounded
by a kelyphitic rim. In contrast to the ultramafic rocks of the
first type, replacement phenomena in the solid phase are absent in
them. These rocks are characterized by: (1) the presence of porphy-
roclasts and a groundmass; (2) gradual transitions from typical
sheared rocks with well-defined parallel-oriented fabric, well-pre-
served rounded grains of garnet and the deformed grains of pyroxene
(Photo 2), which sometimes form 'comet' textures (Bobrievich et al.,
1964), with fine-grained groundmass, into the present pseudo-porphyritic rocks with a recrystallized fine-grained groundmass, as in the
peridotites known in place in the Czech Massif (Fiala, 1965, 1966).
The results of the detailed investigation of the mineral compositions
of the pseudo-porphyritic peridotites from the 'Udachnaya' pipe, have
enabled us to justify their assignment to a special deep-seated type
(see below) and to use a term "pseudoporphyritic" for this group of
xenoliths from the 'Udachnaya' pipe.

In spite of the textural differences and variations in the amounts
of Al_2O_3 and Cr_2O_3, the xenoliths of peridotite composition, assigned
to different types, consist of an identical assemblage of minerals.
These are olivine, ortho- and clinopyroxene, pyrope garnet, and often
a chrome-spinel.

Quantitative-mineralogical investigations, carried out by some
authors on a large number of xenoliths of ultramafic composition
both from Yakutia (the 'Obnazhĕnnaya' pipe) (Milashev et al.,
1963), and South Africa (Rickwood et al., 1969), indicate wide fluctuations in the amounts of the principal rock-forming minerals in
these xenoliths. Thus, in the 'Obnazhĕnnaya' pipe, xenoliths of
pyrope lherzolites are known, which contain very rare grains of
pyroxenes. With decrease in the amount of olivine, there is a
marked increase in the content of pyroxenes, which, along with garnet, are present in all possible combinations (Milashev et al., 1963).

As exemplified by the xenoliths from the South African pipes, it
has been shown that in the peridotites, the amount of garnet (in volume %) does not exceed 19% (Rickwood et al., 1968). Xenoliths, in
which more than 28% of garnet have been recorded, in almost no instance contain olivine or enstatite. In the Yakutian kimberlites,
only the xenoliths of the 'Obnazhĕnnaya' pipe have not been subject
to this pattern, evidently owing to the presence of a large number
of samples with reaction development of garnet and xenoliths of complex composition.

In the porphyric and porphyritic peridotites, the amount of garnet
seldom exceeds 15%, clearly reflecting the features of distribution
of chromium and aluminum in the garnets (Fiala, 1965; Sobolev and
Sobolev, 1967). When considering the features of the m i n e r a l
composition of the peridotites, special attention must be paid to the
garnets and pyroxenes, since the equilibrium between these minerals
depends on temperature and pressure.

Among the garnet-bearing xenoliths of ultramafic composition in
the kimberlite pipes of Yakutia and South Africa, the following asso-

ciations have been identified: (1) pyrope + olivine; (2) pyrope +
olivine + enstatite; (3) pyrope + olivine + enstatite + diopside;
(4) pyrope + olivine + diopside; (5) pyrope + enstatite + diopside;
and (6) pyrope + diopside. Accessory chrome-spinels may be present
in most of the associations.

Lherzolite association 3 is the most widely distributed in the
xenoliths, with wide fluctuations in the relative amount of pyroxenes
and olivine, and also pyrope, but in narrower limits (1-15%). Websterite association 5 has also significant distribution. Association
6, containing a pyrope, analogous in composition to that of the ultramafic rocks, is of frequent occurrence in this series of rocks and is,
along with association 5, a connecting link in composition between the
deep-seated ultramafic rocks and the eclogites.

And finally, the rarest types, identified in isolated cases on the
basis of the compositional features of garnet, include the harzburgite-dunite (1 and 2) and the wehrlite (4) associations. The latter
has been identified in unusual intergrowths from the heavy-fraction
concentrate of the kimberlites (Sobolev, Lavrent'ev et al., 1973a),
and also in the first-discovered xenoliths, containing olivine,
chrome-diopside, and a garnet rich in chromium and calcium (Sobolev
et al., 1973).

Garnet

The amount of calcium component in garnet is the most sensitive
indicator of the nature of the paragenesis of ultramafic rocks. As
an investigation of garnet compositions from ultramafic rocks has
shown, their calcium index is distinguished by exceptional constancy: 13.3 ± 1.9 mol % (Sobolev, 1964a). A similar amount of
calcium has been identified in the garnets, belonging to various
associations of pyrope ultramafic rocks (the lherzolite and websterite types). An analogous amount of calcium has also been demonstrated
for the garnets in xenoliths, described as dunites and olivinites.
This same feature has led to the assumption that a seemingly insignificant amount of pyroxenes is present in even the most olivine-rich
rocks (Sobolev, 1964a). In fact, in such xenoliths from the 'Obnazhĕnnaya' pipe, insignificant amounts of pyroxenes have been identified, which suggests that they should be assigned to the lherzolite
paragenesis (samples 0-800 and 0-802).

Data from the study of garnets of ultramafic rocks, especially from
the xenoliths of kimberlite pipes and kimberlite concentrate (Sobolev,
1964a), have shown a marked predominance of associations containing
two pyroxenes. Later accumulation of factual data has emphasized
these results, but at the same time, has revealed a verified association of garnet-olivine, the pyrope of which is characterized, as follows from the paragenetic diagram for ultramafic rocks (Bobrievich
et al., 1960); a diminished amount of calcium component. The pyrope-
olivine paragenesis has been identified among xenoliths from the
'Udachnaya' pipe (Sobolev, Lavrent'ev et al., 1973a). The composition of a chrome pyrope with diminished amount of CaO is shown in
Table 2 (sample No. Uv-196). On the basis of physical properties, a
few samples have been identified, which belong to this association,
described in greater detail in the group of ultramafic xenoliths of
the 'Udachnaya' pipe.

In the South African pipes, amongst more than 200 ultramafic xenoliths examined, the paragenesis garnet + olivine has been noted

TABLE 3. Chemical Composition of Clinopyroxenes

Oxides	'Obnazhënnaya' Pipe		'Mir' Pipe			'Obnazhënnaya' Pipe			'Mir' Pipe	
	O-145	O-309	M-60	M-99	BM-1050	O-172*	O-467	O-379	M-63	M-201
SiO_2	53,96	54,32	53,82	54,49	54,70	54,9	52,30	53,54	53,80	54,47
TiO_2	0,13	0,15	0,64	0,38	0,58	0,44	0,53	0,29	0,15	0,16
Al_2O_3	6,90	6,88	7,96	2,23	8,14	6,82	7,21	2,17	3,17	1,47
Cr_2O_3	0,14	0,11	0,29	0,22	—	0,56	0,66	0,39	1,38	0,56
Fe_2O_3	1,71	0,98	2,34	0,91	1,37	—	1,09	2,45	2,10	1,98
FeO	0,80	0,65	1,13	0,44	2,07	2,68	1,11	1,64	0,53	0,90
MnO	0,09	0,01	0,08	0,02	0,02	0,03	0,07	0,01	0,05	0,06
MgO	15,69	14,15	12,32	17,07	14,38	12,6	14,37	15,93	16,94	16,25
CaO	17,41	19,29	16,66	21,80	15,98	17,0	19,10	21,39	19,13	22,60
Na_2O	2,87	3,11	4,26	1,98	3,37	3,90	3,12	1,89	2,23	1,08
K_2O	0,04	0,08	0,05	0,01	Сл.	—	0,03	0,03	0,05	—
Total	99,74	99,73	99,53	99,55	100,61	98,93	99,59	99,73	99,53	100.53
Si	1,927	1,943	1,934	1,967	1,933	1,982	1,887	1,949	1,950	1,979
Al^{IV}	0,073	0,057	0,066	0,033	0,067	0,018	0,113	0,051	0,050	0,021
Al^{VI}	0,219	0,235	0,276	0,067	0,272	0,273	0,195	0,045	0,084	0,044
Ti	0,002	0,004	0,017	0,011	0,016	0,011	0,015	0,009	0,004	0,004
Cr	0,004	0,004	0,009	0,009	—	0,017	0,022	0,013	0,039	0,020
Fe^{3+}	0,047	0,026	0,061	0,026	0,037	—	0,030	0,069	0,057	0,052
Fe^{2+}	0,024	0,019	0,035	0,013	0,061	0,082	0,033	0,050	0,015	0,026
Mn	—	—	0,002	—	0,001	—	0,002	—	—	0,002
Mg	0,835	0,754	0,659	0,917	0,758	0,677	0,771	0,863	0,913	0,879
Ca	0,665	0,739	0,642	0,844	0,605	0,657	0,739	0,835	0,741	0,879
Na	0,200	0,215	0,298	0,139	0,231	0,273	0,221	0,135	0,161	0,079
K	0,002	0,004	0,002	—	—	—	—	—	0,001	0,002	—
Total of Cations	3,998	4,000	4,001	4,026	3,981	3,990	4,028	4,020	4,016	3,985
f'	2,8	2,5	5,0	1,4	7,4	10,8	3,7	5,5	1,6	2,9
f	7,8	5,6	12,7	4,1	11,4	10,8	7,6	12,1	7,3	8,2
Ca/Ca+Mg	44,0	49,5	49,3	47,9	44,4	49,3	48,9	49,2	44,8	50,0
Ca/Ca+Mg+Fe	43,6	49,0	48,1	47,7	42,5	46,4	47,9	47,7	41,9	49,3
Cr/Cr+Al	1,4	1,4	2,6	8,3	—	5,5	6,7	11,9	22,5	23,5

Note: A-3: NiO = 0.052, P_2O_5 = 0.015; A-17: NiO = 0.055; P_2O_5 = 0.002; M-201: loss ign = 1.0; 0-466C: loss ign = 0.82; 0-466T: loss ign. = 0.81. Analyses of samples M-114 (after Lutts and Marshintsev, 1963); BM-1050 (afte

in one case only (Mathias et al., 1970) though without definition of the garnet composition. Only the deviation of the garnet composition from the peridotite point may serve as a weighty argument for recognizing the paragenesis garnet + olivine. In the small xenoliths, clinopyroxene may simply not be observed.

'Obnazhënnaya' Pipe				'Mir' Pipe	South Africa								
O-800	O-802	O-466с	O-466т	M-114	E-5*	E-14*	PHN-4*	E-3*	E-11*	MM-4*	M-72*	A-3	A-17
52,68	53,41	52,69	52,37	54,34	55,2	54,0	54,3	55,2	54,2	54,9	54,6	53,90	53,80
0,15	0,53	0,73	0,81	Сл.	0,15	0,39	0,13	0,16	0,41	0,003	0,15	0,20	0,11
2,68	3,52	6,93	6,67	3,45	2,67	2,55	2,39	2,05	3,03	2,93	2,10	2,40	1,80
1,27	3,20	0,30	1,26	0,24	0,71	0,24	0,79	0,85	1,45	1,82	3,15	1,55	1,12
1,33	1,36	1,20	1,27	0,42	—	—	—	—	—	—	—	1,34	1,11
1,20	1,68	1,22	0,83	1,05	4,40	5,94	4,04	3,34	2,59	2,14	2,13	1,28	1,81
0,15	0,04	0,02	0,10	0,04	0,14	0,14	0,14	0,12	0,09	0,08	0,09	0,08	0,10
17,45	14,65	16,05	15,40	18,30	21,9	20,6	20,9	21,2	17,3	16,6	16,1	16,39	17,63
21,37	17,75	17,10	17,18	20,14	14,0	14,5	15,4	16,1	19,5	20,1	19,9	20,57	20,96
1,64	3,39	2,80	3,04	1,37	1,70	1,63	1,52	1,24	1,95	2,31	2,50	1,80	1,26
0,06	0,12	0,06	0,04	0,10	0,02	0,02	0,04	0,05	0,01	0,003	0,01	0,02	0,04
99,98	99,65	99,80	99,92	99,45	100,9	100,0	99,7	100,4	100,6	101,0	100,7	99,60	99,79
1,915	1,946	1,899	1,895	1,956	1,955	1,950	1,960	1,970	1,950	1,970	1,970	1,961	1,956
0,085	0,054	0,101	0,105	0,044	0,045	0,050	0,040	0,030	0,050	0,030	0,030	0,039	0,044
0,030	0,095	0,193	0,180	0,102	0,067	0,059	0,062	0,056	0,078	0,094	0,059	0,063	0,033
0,004	0,016	0,020	0,022	—	0,007	0,011	0,004	0,004	0,011	—	0,004	0,005	0,003
0,037	0,092	0,009	0,036	0,007	0,020	0,007	0,023	0,024	0,041	0,051	0,090	0,045	0,032
0,036	0,039	0,032	0,034	0,011	—	—	—	—	—	—	—	0,037	0,030
0,037	0,053	0,037	0,025	0,032	0,130	0,179	0,122	0,100	0,078	0,064	0,064	0,039	0,055
0,005	—	0,001	0,003	0,001	0,004	0,004	0,004	0,004	0,003	0,002	0,003	0,002	0,003
0,947	0,796	0,862	0,831	0,982	1,155	1,107	1,120	1,130	0,927	0,888	0,865	0,885	0,952
0,833	0,691	0,660	0,666	0,777	0,531	0,559	0,595	0,616	0,752	0,773	0,769	0,801	0,816
0,116	0,232	0,196	0,200	0,096	0,116	0,114	0,106	0,086	0,136	0,160	0,175	0,126	0,089
0,003	0,004	0,003	0,002	0,005	0,001	0,001	0,002	0,002	0,001	—	0,001	0,001	0,002
4,048	4,018	4,013	3,999	4,013	4,031	4,041	4,038	4,022	4,027	4,032	4,030	4,004	4,015
3,8	6,1	4,1	2,9	3,2	10,1	13,9	9,8	8,1	7,8	6,7	6,9	4,2	5,5
7,2	14,1	7,4	6,6	4,2	10,1	13,9	9,8	8,1	7,8	6,7	6,9	7,9	8,2
46,8	46,5	43,4	44,5	44,2	31,5	33,5	34,7	35,4	44,8	46,5	47,0	47,5	46,2
45,8	44,9	42,3	43,8	43,4	29,2	30,3	32,4	33,4	42,8	44,8	45,3	45,6	44,0
24,3	38,2	3,0	11,2	4,6	15,2	6,0	18,4	21,8	24,3	29,1	50,3	30,6	29,4

Frantsesson, 1968); E-5, E-14, E-3, E-11 (after Nixon, 1963); PHN-4, MM-4, M-72 (after Boyd, 1969); A-3, A-17 (after O'Hara and Mercy, 1963).

Experimental data on a study of the system $CaO-MgO-Al_2O_3-SiO_2$ have already provided confirmation of the constancy of the amount of calcium in the garnet of the lherzolite paragenesis based on the a_0 value for the garnet, determined in the products of the experiment (MacGregor, 1964). Later, the result of a direct determination of

22

garnet composition has been obtained in experiments with $\underline{P}$ = 30 kb
and $\underline{T}$ = 1200° in a paragenesis with two pyroxenes (Boyd, 1970), and
the amount of grossular (15.4%) differs only insignificantly from
the average value of the amount of Ca-component which we obtained
in garnets from peridotites (Sobolev, 1964$\underline{a}$). However, in spite of
the constancy of the average composition of the calcium content,
fluctuations in the amount of Ca-component of from 10% to 18% have
been identified in a number of cases in garnets of natural lherzolites and websterites. As information has accumulated subsequently,
the limits of variation have expanded still farther from 9% to 21%.

When considering the question of the use of the results of experiments in the above-noted four-component system in the study of
natural ultramafic parageneses, many authors have pointed out the
possibility of the effect of additional components, involved in the
composition of the rocks: FeO, Fe_2O_3, Cr_2O_3, Na_2O (MacGregor, 1970;
Sobolev, 1970; Boyd, 1970). In fact, the composition of the garnet
reacts sensitively to the effect of added components, and this may be
calculated in individual cases.

In this respect, the effect of increased amounts of Na_2O in pyroxenes, in individual examples of the peridotite association, exceeds
by twice or even more the normal 'background' amount of Na_2O in the
clinopyroxenes of the peridotites (1-2 wt %). Such increased amounts
of Na_2O, as substantiated by Banno (1967) for the paragenesis garnet
+ clinopyroxene + enstatite, lowers the amount of Ca in the garnet,
as established for the kyanite eclogites (Sobolev, Zyuzin and Kuznetsova, 1966). In our collection, samples from different pipes have
been examined, which clearly emphasize such a feature, with an amount
of Ca-component in the garnets of up to 9% (see Table 1, Samples M-60,
0-172).

In the garnets with variable Cr_2O_3 content, a positive correlation
has been defined between the amounts of Ca- and Cr-components (Sobolev,
Lavrent'ev et al., 1969). This correlation appears especially clearly
in garnets, containing more than 3-4% Cr_2O_3, since in garnets with a
smaller amount of chromium, the picture is often complicated by the
additional effect of Na_2O (see Table 1). The established relationships between the amount of calcium and chromium has also been emphasized for pyropes from lherzolites, which outcrop in the Czech Massif
and Norway, the features of which have been considered by us along
with those of the garnets in the appropriate parageneses of xenoliths.

As compared with the effect of Cr_2O_3 and Na_2O, the amount of which
fluctuates quite widely in the minerals of the peridotites, the effect
of the increased amount of iron on the composition of garnet can almost be ignored, this being associated with the relative constancy of
the amount of iron in the peridotites.

Judging by the composition of the garnet, which most clearly
reacts to the change in iron content, its iron index changes little
and varies within a narrow range: from 15% to 25% for xenoliths,
very rarely increasing up to 30-35% in the Norwegian peridotites.
It is possible to estimate the effect of the amount of FeO, by comparing the garnet compositions in the peridotites and the bipyroxene
paragenesis of the granulite facies, for which a positive correlation for the amounts of iron and calcium has been established
(Sobolev, 1964$\underline{a}$).

The amount of Fe_2O_3, as a rule, is very small and constant in
peridotites, and in the garnets, the amount of andradite component
exceptionally rarely exceeds 5 mol %, and its effect may be ignored
in practice, like the effect of a very small amount of manganese.

TABLE 4. Chemical Composition of Minerals from an Ilmenite-Bearing Peridotite from the 'Mir' Pipe (A-147) and a Peridotite from the Czech Massif (T7/272)

Oxides	Garnets A-147	Garnets T7/272	Clinopyroxene A-147	Clinopyroxene T7/272	Olivine A-147	Ilmenite A-147
SiO_2	42,3	41,5	55,8	53,7	40,4	0,19
TiO_2	2,38	0,11	0,59	0,01	0,08	49,2
Al_2O_3	18,4	17,0	2,52	1,07	—	0,68
Cr_2O_3	1,90	7,62	0,51	1,33	0,03	1,28
Fe_2O_3	—	—	—	—	—	7,5
FeO	11,2	7,26	6,50	1,92	14,3	31,5
MnO	0,27	0,28	0,19	0,07	0,10	0,25
MgO	19,3	19,1	20,2	17,2	46,2	8,86
CaO	5,70	6,58	12,2	21,8	0,03	
Na_2O	0,19	—	2,10	0,96		
Total	101,64	99,45	100,61	98,06	101,17	99,46
Si	3,029	3,036	1,988	1,983	0,998	0,005
Al^{IV}		—	0,012	0,017		
Ti	0,128	0,004	0,016	—	0,001	0,906
Al^{VI}	1,549	1,468	0,094	0,029		0,200
Cr	0,108	0,439	0,015	0,040	0,001	0,025
Fe^{3+}	0,215	0,089	—	—	—	0,069
Fe^{3+}	—	—	—	—	—	0,069
Fe^{2+}	0,456	0,355	0,194	0,058	0,295	0,645
Mn	0,016	0,018	0,006	0,002	0,002	0,005
Mg	2,061	2,083	1,072	0,947	1,702	0,323
Ca	0,439	0,514	0,467	0,863	0,001	—
Na	0,027	—	0,145	0,071	—	—
f	24,6	17,5	15,3	5,8	14,8	66,5
Ca/Ca+Mg	—	—	30,3	47,7	—	—
Ca/Ca+Mg+Fe	5,6	17,3	26,9	46,1	—	—
Cr/Cr+Al	5,4	22,0	12,4	46,5	—	—

Note. For the garnets of samples A-147 and T7/272 respectively, pyrope = 61.7, 60.7; almandine = 15.2, 12.0; spessartine = 0.5, 0.6; andradite = 9.2, 4.5; Ti-andradite = 6.4, 0.2; knorringite = 5.4, 9.4; uvarovite for T7/272 = 12.6; koharite for A-147 = 1.6.

In the magnesian garnets of the pyrope peridotites, a variable amount of titanium has also been identified (Sobolev, Lavrent'ev et al., 1973a), and in the chrome-rich pyropes, up to 1.56% TiO_2 has been found (see §20). A still greater amount of TiO_2 (2.38%) has been identified in a garnet from a xenolith of ilmenite peridotite from the 'Mir' pipe (Table 4).

Pyroxenes

The c l i n o p y r o x e n e s of the pyrope peridotites consist of diopside, containing a certain amount of Na_2O and Cr_2O_3, and also an isomorphous amount of enstatitic solid solution. The latter is of importance in determining the equilibrium temperature of the peridotites (see §22). As a result of detailed investigations in recent years, wide variations in the composition of the clinopyroxenes have been identified depending on the amount of the magnesian solid solution. The value of the Ca/Ca + Mg ratio varies from 50% (corresponding to a pure diopside) to 30% in one of the highest-temperature natural parageneses of peridotites known, in sample A-147, an ilmenite peridotite from the 'Mir' pipe (see Table 4). Besides these extreme values, all intermediate values are known, corresponding to a significant variation in the temperature conditions of crystallization of the deep-seated peridotites containing pyrope.

The e n s t a t i t e of the peridotites contains very little in the way of additives (see §14). The amount of calcium (a solid solution of diopside similar to the features of composition of clinopyroxene) is a clear indicator of the temperature. In analyses of enstatites, devoid of mechanical additives, the amount of Ca, as a rule, is negatively correlated with the value of the Ca/Ca + Mg ratio of the associated clinopyroxenes. An isomorphous amount of Na_2O has also been identified in the enstatites, the entry of which is most probably associated with increased pressure (see §14), and also in the diamond-bearing associations, with the nature of the paragenesis (with or without clinopyroxene).

Other Minerals

The o l i v i n e belongs to the most magnesian variety, containing 90-92% of forsterite. In some of the ilmenite-bearing peridotites, the iron index of the olivine increases up to 13% fayalite.

The composition of the c h r o m e - s p i n e l s reflects a general trend, typical of the ultramafic rocks: from enrichment in chromium in the dunite associations to a decrease in its amount with simultaneous increase in the amount of Al_2O_3 as the paragenesis of dunite → lherzolite → websterite changes, that is, with increase in the amount of SiO_2 and Al_2O_3 in the rocks. At the same time, there are marked differences in the chrome-spinels which belong to the series $(Mg,Fe)Al_2O_4$-$(Mg,Fe)Cr_2O_4$ with approximately equal amounts of Fe^{3+}, from certain peridotites. This applies to the abnormal chromites, containing up to 6.5% TiO_2 (see §18), and also the chrome-spinels with a substantial amount of the magnetite component (up to 26 mol %), which belong to the series of solid solutions of magnetite-chromite with very small amounts of alumina (see §2).

I l m e n i t e has been found in certain xenoliths of peridotites in significant amount. Thus, in sample A-147 from the 'Mir' pipe, which consists of an association of subcalcic clinopyroxene with pyrope, enstatite, olivine, and ilmenite (Ponomarenko et al., 1972), the amount of TiO_2 reaches 10%, corresponding to amount of about 15% ilmenite (see Table 4).

In a number of samples of xenoliths of pyrope peridotites from the 'Obnazhennaya' pipe, s u l f i d e segregations have been found in the fresh samples, containing a high-temperature association of chalcopyrite + pyrrhotite + pentlandite (Vakhrushev and Sobolev, 1971).

The Pyrope Peridotites of the 'Udachnaya' Pipe

The pyrope peridotites of the 'Udachnaya' pipe have been described as 'porphyritic peridotites' (Bobrievich et al., 1959a, 1964; Sarsadskikh et al., 1960) and have been allocated to a special type of peridotite on the basis of textural features, that is, on the presence of porphyritic oclasts of pyrope, enstatite, olivine, diopside, and groundmass (olivine) and gradual transitions to recrystallized pseudoporphyritic rocks. The compositions of certain minerals, especially the garnets, in these rocks, have been studied by Sarsadskikh et al. (1960). In addition to some chemical analyses of garnets with small amounts of Cr_2O_3, these authors have presented data on the values of n for certain garnets in the peridotites from this pipe, including an unusually high n value of 1.780. By analogy with the garnet composition for the ultramafic rocks (Sobolev, 1964a), which contain a low and practically constant 'background' amount of the almandine component, we have assumed that such a high n value for the garnet in the peridotite may be the result of a trace of Cr_2O_3 alone.

The nature of the overall distribution pattern for Cr and Al in the garnets and the surrounding peridotites (Fiala, 1965; Sobolev and Sobolev, 1967) indicates that, with a particularly large amount of Cr_2O_3 in the garnet, the amount of garnet itself in the rock must be very small, and its color must approach green in daylight (see Chapter VIII); in other words, the xenoliths of peridotites with chrome-rich garnets are difficult to diagnose visually.

As a result of our search for such xenoliths in the 'Udachnaya' pipe, making allowances for the possible features indicated, we have collected a representative series of samples from completely serpentinized to fresh. Most of the xenoliths found are analogous to the petrographically well-defined porphyritic peridotites. Besides this well-known type of peridotite, xenoliths and their fragments have been found with idiomorphic garnets, located within the large olivine grains or as pseudomorphs after olivine. Optical investigations have demonstrated the presence of garnets with n = 1.782.

Bulk determinations of refractive indices of about 300 garnets from the collection of xenoliths indicate very wide variations in the n values: from 1.734 up to 1.796 (Sobolev, Lavrent'ev et al., 1973a). On the basis of these data, a histogram has been constructed (Fig. 3, I). For comparison, a histogram of n values for garnets from ultramafic xenoliths (III) is presented, after Pankratov (1960). In addition, from the literature data, we have constructed a histogram for the n values for garnets from peridotite xenoliths in the South African kimberlites (II) (Mathias and Rickwood, 1969; Rickwood et al., 1968). The results of statistical treatment of the data on the n values are presented in Table 5. The double-peak histograms for the garnets from the peridotites in the 'Udachnaya' pipe and the Yakutian peridotites show an approximate coincidence in the left-hand maxima, similar also to the maximum for the garnets in the South African peridotites. For the double-peak histogram of the 'Udachnaya' pipe, the null hypothesis for normal distribution has been rejected on the basis of Pearson's χ^2 criteria (χ^2 = 51.9; χ^2_{05} = 23.7).

Clear differences have also been identified in the width of the intervals of variations of n for garnets from the peridotites of the 'Udachnaya' pipe (0.062), the Yakutian pipes (0.035), and those from South Africa (0.036). For histogram I of Figure 3, two distributions may at least be assumed, which is partly explained by the complex nature of the sampling, in which two different types of xenoliths are present.

A comparison of the data demonstrates that the garnets from the peridotites in the 'Udachnaya' pipe, are characterized on average by higher n values than those for the general sampling of garnets from

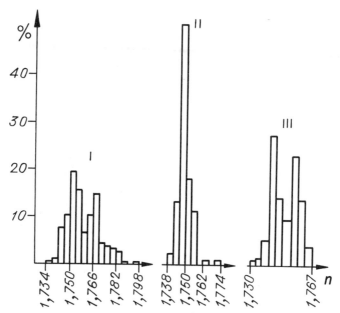

Fig. 3. Histograms refractive index values for magnesian garnets from peridotite xenoliths: 'Udachnaya' pipe (I), South African pipes (II), and Yakutian pipes (III)

the Yakutian peridotites, and also from Africa. In addition, with the aid of our composition-properties diagram for the magnesian garnets (see §12), we have made an estimate of the compositional features of more than 50 garnets, for which the a_0 values have also been determined.

Already, as a result of a study of physical properties, it has been shown that the great majority of the garnets belong to the lherzolite paragenesis with increased amounts of CaO and Cr_2O_3 (Fig. 4). At the same time, these garnets are clearly distinguished in composition (amount of CaO) from the chromium-rich pyropes from diamonds, in spite of the similar amounts of Cr_2O_3 and n values. Nevertheless, on the basis of physical properties, a decreased amount of CaO has been identified in certain garnets, which allies them with the garnet compositions from diamonds (see Fig. 4). However, it must be stressed that all six samples, in which garnets with decreased amounts of CaO have been found, are markedly distinguished on the basis of textural features from the pseudoporphyritic peridotites, and clinopyroxene is absent from them. They are more like the diamond-bearing peridotites of the 'Aikhal' pipe (V. S. Sobolev, Nai et al., 1969).

Thus, on the basis of a study of the physical properties and composition of the garnets, we may conclude that there are clear differences between the pseudoporphyritic peridotites of the 'Udachnaya' pipe and the diamond-bearing peridotites (dunites) of the 'Aikhal' pipe, in which respect there are no grounds for regarding these peridotites as a possible source of the diamonds in the 'Udachnaya' pipe (Sarsadskikh et al., 1960). The xenoliths with garnet, similar in composition to that enclosed in the diamonds, belong to a different type, clearly distinguishable from the pseudoporphyritic peridotites.

TABLE 5. Variations in n Values for Garnets from Peridotite Xenoliths in Kimberlite Pipes

Sampling locality for peridotites	Number of samples	Limits of variation of n for garnets	$\bar{x}$	S	Confidence intervals for x
'Udachnaya' kimberlite pipe	297	1.734–1.796	1.760	0.011	1.756–1.764
South African kimberlites	88	1.737–1.773	1.760	0.004	1.748–1.750
Yakutian kimberlites	96	1.732–1.767	1.751	0.008	1.749–1.753

The results of a study of the chemical composition of a representative series of garnets, selected from amongst a wide range of composition with n ranging from 1.755 to 1.796 (see Table 2), have emphasized the established features and have enabled us to provide a more precise estimate of the variations in the amounts of CaO in garnets from lherzolites: from 4.5% up to 7.7%. Such wide variations in the amount of the calcium component, seemingly contradict the experimental data on the constancy of the calcium index of the peridotitic garnets. However, a clear positive correlation has been established here between the amounts of CaO and Cr_2O_3 (N. V. Sobolev, Lavrent'ev et al., 1969) at a high $\rho = 0.90$, significant at 95% probability. The position of the regression lines have been determined by the equations:

$$Cr = 2.534 \; Ca - 187.7; \quad Ca = 0.333 \; Cr + 89.48.$$

The identification of this correlation is important not only for indicating the cause of the increase in the calcium index of the garnet in natural lherzolites, but also for assessing the possible paragenesis of both the individual chrome-rich garnets from kimberlites, and also inclusions in diamonds (see §9 and Table 38).

A constant amount of TiO_2 has been identified along with the large amount of Cr_2O_3 in the series of garnets studied, and special investigations have enabled us to treat this amount up to 1.56% as isomorphous (see §12).

The compositions of 14 clinopyroxenes studied (Table 6) are characterized in almost all the samples by a lowered calcium index (Ca/Ca + Mg = 38–42), which indicates the high-temperature nature of the entire series of rocks (see §§13 and 22). In accordance with the high chromium index of the garnets, the chromium index of these pyroxenes is also markedly increased, as compared with that of the pyroxenes of granular peridotites, owing to decrease in the amount of aluminum.

The compositions of the enstatites, studied with the aid of the microprobe (Table 7), confirm the features of distribution of the additional components found in the clinopyroxenes. The enstatites are also characterized by increased amounts of CaO and Cr_2O_3. In addition, a stable amount of Na_2O (0.12–0.25%) has been identified, which clearly distinguishes them from the enstatites of other pyrope peridotites (see §14).

The olivine is not distinguished in its iron index from that of the rocks described above, with the exception of a certain addition of chromium (up to 0.08% Cr_2O_3; see §15). In only one case in a rock showing few traces of shearing (Ud-101), has an increased iron index been identified for the olivine of the groundmass by 2% as compared with that of the phenocrysts.

Our investigations have also emphasized the presence of chromite and ilmenite in individual specimens (Sarsadskikh et al., 1960). In

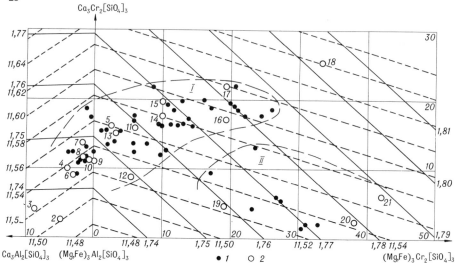

Fig. 4. Relationships of compostional points for garnets from peridotites (1) of the 'Udachnaya' pipe, plotted on the basis on n and a_0 values on a composition-properties diagram for magnesian garnets: field I)   garnets with larger amount of Ca (from lherzolites); field II) garnets with decreased amount of Ca (from dunites and harzburgites);   2)   points for reference compositions of garnets, used for checking diagram.

one of the samples, lacking, it is true, clear textural features of the pseudoporphyritic peridotites, graphite has been found (Spec. No. Ud-111).

The chromites, associated with titanium-rich garnets are also characterized by an increased amount of TiO_2.   In two samples (Ud-126 and Ud-161), 6.45% and 4.3% have been found respectively (see §18).   An additional feature of these chromites is the relatively high chrome-index (52 and 61) and the low iron index (33 and 45).

The amount of the ulvöspinel component (15.6% and 11.5%) is the greatest, reliably identified in chromites of terrestrial origin (Chukhrov, 1967; Deer et al., 1964; Haggerty et al., 1970).   It is evident that this feature, like the solid solution of titanium in the chrome pyropes, may be explained by the especially high-temperature conditions of crystallization and the appropriate composition of the melt.

Thus, this series of peridotitic xenoliths from the 'Udachnaya' pipe is characterized by a number of stable features of mineral composition: the increased role of Cr_2O_3 in the coexisting silicates (garnets and pyroxenes); the decreased calcium index of the clinopyroxenes; a trace of TiO_2 in the garnets; a trace of Na_2O in the enstatites; and a trace of chromium in the olivines.   These features not only indicate the high-temperature nature of the xenoliths examined, but also their greater depth of origin as compared with that of the normal, widely distributed granular peridotitic xenoliths.

On the basis of the data obtained, these peridotites must be assigned, from the conditions of formation, to rocks intermediate between the diamond-bearing types and those formed at shallower depths.

TABLE 6. Chemical Composition of Clinopyroxenes From Peridotite Xenoliths of the 'Udachnaya' Pipe

Oxides	Ud-96*	Ud-93	Ud-40*	Ud-3	Ud-111*	Ud-94*	Ud-5*	Ud-20*	Ud-91*	Ud-18*	209/576	Uv-161*	Ud-183*	Ud-92*
SiO_2	54.9	54,32	54,4	54,05	54,5	54,3	55,5	54,2	54,5	54,9	55,7	55,7	56,3	54,9
TiO_2	0,34	0,37	0,41	0,15	0,00	0,34	0,11	0,17	0,17	0,26	0,19	0,26	0,23	0,04
Al_2O_3	1,63	2,28	1,57	2,66	1,01	1,46	1,11	1,34	1,36	1,47	1,14	1,23	0,95	0,82
Cr_2O_3	0,91	2,29	0,88	1,61	1,08	1,59	1,10	1,78	1,44	2,00	1,85	1,98	0,94	1,08
Fe_2O_3	—	1,79	—	0,90	—	—	—	—	—	—	—	—	—	—
FeO	3,51	1,48	3,12	1,20	1,67	2,95	3,07	3,35	3,21	3,32	3,13	3,25	2,44	2,96
MnO	0,10	0,05	0,07	сл	0,00	0,05	0,11	—	—	—	0,08	0,05	0,12	—
MgO	18,3	18,70	18,1	19,11	17,7	18,3	19,5	18,8	18,7	18,4	19,8	19,4	19,4	19,5
CaO	17.4	16,82	17,3	18,30	21,6	17,7	17,5	17,4	18,0	16,9	17,0	16,8	19,3	19,2
Na_2O	1,52	2,13	1,68	1,79	0,84	1,84	0,98	1,47	1,48	1,66	1,41	1,50	1,19	0,91
K_2O	0.00	0,09	0,05	0,25	0,08	0,03	0,04	—	—	—	0,04	0,04	0.02	—
Total	98,61	100,32	97,58	100,02	98,51	98,57	98,98	98.51	98,86	98,91	100,34	100,21	100,89	99,31
Si	2.001	1,947	2.001	1,942	1,996	1,982	2,009	1,982	1,984	1,994	2,019	1,994	2,003	1,992
Al^{IV}	—	0.053	—	0,058	0,004	0,018	—	0,018	0,016	0,006	—	0,006	—	0.008
Al^{VI}	0.070	0,046	0,071	0,054	0,040	0,048	0,048	0,039	0,045	0,059	0.048	0,046	0,043	0,027
Ti	0,009	0,010	0,011	0,004	0,000	0,009	0,002	0,004	0,004	0,004	0.007	0,009	0,006	0,001
Cr	0,026	0,065	0,027	0,047	0,031	0,048	0,030	0,053	0,042	0,057	0,052	0,056	0,026	0,031
Fe^{3+}	—	0,047	—	0,026	—	—	—	—	—	—	—	—	—	—
Fe^{2+}	0,107	0,045	0,095	0,037	0,051	0,090	0,094	0.103	0,098	0,100	0,096	0,099	0.073	0,089
Mn	0.002	0,002	0,002	—	0,000	0,002	0,002	—	—	—	0.002	0,002	0,002	—
Mg	0,994	0,999	0,993	1,023	0,966	0,996	1,052	1,024	1,015	0,995	1,069	1,035	1.028	1,048
Ca	0,679	0,646	0,681	0,703	0,847	0,693	0,678	0,681	0,702	0,657	0,660	0,645	0,735	0,745
Na	0,105	0,151	0,119	0,124	0,062	0,132	0,070	0,105	0,105	0,118	0,100	0,103	0,081	0.065
K	0,000	0,004	0,002	0,008	0,005	0,002	0,000	—	—	—	0.002	0,002	0,001	—
Total of cations	3,993	4.015	4,002	4.026	4.002	4.020	3.985	4.009	4,011	3,990	4,055	3,997	3,998	4,006
f	9,7	8.4	8.7	5,8	5.0	8.3	8,2	9,1	8,8	9.1	8.2	9.6	6.6	7,8
Ca/Ca+Mg	40.6	39,3	40.7	40,7	46,7	41,0	39.2	39,9	40,9	39,8	38.2	38,4	41,7	41,6
Ca/Ca+Mg+Fe	38,1	37,2	38.5	40,0	45,4	39,0	37,2	37,7	38.7	37,5	36.2	36,3	40.0	39,6
Cr/Cr+Al	27,1	39,6	27,6	29,3	41,3	42,1	38.5	48,2	40.8	46,7	52.0	52,0	37,7	47,0

It must be stressed again that a clear association has been
observed between the porphyritic and porphyroclastic textures in these
sheared and pseudoporphyritic rocks, as finely illustrated in the work
of Bobrievich et al. (1964). In a description of analogous textures
in certain ultramafic xenoliths from the kimberlites of Lesotho, Boyd
(1973) has noted that the presence of garnet in the form of porphyritic segregations is dependent on the fact that this mineral is
afféctéd least of all by shearing. All the olivine has been crushed
and recrystallized. Clear traces of deformation have also been
identified in the diopside. The particularly high-temperature and
deep-seated nature of such lherzolite xenoliths, assessed by means

TABLE 7. Chemical Properties of Enstatites

Oxides	'Obnazhënnaya' Pipe			'Mir' Pipe				
	O-172*	O-145*	O-802*	M-60*	Ud-3*	Ud-111*	Ud-40*	Ud-5*
SiO_2	57,8	58,6	57,0	56,2	58,2	59,0	58,8	58,5
TiO_2	0,04	—	—	0,04	—	—	0,17	0,05
Al_2O_3	1,11	0,90	0,86	0,51	0,98	0,33	0,61	0,65
Cr_2O_3	0,13	0,01	0,38	0,03	0,39	0,25	0,27	0,29
Fe_2O_3	—	—	—	—	—	—	—	—
FeO	5,98	5,70	5,25	7,18	4,67	4,64	5,53	4,95
MnO	0,04	0,05	0,11	0,08	0,11	0,13	0,11	—
MgO	35,9	35,9	35,2	34,7	35,7	36,3	35,5	35,8
CaO	0,08	0,19	0,27	0,06	0,40	0,39	0,67	1,03
Na_2O	0,07	0,07	0,17	0,06	0,07	0,04	0,19	0,17
Total	101,15	101,42	99,24	98,86	100,52	101,08	101,85	101,44
Si	1,966	1,982	1,973	1,968	1,981	1,996	1,985	1,979
Al^{IV}	0,017	0,016	0,017	0,011	0,019	0,004	0,012	0,013
Ti	0,002	—	—	0,001	—	—	0,004	0,002
Al^{VI}	0,017	0,020	0,018	0,010	0,020	0,009	0,012	0,013
Cr	0,004	0,000	0,010	0,001	0,011	0,006	0,007	0,008
Fe^{+3}	—	—	—	—	—	—	—	—
Fe^{+2}	0,172	0,161	0,152	0,210	0,133	0,130	0,156	0,140
Mn	0,001	0,002	0,003	0,002	0,003	0,003	0,003	—
Mg	1,818	1,810	1,815	1,812	1,810	1,830	1,784	1,804
Ca	0,004	0,007	0,010	0,002	0,015	0,014	0,024	0,037
Na	0,004	0,004	0,011	0,004	0,004	0,002	0,013	0,011
Total of Cations	4,015	4,002	4,009	4,021	3,996	3,994	4,000	4,007
f	8,6	8,2	7,7	10,4	6,8	6,6	8,0	7,2
Ca/Ca+Mg	0,2	0,4	0,5	0,1	0,8	0,8	1,3	2,0
Ca/Ca+Mg+Fe	0,2	0,4	0,5	0,1	0,8	0,7	1,2	1,9
Cr/Cr+Al	10,5	1,0	22,2	4,5	22,0	31,6	23,1	23,5

Note. K_2O has been determined for BLT, 0.01; E-2, 0.08; E-3, 0.03; E-8, traces; KP-1, 0.16; P_2O_5 for BLT,

of the compositional features of the coexisting pyroxenes and garnet (Wood and Banno, 1973) has led Boyd to the conclusion that the shearing of such rocks is associated with the displacement of plates and the release of heat, which caused a rise in temperature and remelting of the rocks to a level where diamonds are stable. However, from this point of view, it is difficult to explain the differences in the iron index of the olivines in the phenocrysts and in the groundmass, which we have identified in one sample of pseudoporphyritic peridotite of the 'Udachnaya' pipe. This problem is arguable and deserves further

'Udachnaya' Pipe				South Africa						Czechoslovakia	
Ud-18*	Uv-161*	Ud-92*	209,576*	BLT-1	BKM-1	E-2	E-3	E-8	E-9	KP-1	T7,272*
58,6	58,0	58,4	57,2	57,02	57,79	58,48	57,11	56,61	57,70	54,13	58,9
0,19	0,15	0,04	0,13	0,03	0,007	Сл.	0,13	—	Сл.	0,12	—
0,54	0,50	0,46	0,60	1,07	0,79	0,88	1,00	0,86	0,37	1,62	0,90
0,43	0,50	0,36	0,45	0,23	0,23	0,25	0,35	0,17	0,34	0,21	0,41
—	—	—	—	0,72	—	0,72	1,82	1,35	1,18	0,85	—
5,33	5,50	4,93	4,98	3,76	3,96	3,93	3,61	3,73	3,79	5,70	5,22
0,13	0,13	0,08	0,15	0,11	0,10	0,02	0,12	0,10	0,09	0,14	0,10
35,6	34,7	35,4	34,7	35,86	36,36	34,71	34,02	35,90	35,55	34,55	35,7
0,86	0,80	0,85	0,98	0,82	0,53	0,50	1,50	0,70	0,55	0,57	0,33
0,24	0,29	0,12	0,20	0,09	—	0,23	0,35	0,09	0,24	0,16	0,07
101,92	100,60	100,64	99,39	99,74	100,22	100,01	100,32	99,95	100,24	100,11	101,63
1,978	1,985	1,991	1,980	1,960	1,976	2,003	1,965	1,952	1,978	1,913	1,986
0,011	0,010	0,009	0,012	0,022	0,017	—	0,021	0,018	0,008	0,034	0,014
0,006	0,004	0,001	0,002	0,001	0,000	0,000	0,004	—	0,000	0,003	—
0,011	0,010	0,009	0,012	0,021	0,016	0,035	0,020	0,017	0,007	0,034	0,022
0,011	0,014	0,010	0,012	0,006	0,006	0,007	0,010	0,005	0,009	0,006	0,011
—	—	—	—	0,017	—	0,016	0,046	0,037	0,033	0,025	—
0,150	0,158	0,139	0,146	0,109	0,115	0,111	0,103	0,108	0,109	0,168	0,146
0,003	0,003	0,002	0,002	0,002	0,002	0,000	0,002	0,002	0,002	0,004	0,003
1,791	1,771	1,798	1,790	1,836	1,852	1,770	1,744	1,844	1,818	1,820	1,794
0,031	0,029	0,031	0,038	0,029	0,021	0,019	0,056	0,025	0,021	0,021	0,012
0,016	0,019	0,008	0,012	0,006	—	0,015	0,023	0,006	0,016	0,011	0,004
4,008	4,003	3,998	4,006	4,009	4,005	3,976	3,994	4,014	4,001	4,039	3,992
7,7	8,2	7,2	7,5	6,4	5,8	6,7	7,9	7,3	7,2	9,6	7,5
1,7	1,6	1,7	2,1	1,6	1,1	1,1	3,1	1,3	1,1	1,1	0,7
1,6	1,5	1,6	2,0	1,5	1,1	1,1	2,9	1,3	1,1	1,1	0,7
33,3	41,2	35,7	33,3	12,2	15,4	16,7	19,6	12,5	37,5	8,1	23,4

0.02; H_2O^+ for BKM-1, 0.32; E-3, 0.28; E-8, 0.44; E-9, 0.43; KP-1, 1.17; H_2O for BKM-1, 0.04; E-2, 0.21; KP-1, 0.73. Analyses of samples BKM-1, after Boyd, 1969; E-2, E-3, E-8, and E-9, After Nixon et al., 1963.

study. Here we must note: (1) the existence in Yakutia of sheared and pseudoporphyritic rocks, also formed at a shallower depth (ilmenite peridotites); and (2) the presence of typical pseudoporphyritic textures in the Czechoslovakian garnet peridotites, which occur in the crust. Therefore, we must compare the possibility and probability of the formation of such rocks as the result of vertical, and not horizontal displacements, in a fault zone. It is also necessary to note that the mineral paragenesis in the diamonds themselves, is considerably more deep-seated and different in nature (see Chapter

4), with a predominance of harzburgite-dunites, and not lherzolites. If the shearing had been associated with horizontal displacements, then it must have occurred at a higher level than the formation of the kimberlitic foci.

§2. MINERALS OF THE ULTRAMAFIC ASSOCIATION FROM A KIMBERLITE CONCENTRATE

It is known that in concentrates, extracted from kimberlites, there are a large number of fragments of mineral grains of disintegrated xenoliths, and also those of minerals that crystallized during the process of evolution of the kimberlites. In most cases, difficulties have arisen during identification of the paragenesis of individual grains, especially in the case of the peculiar composition of minerals, very rare in xenoliths, which it is possible to assign to grains of green garnets, rich in Ca and Cr, found in kimberlites from a series of pipes. In only one case previously was it possible to identify the paragenesis of similar garnets (§4) in unusual chrome-rich kyanite eclogites and grospydites. But in most of the identified samples (Table 8), the Ca-Cr garnets are distinguished in composition from those of the chrome-bearing eclogites, mainly by their lower iron index. Isolated finds of unusual peridotites with rounded segregations of green garnets in olivine (see Samples Nos. D-332 and S-1 in Table 8) have enabled us to determine the nature of the paragenesis of these unusual garnets (Sobolev, Lavrent'ev et al., 1973a). An analogous wehrlitic type of paragenesis was also defined earlier, based on a study of the paragenesis of calcium- and chromium-rich garnet, olivine, and chrome-diopside from a diamond (Sobolev, Bartoshinsky et al., 1970).
Allowing for the fragmentary nature of the literature data on unusual garnets, rich in chromium, from kimberlites in general, we are faced not only with the problem of searching for garnets rich in calcium and chromium, but also of assessing the variations in composition of the chrome-rich garnets, characterized by variable calcic content.
In the concentrates of kimberlites from the 'Mir', 'Udachnaya', and 'Dal'nyaya' pipes, garnets have been collected with the most intense coloration (effect of chromium) and varying degree of manifestation of the alexandrite effect, which characterizes the combined influence of calcium and chromium (Sobolev, 1971). These garnets have been analyzed for Cr and Ca (the data for the three pipes are shown in Figure 5). On the basis of information obtained from the entire series of compositions, garnets have been selected with varying amount of chromium and calcium and complete analyses have been made for them (see Table 8). In this table, only one analysis (K-47) has been taken from the literature data. Until recently, it had been assigned to the most chrome-rich magnesian garnet (Nixon and Hornung, 1968). These garnets (mainly from concentrates and less frequently from xenoliths) contain a variable amount of CaO (from 0.69 to 26.0% = 1.7 − 72% Ca component) and up to 18.9% Cr_2O_3. All the garnets form a continuous series of compositions up to an amount of more than 50% chromium and calcium components.
The variation in the amount of calcium is controlled by the difference in the parageneses, amongst which there is marked predominance of the lherzolitic type. The results of the investigation have emphasized the possibility of finding garnets poor in calcium (i.e.

completely analogous to those enclosed in diamonds) in the concentrates of the diamond-bearing kimberlite pipes (Sobolev, Lavrent'ev et al., 1973a).

The data presented may be regarded as sufficiently representative (together with the results of partial analyses) to decide on the possible limit to the amount of chromium in natural magnesian garnets from kimberlites. It is likely that the limiting amount of knorringite, of the order of 50 mol %, depends on maximum pressures, characteristic during diamond-formation (Sobolev, Lavrent'ev et al., 1973).

The importance of the amount of knorringite as an indicator of pressure has also been emphasized experimentally (with P = 30 kb and T = 1200°C, it has been possible to dissolve only up to 10 mol % of knorringite in pyrope--Malinovsky et al., 1973). At P = 50 kb and T = 1200°C pyrope contained up to 25% mol of knorringite (Malinovsky, Doroshev, 1975). Data on natural compositions and results of the experiments cited do not support the suggested possibility of synthesizing knorringite at P = 37 kb (Bykova and Genshaft, 1972). Judging by the physical properties, the garnet synthesized by the latter authors, is similar in composition to uvarovite. It has also not been possible to support the preliminary statement about the synthesis of a magnesian-chrome garnet at P = 20 kb (Coes, 1955).

Important information on the features of the paragenesis and fluctuations in the composition of the chrome-bearing minerals of the kimberlites is contained in unusual intergrowths of chromite-pyroxene-garnet composition, found in the 'Mir' pipe (Sobolev, Khar'kiv et al., 1973). Serpentine-like products are sometimes present in these intergrowths, which are, in all probability, altered olivine. The intergrowths are distinguished by very small dimensions, and their detailed investigations are only possible using the microprobe.

We have studied the compositions of the associated minerals in five such intergrowths (see Table 9), where, besides garnet and a chrome-spinel, clinopyroxenes have been identified. There is also Sample No. MS-59, in which pyroxene is absent.

There is considerable interest in the compositions of garnets with a diminished amount of calcium, discovered in a paragenesis with clinopyroxenes. They include Samples Nos. M-66, M-40, and M-56, containing respectively 4.01%, 3.60%, and 4.97% CaO, along with an amount of about 10% Cr_2O_3. A typical feature of this paragenesis is the large amount of Na_2O in the associated pyroxenes (more than 5 wt %, or about 40% $Na(Al,Cr)Si_2O_6$).

Such a paragenesis, as emphasized below, is an alternative association of a calcium-poor garnet with olivine, and in the case of a still greater amount of Na_2O in the pyroxene, the amount of CaO in the garnet falls to 2.45% (Sample No. B-1, intergrowth with diamond, §10). This relationship, demonstrated for the grospydites, and the kyanite and enstatite eclogites (Sobolev, Zyuzin and Kuznetsova, 1966; Banno, 1967; Sobolev et al., 1968) is of considerable importance in assessing the effect of sodium in the deep-seated associations of kimberlites (see §6).

Besides the calcium-poor garnets, specimens have been found in individual cases in the intergrowths mentioned, completely analogous to the above Ca-Cr garnets from the concentrate, for example, in Sample No. MZ-2.

Valuable additional information on the features of these intergrowths is contained in the composition of the associated chromites. A series of compositions of chromites from 10 garnet-pyroxene inter-

TABLE 8. Chemical Composition of Chrome-Rich Garnets From Kimberlite Concentrates and Some Peridotite Xenoliths

Oxides	Calcium-poor chrome-pyropes								Pyropes with moderate and large amounts of Calcium								
	u-50	M-73	M-24	u-20	D-22	u-19	D-23	D-140	Ms-59	Ms-27	Ms-29	Ms-35	Ms-28	u-35	D-70	D-72	D-48
SiO_2	42,0	42,7	41,4	41,9	42,8	41,8	42,2	41,1	38,8	39,9	38,1	40,4	40,8	41,2	40,7	40,2	39,0
TiO_2	0,15	0,02	0,01	0,07	0,07	0,04	0,10	0,11	0,57	0,21	0,27	0,04	0,05	0,24	0,09	0,28	0,34
Al_2O_3	18,7	18,8	17,6	16,5	16,4	15,2	14,5	13,0	13,1	13,5	13,2	13,2	12,9	12,6	10,7	11,7	9,64
Cr_2O_3	6,56	7,24	8,76	10,2	10,7	12,1	12,8	13,7	10,0	11,4	11,7	12,7	13,8	15,0	15,6	15,7	16,3
FeO	6,50	7,12	7,26	6,71	6,86	6,74	6,58	6,17	8,23	8,35	8,60	6,75	7,33	6,45	7,35	6,99	6,55
MnO	0,35	0,41	0,43	0,22	0,25	0,25	0,22	0,21	0,41	0,52	0,43	0,46	0,51	0,44	0,27	0,28	0,22
MgO	22,3	22,4	23,5	22,5	22,2	23,2	21,3	22,1	11,4	13,4	11,7	16,9	18,4	19,2	17,2	17,7	12,9
CaO	2,33	1,43	0,69	1,22	2,00	0,98	3,38	3,00	15,4	13,0	14,2	9,20	7,38	6,10	8,79	7,84	14,3
Total	98,89	100,12	99,65	99,32	101,28	100,31	101,08	99,39	97,91	100,28	98,00	99,65	101,17	101,23	100,70	100,69	99,25
Si	3,024	3,037	2,984	3,035	3,047	3,016	3,048	3,024	3,019	3,015	2,968	3,025	3,007	3,007	3,042	2,996	3,020
Ti	0,009	0,001	—	0,004	0,004	0,002	0,004	0,004	0,019	0,014	0,019	0,002	0,002	0,013	0,005	0,018	0,019
Al	1,586	1,583	1,498	1,405	1,378	1,291	1,237	1,132	1,196	1,199	1,208	1,161	1,125	1,087	0,943	1,027	0,875
Cr	0,374	0,408	0,499	0,583	0,599	0,689	0,729	0,796	0,617	0,681	0,721	0,756	0,806	0,868	0,923	0,925	0,998
Fe^{3+}	0,031	0,008	0,003	0,008	0,019	0,018	0,030	0,068	0,152	0,106	0,052	0,081	0,067	0,032	0,129	0,030	0,108
Fe^{2+}	0,363	0,416	0,434	0,396	0,388	0,389	0,396	0,312	0,385	0,425	0,510	0,342	0,385	0,362	0,334	0,404	0,315
Mn	0,022	0,026	0,026	0,013	0,017	0,017	0,013	0,013	0,028	0,032	0,028	0,032	0,031	0,026	0,018	0,018	0,014
Mg	2,396	2,379	2,521	2,427	2,354	2,488	2,288	2,423	1,322	1,507	1,358	1,886	2,019	2,086	1,918	1,966	1,489
Ca	0,182	0,111	0,052	0,096	0,154	0,074	0,261	0,234	1,285	1,053	1,184	0,738	0,584	0,478	0,710	0,627	1,187
Total	7,987	7,970	8,017	7,971	7,964	7,986	7,983	8,010	8,039	8,032	8,048	8,023	8,026	7,959	8,022	8,011	8,025
Pyrope	66,2	64,1	59,6	56,3	54,9	50,8	48,8	45,7	43,9	44,7	42,8	45,5	42,4	41,1	35,3	37,3	32,7
Almandine	12,3	14,2	14,3	13,5	13,3	13,1	12,6	10,5	12,7	14,1	16,6	11,4	12,8	12,3	11,2	13,4	10,5
Spessartine	0,7	0,9	0,9	0,4	0,6	0,6	0,4	0,4	0,9	1,1	0,9	1,1	1,0	0,9	0,6	0,6	0,5
Grossular	—	—	—	—	—	—	—	—	2,2	5,3	2,6	4,1	3,4	1,6	6,4	1,5	5,4
Andradite	1,6	0,4	0,2	0,4	1,0	0,9	1,5	3,4	7,6	0,7	1,0	0,1	0,1	0,7	0,3	0,9	1,0
Ti-Andradite	0,5	—	—	0,2	0,2	0,1	0,2	0,2	1,8	—	—	—	—	—	—	—	—
Uvarovite	4,0	3,4	1,5	2,7	4,1	1,5	7,2	4,2	30,9	28,9	34,8	20,4	15,8	13,9	17,1	18,4	33,1
Knorringite	14,7	17,0	23,5	26,5	25,9	33,0	29,3	35,6	—	5,2	1,3	17,4	24,5	29,5	29,1	27,9	16,8
f	14,1	15,1	14,8	14,3	14,7	14,1	14,8	13,6	—	26,1	29,3	18,3	18,3	15,9	19,4	18,1	22,1
Mg-Component	80,9	81,1	83,1	82,8	80,8	83,8	78,1	81,3	29,0	49,9	44,1	62,9	66,9	70,7	64,4	65,2	49,5
Ca-Component	6,1	3,8	1,7	3,3	5,3	2,5	8,9	7,8	42,5	34,9	38,4	24,6	19,3	16,2	23,8	20,8	39,5
Cr-component	18,7	20,4	25,0	29,2	30,0	34,5	36,5	39,8	30,9	34,1	36,1	37,8	40,3	43,4	46,2	46,3	49,9

(table 8)

	Pyropes with moderate and large amounts of calcium					Chromium-rich Ca-Garnets											
						from concentrate										from xenoliths	
Oxides	D-45	D-19	D-40	K-47	D-80	U-41	U-27	D-90	U-3	U-29	U-18	D-336	D-7	U-30	U-15	D-332	S-1
SiO_2	38,8	39,0	38,9	39,92	39,3	39,9	39,9	38,4	38,7	39,4	39,7	39,2	38,8	40,1	38,8	40,3	39,2
TiO_2	0,39	0,54	0,36	0,11	0,15	0,25	0,19	0,47	0,31	0,27	0,31	0,35	0,31	0,23	0,23	0,44	0,18
Al_2O_3	8,48	9,08	8,50	9,74	8,34	16,4	14,9	11,8	13,3	12,8	14,0	11,9	11,9	13,0	7,46	13,9	10,4
Cr_2O_3	17,0	17,2	17,3	17,47	18,9	7,76	10,2	10,5	11,3	11,7	11,8	11,9	12,0	13,1	18,7	9,89	14,9
FeO	6,94	7,57	7,44	7,62	6,62	4,74	5,72	6,25	4,56	4,83	5,20	6,43	4,59	4,85	6,48	7,61	5,38
MnO	0,28	0,38	0,33	0,60	0,33	0,29	0,40	0,21	0,34	0,34	0,42	0,30	0,14	0,27	0,25	0,36	0,34
MgO	13,2	14,7	13,3	16,97	15,2	10,5	9,69	7,71	7,34	6,36	10,2	9,90	5,86	9,12	10,3	13,2	9,46
CaO	13,2	11,7	13,9	8,14	11,3	18,8	20,2	23,3	23,9	25,4	19,2	19,2	26,0	21,0	19,8	14,1	19,0
Total	98,29	100,17	100,03	100,68	100,14	98,64	101,20	98,64	99,75	101,10	100,9	98,98	99,60	101,67	100,02	99,80	98,86
Si	3,023	2,994	3,009	3,006	3,012	3,022	2,997	3,017	2,990	3,014	2,995	3,038	3,027	3,016	2,992	3,045	3,051
Ti	0,024	0,028	0,023	0,005	0,009	0,014	0,014	0,028	0,019	0,018	0,018	0,023	0,019	0,023	0,014	0,023	0,014
Al	0,783	0,821	0,772	0,869	0,754	1,465	1,318	1,095	1,207	1,158	1,242	1,072	1,092	1,157	0,685	1,234	0,955
Cr	1,053	1,044	1,058	1,041	1,145	0,464	0,605	0,653	0,687	0,707	0,707	0,730	0,740	0,778	1,139	0,591	0,917
Fe^{3+}	0,140	0,107	0,147	0,072	0,092	0,053	0,063	0,224	0,087	0,107	0,033	0,175	0,149	0,051	0,162	0,152	0,114
Fe^{2+}	0,312	0,377	0,332	0,412	0,332	0,247	0,294	0,187	0,210	0,191	0,293	0,240	0,151	0,256	0,255	0,329	0,237
Mn	0,019	0,028	0,023	0,036	0,018	0,018	0,027	0,014	0,023	0,023	0,027	0,019	0,009	0,018	0,019	0,023	0,023
Mg	1,540	1,684	1,535	1,910	1,736	1,183	1,088	0,902	0,845	0,726	1,147	1,146	0,679	1,022	1,181	1,484	1,100
Ca	1,106	0,964	1,153	0,647	0,930	1,524	1,625	1,960	1,978	2,081	1,550	1,594	2,174	1,691	1,635	1,139	1,587
Total	8,000	8,047	8,052	7,998	8,028	7,994	8,031	8,080	8,046	8,035	8,012	8,037	8,040	8,003	8,082	8,020	8,002
Pyrope	28,0	27,9	26,9	28,8	26,0	39,8	35,8	29,4	27,6	24,0	38,0	38,2	22,5	34,2	25,3	49,8	37,3
Almandine	10,5	12,3	10,9	13,7	11,0	8,3	9,7	6,1	6,9	6,3	9,7	8,0	5,0	8,6	8,3	11,1	8,0
Spessartine	0,6	0,9	0,8	1,2	0,6	0,8	0,9	0,5	0,8	0,8	0,9	0,6	0,3	0,6	0,6	0,8	0,8
Grossular	—	—	—	—	—	24,5	19,4	18,7	25,0	26,7	13,4	6,8	26,8	14,5	8,1	—	1,6
Andradite	7,0	5,3	7,3	3,6	4,6	2,7	3,2	11,2	4,3	5,9	1,7	8,7	7,4	2,5	0,7	7,6	5,7
Ti-andradite	1,2	1,4	1,2	0,3	0,5	0,9	0,7	1,4	1,0	0,9	0,9	1,2	1,0	0,7	—	1,1	0,7
Uvarovite	29,0	24,9	29,4	17,6	25,7	23,2	30,3	32,7	34,4	35,4	35,4	36,5	37,0	38,9	44,1	29,6	45,9
Knorringite	23,7	27,3	23,5	34,8	31,6	—	—	—	—	—	—	—	—	—	12,9	—	—
{	22,7	22,3	23,8	20,2	19,6	20,2	22,6	31,3	26,0	29,8	22,1	26,6	30,6	23,1	26,1	24,5	24,2
Mg-Component	51,7	55,2	50,4	63,6	57,8	39,8	35,9	29,4	27,6	24,0	38,0	38,2	22,5	34,2	38,2	49,8	37,2
Ca-Component	37,2	31,6	37,9	21,5	30,6	51,3	53,6	64,0	64,7	68,9	51,4	53,2	72,2	56,6	52,9	38,3	53,9
Cr-Component	52,7	52,2	52,9	52,4	57,3	23,2	30,3	32,7	34,4	35,4	35,4	36,5	37,0	38,9	57,0	29,6	45,9

TABLE 9. Chemical Composition of Garnets and Clinopyroxenes From Intergrowths in Kimberlites From the 'Mir' Pipe

Garnets

Oxides	M-66	M-40	Mz-2	Ms-59	M-67	M-56
SiO_2	40,7	41,4	40,6	38,8	40,0	40,4
TiO_2	0,08	0,08	0,49	0,57	0,34	0,16
Al_2O_3	17,2	17,2	14,7	13,1	15,2	16,0
Cr_2O_3	8,69	9,42	9,61	10,0	10,1	10,2
FeO	7,57	7,35	8,32	8,23	8,00	7,97
MnO	—	—	0,24	0,41	—	—
MgO	22,1	22,4	14,1	11,4	18,0	20,2
CaO	4,01	3,60	9,76	15,4	7,03	4,97
Na_2O	—	—	0,10	—	—	—
Total	100,35	101,45	97,92	97,91	98,67	99,90
Si	2,945	2,954	3,084	3,019	2,983	2,963
Ti	0,004	0,004	0,027	0,035	0,018	0,009
Al	1,470	1,449	1,314	1,196	1,335	1,384
Cr	0,496	0,532	0,575	0,617	0,591	0,591
Fe^{3+}	0,030	0,015	0,084	0,152	0,056	0,016
Fe^{2+}	0,427	0,422	0,445	0,485	0,431	0,473
Mn	—	—	0,014	0,028	—	—
Mg	2,384	2,392	1,597	1,322	1,998	2,209
Ca	0,309	0,274	0,794	1,285	0,609	0,392
Na	—	—	0,018	—	—	—
Pyrope	59,9	58,8	49,6	43,9	52,2	53,1
Almandine	13,6	13,6	15,5	12,7	14,5	15,3
Spessartine	—	—	0,5	0,9	—	—
Grossular	—	—	—	2,2	—	—
Andradite	1,5	0,8	4,2	7,6	2,8	0,8

Clinopyroxenes

Oxides	M-66	M-40	Mz-2	M-67	M-56
SiO_2	54,3	56,6	55,0	55,0	54,9
TiO_2	0,04	0,07	0,15	0,15	0,17
Al_2O_3	3,71	4,42	1,90	2,73	4,10
Cr_2O_3	6,27	6,44	3,03	3,29	5,80
FeO	2,51	2,61	2,61	2,34	2,84
MnO	—	0,03	0,04	—	—
MgO	12,1	12,4	14,5	14,9	12,3
CaO	14,0	12,6	18,1	18,4	13,6
Na_2O	5,21	5,48	2,87	2,98	5,42
K_2O	—	—	0,02	0,04	—
Total	98,14	100,65	98,22	99,83	99,13
Si	2,002	2,016	2,021	1,993	2,001
Al^{IV}	—	—	—	0,007	—
Al^{VI}	0,159	0,188	0,084	0,106	0,175
Ti	0,001	0,002	0,004	0,004	0,004
Cr	0,182	0,182	0,088	0,094	0,166
Fe^{3+}	0,030	—	0,032	0,014	0,032
Fe^{2+}	0,048	0,077	0,048	0,056	0,053
Mn	—	0,001	0,001	—	—
Mg	0,664	0,659	0,795	0,806	0,668
Ca	0,554	0,482	0,713	0,714	0,532
Na	0,372	0,378	0,208	0,209	0,381
K	—	—	0,001	0,002	—
Total of cations	4,012	3,985	3,995	4,005	4,012

(table 9)

	f'	f	Ca/Ca+Mg	Ca/Ca+Mg+Fe	Cr/Cr+Al
	6,7	10,5	45,5	43,8	53,4
	10,5	10,5	42,2	39,6	49,2
	5,7	9,1	17,3	45,8	51,2
	6,5	8,0	17,0	15,3	45,4
	7,4	11,3	44,5	42,5	48,7

Ti-Andradite	0,2	0,2	1,4	1,8	0,9	0,5
Uvarovite	8,5	8,3	22,7	30,9	16,3	11,7
Knorringite	16,3	18,3	6,1	—	13,3	17,9
f	16,1	15,4	24,9	29,0	20,0	18,1
Mg-Component	76,2	77,1	55,7	43,9	65,5	71,7
Ca-component	10,2	9,3	27,6	42,5	20,0	13,0
Cr-component	24,8	26,6	28,8	30,9	29,6	29,6

growths from the 'Mir' pipe is given in Table 10. These compositions represent a continuuous series from the Cr_2O_3-rich to less chromiferous types, with fluctuations in the Cr/R^{3+} ratio of from 79% to 53%. On the basis of the iron index, they are very similar (44-51%), with the exception of Sample No. MS-1/1, which is characterized by an exceptionally large amount of Fe^{3+}, approximating in composition the recently discovered new mineral species, donathite (Seeliger and Mücke, 1969), which is a chrome-spinel of composition intermediate between chromite and magnetite.

The high content of Fe^{3+} and the appropriately increased value of the oxidation coefficient (K_0 = 25-43%), along with the increased chrome-index, are typical features of the chromites from intergrowths, which distinguish them from the chromites in diamonds, and from the minerals of the $(Mg,Fe)Al_2O_4-(Mg,Fe)Cr_2O_4$ series, which are widely represented in xenoliths of ultramafic rocks of varying composition. Therefore, in the chromites from intergrowths, the low content of Al_2O_3 (5.13-8.10%) cannot be regarded as a feature of their possible paragenesis with the diamond. In these chromites, as the chrome-index falls, the composition becomes more complex owing to the increasing role of Fe^{3+}, and also the addition of titanium (see §18).

Thus, a study of the composition of the associated minerals of the intergrowths has identified a number of features, which markedly distinguish them from the composition of minerals from the xenoliths of different peridotites. Primarily, these are the very high chrome-index of the associations and the correspondingly high chrome-index of all the associated minerals up to the sodium-rich pyroxenes. Such a mineral composition is completely unusual for the most common paragenesis (lherzolitic) in the deep-seated rocks, and these intergrowths may be sufficiently well-defined to assign them to the specially deep-seated lherzolites, the more so since individual occurrences of unusual lherzolites with mineral compositions, similar to those studied, are known (Sobolev, Khar'kiv et al., 1973).

Zoned Garnets from Kimberlite

In the heavy concentrate from the 'Mir' pipe, a grain of zoned chrome-bearing garnet has been found (Sobolev, Khar'kiv et al., 1973a). The grain is almost spherical in shape, measures 5.0 mm across, and is covered with a very thin selvedge (0.5-0.1 mm) of dark-gray color. The central portion of the grain is almost without cracks, and contains rare inclusions of chrome-spinel in the form of equant and acicular grains (Photo 3). The peripheral zone of the grain is cracked, and the number of inclusions of chrome-spinel markedly increases here. The central zone of the grain is colored violet-red ($n = ca$ 1.785), and the outer zone has a greenish-gray color ($n = ca$ 1.792). In the peripheral zone of the garnet, there is a pseudomorph of serpentine after olivine, about 0.7 mm across.

A preparation for examination under the microprobe (polished section) was made from one half of the grain. In this case, the zonation appeared particularly clearly, having a concentric nature, and with the outer zone a little asymmetric. A quantitative microanalysis of the $K\alpha$-lines of radiation of individual elements for the sample and standards were determined at points 40μ apart. The results of the analysis in wt % are shown in Figures 6 and 7, and Table 11. The results of the investigation have emphasized the presence of zonation

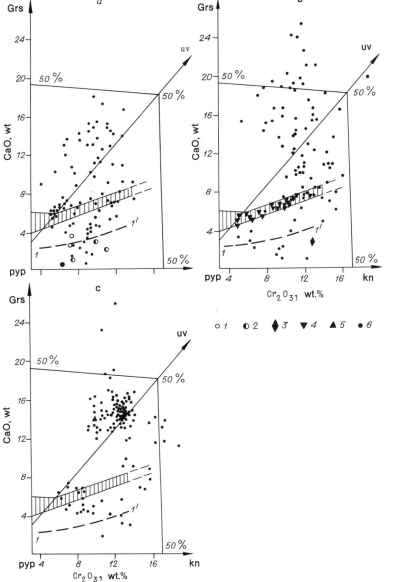

Fig. 5. Features of the amounts of Cr_2O_3 and CaO in chrome-rich garnets from the kimberlitic pipes: 'Mir' (a), 'Udachnaya' (b), and 'Dal'
nyaya' (c) (after Sobolev, Lavrent'ev *et al.*, 1973), Garnets taken
from diamonds (*1*), intergrowths with diamonds (*2*), dunites (*3*),
pseudo porphyritics lherzolites from the 'Udachnaya' pipe (*4*),
wehrlites (*5*), and kimberlite concentrate (*6*). Continuous lines
frame the compositional fields of grants with $f = 20\%$, containing
up to 50% of calcic and chrome componets; the compositional field
for garnets from the paragenesis with two pyroxenes is cross-hatched; the lower possible boundary for the amount of calcium in the
garnet from a paragenesis with sodic clinopyroxenes is denoted by
the line $l - l'$. (pyp = pyrope, grs = grossular, uv = uvarovite,
and kn = knorringite).

TABLE 10.   Chemical Composition of Chromites from Intergrowths With Garnet and Pyroxene in the 'Mir' Pipe

Oxides	M-66	M-40	Ms-59	M-67	M-56	Ms-1/1	Ms-12/1	Ms-7/1	Ms-6/1	Ms-9/1
SiO_2	0,10	0,09	0,23	0,11	0,11	0,10	0,20	0,08	0,08	0,11
TiO_2	0,12	0,27	1,80	1,51	0,87	3,80	0,56	2,06	2,34	2,08
Al_2O_3	6,23	6,21	7,85	7,04	5,79	5,13	6,17	7,62	8,10	7,98
Cr_2O_3	59,5	60,5	46,8	53,5	58,0	38,4	58,0	50,3	48,7	49,8
Fe_2O_3	6,60	6,1	13,7	9,30	8,2	20,3	6,5	11,3	11,3	11,0
FeO	16,8	16,7	16,5	17,3	17,6	25,3	18,2	18,4	19,0	18,2
MnO	0,30	0,34	0,29	0,32	0,35	0,40	0,33	0,34	0,36	0,37
MgO	10,5	10,8	10,6	11,0	10,6	6,91	9,68	10,8	10,5	10,9
Total	100,2	100,9	97,77	100,08	101,5	100,34	99,7	101,0	100,4	100,5
Si	0,004	0,003	0,008	0,004	0,004	0,004	0,006	0,003	0,003	0,004
Ti	0,003	0,007	0,047	0,038	0,022	0,100	0,014	0,052	0,058	0,052
Al	0,246	0,243	0,314	0,277	0,227	0,209	0,246	0,298	0,317	0,311
Cr	1,577	1,587	1,254	1,410	1,521	1,057	1,553	1,314	1,280	1,306
Fe^{3+}	0,167	0,152	0,350	0,232	0,203	0,529	0,167	0,282	0,284	0,274
Fe^{2+}	0,471	0,463	0,468	0,483	0,488	0,732	0,514	0,508	0,528	0,504
Mn	0,008	0,010	0,008	0,008	0,010	0,015	0,010	0,010	0,010	0,010
Mg	0,524	0,534	0,536	0,546	0,524	0,356	0,488	0,532	0,520	0,538
f	47,2	46,1	44,0	44,9	47,1	64,0	50,6	46,2	47,5	45,7
Cr-component	78,9	79,4	62,7	70,5	76,2	52,9	77,7	65,7	64,0	65,3
Al-component	12,3	12,2	15,7	13,9	11,4	10,5	12,3	14,9	15,9	15,6
Ulvöspinel	0,3	0,7	4,7	3,8	2,2	10,0	1,4	5,2	5,8	5,2
K_0	26,1	24,7	42,8	32,5	29,5	42,0	24,3	35,5	34,9	35,2

Note: Iron has been divided after recalculation in formula.

with a clearly defined internal zone with a decreased content of CaO (about 7%) and an outer zone, with an increased content (11-14%), with significantly narrower fluctuations in the amount of Cr_2O_3 (7-10%) in both zones.

The clearest zonation has been manifested in the content of CaO, MgO, Al_2O_3, and Cr_2O_3 (see Fig. 6).   In regard to the amount of FeO, its clear minimum corresponds to the central zone of the garnet, where the amount of titanium is also decreased.

Within the peripheral zone, there is a cracked area ('crush zone') in which fragments of grains of garnet, and clinopyroxene have been found.   An analysis of the fragments has demonstrated a composition identical to that of the outer zone (see Table 11, garnet a).   The chrome-bearing clinopyroxene has been enriched in Na_2O and titanium.

The composition of the chrome-spinels is of interest.   The grains studied are characterized by variable amounts of Al_2O_3 and Cr_2O_3, and also an increased amount of titanium and similar values of K_0 = 26-30% (see Table 11).   A grain with zoned structure (see Photo 2) has been found with a titanium-rich inner zone (see Table 12, chromite a), containing very little Al_2O_3.   The composition of this zone is most similar to that of the chromites from the above-described intergrowths.   The peripheral zone of the grain has been thrice enriched in Al_2O_3, with the same content of Cr_2O_3, along with a decrease in the amount of titanium.

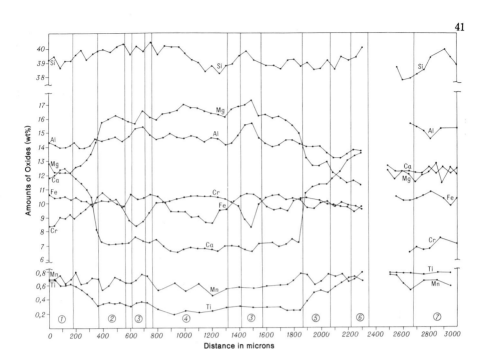

Fig. 6. Profile through grain of zoned garnet M-49 based on the
content of the principal oxides. Numbers in circles indicate
zones of garnets, the compositions of which are shown in Table 11.

On the basis of these results, seven garnet zones of different composition have been recognized, with ranges shown in Figure 6. The
compositions of these zones are shown in Table 11. The average position of the compositions of the zones recognized on the composite
diagram for the chrome-bearing garnets (Fig. 7) indicates their similarity to the above-described garnets, rich in Ca and Cr, from the
concentrate. The presence of a pseudomorph after olivine suggests
an analogous paragenesis (wehrlitic).

Another zoned garnet, found in the 'Mir' pipe (Sample M-41), is
distinguished from the above-described both in outward appearance
and by the numerous inclusions of chrome-spinels and ilmenite
(Sobolev et al., 1975). This is a large rounded grain, about 7 mm
in diameter, covered with a kelyphitic rim. The inner zones of the
garnet are violet-colored, with the color towards the outer zones
becoming less intense as far as orange. The chrome-spinels of the
inner zones consist of acicular inclusions, and the ilmenite, of
rounded grains.

The change in composition of the inner zones of the garnet (M-41)
(Table 12) is similar in nature to that of zones 2, 3, 4, and 5 of
Sample M-49 (see Fig. 7). However, in the outer zones of Sample M-41,
there is a clear lowering in the amount of Cr_2O_3 and CaO. The unzoned
chromites of Sample M-41 are equivalent in composition to that of
chromite a (see Table 11) of the first generation from Sample M-49.
In the outer zones, the chromite is replaced by ilmenite, and in one
of the zones, chromite and ilmenite coexist in the garnet, which contains 5.7-6.0% Cr_2O_3. The ilmenite of this zone is one of those most
enriched in Cr_2O_3 (10.7%) (see Table 12), known to have formed under

42

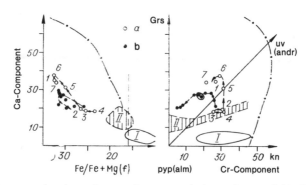

Fig. 7. Diagram showing change in composition of zoned Mg-Fe garnets
a) Sample M-49 (zone number corresponds to that in Table 11); b)
Sample M-41 (composition of zones in which chromite and ilmenite coexist, surrounded by continuous line); field I) compositions of
garnets of dunitic paragenesis from diamonds; field II) compositions
of garnets of lherzolitic paragenesis: vertical broken line separates magnesian garnets from diamonds (right) from xenolith-lherzo-
lites; boundaries of compositional field mostly of natural chrome-
bearing garnets shown by cross-hatched lines.

natural conditions. In the less chromiferous zones, the garnet contains only inclusions of ilmenite with gradually decreasing amounts
of Cr_2O_3, and finally, the outermost zone consists of an association
of garnet, clinopyroxene, and ilmenite, with garnet of the same composition as in the paragenesis, often identified in kimberlites for
individual discrete garnet grains and in intergrowths with ilmenite.
The range of iron index, which is especially wide for the different
zones of Sample M-49 (from 24% to 33%), is narrower for Sample M-41
(from 26% to 31.5%), and for the latter, the iron index in the outer
zones is even decreased. Olivine, which is possible in paragenesis
with both garnets, must be defined by a change in iron index approximately in the range of 10-15%.

   The described zoned garnets belong in their features to the group
very rare in the kimberlite concentrates. They are markedly distinguished from the garnets from xenoliths of pyrope peridotites and
garnets, included in diamonds, for which, in all known cases, a relatively uniform composition and the absence of marked zonation have
been identified. These differences are expressed primarily by the
presence of a zonation, and also by a markedly increased iron-index,
even in the cores of the zoned garnets, which seemingly are similar
in the content of CaO and Cr_2O_3 to the composition of garnets from
lherzolites. Such increased iron-indices suggest that the central
zones of the garnets studied crystallized sufficiently later than
the more magnesian garnets and coexisting olivines. The magnesian
garnets of lherzolite paragenesis predominante in the kimberlite concentrates of many Yakutian pipes over garnets of other types, as with
the most magnesian olivines also. However, even this type of garnet
also is more iron-rich than the chrome-pyropes, which are known as
inclusions in diamonds and as a component part of the diamond-bearing dunites (see Fig. 7).

   Consequently, these zoned garnets may only reflect a certain
intermediate stage of evolution in the kimberlite melt, and for the
second sample (M-41), it may be emphasized that here, even the final

phases of such evolution or close to them are defined, because in
the last phases it is more likely that the non-chrome orange garnet
which is close to garnet megacrysts in African and Yakutian kimberlites in composition would crystallize in a paragenesis with ilmenite.

§3. ECLOGITES

The eclogites of the kimberlite pipes are coarse-grained rocks which
consist mainly of garnet and clinopyroxene only, with accessory rutile.
Besides the most common two-mineral eclogites, corundum types are known,
and also kyanite eclogites and grospydites (see §4).

During descriptions of eclogite xenoliths from the South African
pipes, a tendency was earlier attempted to separate them on the basis
of the classification of the Norwegian eclogites (Eskola, 1941), into
'magmatic' and 'metamorphic' types (Wagner, 1928). The latter include
the xenoliths, containing minerals atypical of the kimberlites, such
as kyanite, scapolite, and ferriferous garnet. In this way, two different groups of rocks were mixed together, the eclogite-like rocks
of shallow depth, taken from the crust (Bobrievich and Sobolev, 1957),
and the rocks of significantly deeper origins with ferriferous garnet,
which even contain diamonds in individual cases (Bobrievich et al.,
1959b). A start towards a clear separation of the eclogite-like rocks
(in a number of cases containing plagioclase) and the deep-seated
eclogites, was proposed in the works of the above authors, based
on information from the Yakutian kimberlites.

Following the classification of Eskola (1921), who had separated
the pyrope peridotites, containing an association of olivine, garnet,
diopside, and enstatite, minerals of similar composition in all the
rocks, in spite of very wide variations in the quantitative ratio (up
to the appearance of garnet-pyroxene rocks), we shall consider the
analogous associations of kimberlites individually.

The eclogite xenoliths are characterized by morphologic features,
analogous to those of all the deepest-seated mantle rocks of the kimberlites (a rounded smooth surface and varied dimensions, from a few
centimeters up to 20-25 cm (see Photo 1).

The quantitative ratios of the principal rock-forming minerals
(garnet and pyroxene) are extremely varied. Calculation of the
amounts of minerals (in volume %) for a large series of eclogite xenoliths in the South African pipes (Rickwood et al., 1968) has shown
very wide fluctuations in the amount of garnet (from 28% to 80%). A
break has been identified between the amount of garnet in the peridotites and the eclogites (from 19% to 28%) and recognition of different types of rocks has been proposed. However, this pattern is not
general, because even wider fluctuations in the amount of garnet
(from 10% to 80%) have also been demonstrated in the South African
eclogites (MacGregor and Carter, 1970), and calculations for xenoliths from the 'Obnazhennaya' pipe indicate the presence of an almost
continuous series of compositions (based on amount of garnet) between
the peridotites and eclogites (Milashev, 1960; Milashev et al., 1963).
In the olivine-bearing rocks of this pipe, sometimes more than 20% of
garnet has been identified (cases that are exceptionally rare for the
South African xenoliths).

These comparisons demonstrate that the quantitative-mineralogical
composition cannot be used as a basis for separating the peridotites
and eclogites, but may be taken into account only as a supplementary
feature after separating the xenoliths on the basis of paragenesis

TABLE 11. Chemical Composition of Individual Zones of a Garnet from the 'Mir' Pipe and Inclusions in It (Sample No M-49).

| Oxides | Zone of garnet | | | | | | | Inclusions in garnet | | | | | | |
	1	2	3	4	5	6	7	garnet a	pyroxene b	chromites x	chromites y	chromites z	chromites from zone 5 a	chromites from zone 5 b
SiO_2	39,1	40,1	40,3	39,5	38,8	39,7	39,2	39,4	50,7	0,17	0,19	0,18	0,21	0,19
TiO_2	0,70	0,35	0,31	0,24	0,52	0,76	0,76	0,60	1,10	2,21	2,39	2,24	4,53	2,00
Al_2O_3	14,1	14,5	15,5	14,6	13,9	13,8	15,3	16,4	6,11	16,3	23,1	14,0	5,08	17,8
Cr_2O_3	8,70	10,1	8,44	10,3	10,2	9,54	7,0	6,57	2,64	38,5	32,5	40,3	36,5	36,6
Fe_2O_3	—	—	—	—	—	—	—	—	—	9,90	8,36	10,2	18,5	8,89
FeO	10,4	10,2	10,1	9,24	9,81	9,77	10,4	11,0	4,00	22,8	21,9	21,5	23,1	22,2
MnO	0,66	0,64	0,66	0,56	0,67	0,68	0,64	0,93	0,19	0,42	0,52	0,38	0,30	0,31
MgO	12,0	16,0	16,5	16,6	12,9	11,5	12,3	12,4	12,2	8,45	9,77	8,88	7,97	8,42
CaO	12,4	7,15	6,90	6,73	11,3	13,6	12,4	11,2	18,3	0,00	0,00	0,00	0,00	0,00
Na_2O	—	—	—	—	—	—	—	—	2,55	—	—	—	—	—
Total	98,06	99,04	98,61	97,77	98,10	99,35	98,00	98,50	97,79	98,75	98,7	97,69	96,13	96,34
Si	3,029	3,026	3,032	3,009	3,005	3,040	3,019	3,010	1,892	0,006	0,006	0,006	0,006	0,006
Al^{IV}	—	—	—	—	—	—	—	—	0,108	—	—	—	—	—
Ti	0,042	0,023	0,018	0,014	0,028	0,046	0,046	0,037	0,031	0,055	0,057	0,056	0,123	0,050
Al^{VI}	1,284	1,288	1,374	1,310	1,265	1,242	1,389	1,478	0,162	0,634	0,870	0,552	0,215	0,702
Cr	0,532	0,612	0,506	0,621	0,623	0,577	0,426	0,396	0,076	1,002	0,820	1,068	1,032	0,966
Fe^{3+}	0,142	0,077	0,102	0,055	0,084	0,135	0,139	0,089	—	0,246	0,199	0,258	0,499	0,225
Fe^{2+}	0,533	0,567	0,535	0,531	0,549	0,491	0,532	0,613	0,126	0,628	0,582	0,603	0,690	0,620
Mn	0,042	0,041	0,041	0,037	0,042	0,046	0,042	0,060	0,007	0,012	0,013	0,010	0,009	0,012
Mg	1,387	1,801	1,848	1,887	1,488	1,311	1,412	1,414	0,679	0,414	0,464	0,445	0,426	0,417
Ca	1,028	0,581	0,557	0,550	0,940	1,118	1,023	0,918	0,731	0,000	0,000	0,000	0,000	0,000
Na	—	—	—	—	—	—	—	—	0,184	—	—	—	—	—
Pyrope	45,0	49,0	55,4	50,1	49,2	44,2	46,9	47,1	—	—	—	—	—	—
Almandine	17,8	19,0	17,9	17,7	18,2	16,3	17,7	20,4	—	—	—	—	—	—
Spessartine	1,4	1,4	1,4	1,2	1,4	1,5	1,4	2,0	—	—	—	—	—	—

Grossular	—	—	—	—	—	—	3,4	4,4	—	—	—	—	—	—
Andradite	7,1	3,8	5,1	2,8	4,2	6,8	7,0	4,5	—	—	—	—	—	—
Ti-andradite	2,1	1,2	0,9	0,7	1,4	2,3	2,3	1,8	—	—	—	—	—	—
Üvarovite	25,2	14,4	12,7	14,8	25,5	28,6	21,3	19,8	—	—	—	—	—	—
Knorringite	1,4	11,2	6,6	12,8	0,1	0,3	—	—	—	—	—	—	—	—
f	32,7	26,3	25,6	23,7	29,8	32,3	32,2	33,2	15,7	58,1	53,1	55,1	57,1	57,8
Ca-component	34,4	19,4	18,7	18,3	31,1	37,7	34,0	30,5	47,6	—	—	—	—	—
Cr-component	26,6	30,6	25,3	31,1	31,2	28,9	21,3	19,8	22,0	50,1	41,0	53,4	51,6	48,3
Al-component	—	—	—	—	—	—	—	—	—	21,7	43,5	27,6	10,8	35,1
Ulvöspinel	—	—	—	—	—	—	—	—	—	5,5	5,7	5,6	12,3	5,0

TABLE 12. Chemical Composition of Individual Zones of a Garnet from Sample No. M-41 from the 'Mir' Pipe and of Mineral Inclusions in It

Oxides	Zones of garnet										Chromites				Ilmenites				
	1	2	3	4	5	6	7	8	9	10	1	4	5	6	a	6	8	9	10
SiO_2	40,3	39,4	39,8	40,4	39,9	41,2	40,6	40,4	41,1	41,1	0,18	0,16	0,23	0,10	0,09	0,10	0,17	0,10	0,14
TiO_2	0,34	0,67	0,67	0,67	0,64	0,61	0,61	0,67	0,46	0,70	3,10	4,45	4,43	35,5	38,2	40,0	42,2	44,2	43,3
Al_2O_3	15,5	15,6	15,4	16,1	16,7	17,0	17,9	18,8	19,9	20,8	5,60	4,98	13,0	0,83	0,79	0,81	0,70	0,73	0,61
Cr_2O_3	8,87	8,41	7,90	6,80	6,0	5,71	4,83	2,69	2,31	1,61	41,6	37,7	30,7	10,7	8,26	7,15	4,76	4,84	4,11
Fe_2O_3	—	—	—	—	—	—	—	—	—	—	16,7	19,0	18,9	27,4	24,6	22,6	20,2	16,7	19,2
FeO	11,2	10,8	11,0	11,1	10,9	11,0	11,5	12,1	11,1	11,5	22,0	23,8	23,6	22,3	24,6	25,3	23,8	27,5	24,2
MnO	0,56	0,63	0,63	0,63	0,48	0,46	0,49	0,59	0,46	0,50	0,32	0,31	0,31	0,10	0,09	0,14	0,20	0,15	0,24
MgO	15,1	14,1	13,3	13,5	13,5	14,2	14,1	14,7	16,1	15,3	8,31	8,00	9,39	5,38	5,44	5,93	7,93	6,83	8,19
CaO	7,83	9,09	10,4	10,3	9,48	10,1	10,3	8,03	7,96	7,80	0,08	0,05	—	—	—	—	—	—	—
Total	99,70	98,70	99,10	99,50	97,60	100,28	100,33	97,98	99,39	99,31	97,89	98,45	100,53	102,11	102,07	102,03	99,96	101,05	100,05
Si	3,028	2,999	3,024	3,042	3,042	3,056	3,012	3,041	3,025	3,022	0,006	0,006	0,006	0,002	0,003	0,003	0,004	0,002	0,003
Ti	0,018	0,041	0,041	0,041	0,037	0,036	0,036	0,041	0,027	0,040	0,082	0,118	0,110	0,624	0,687	0,716	0,759	0,792	0,777
Al	1,372	1,399	1,375	1,430	1,503	1,488	1,568	1,665	1,725	1,803	0,230	0,206	0,513	0,022	0,023	0,024	0,020	0,021	0,017
Cr	0,527	0,566	0,475	0,405	0,367	0,335	0,285	0,163	0,133	0,097	1,151	1,044	0,810	0,198	0,156	0,134	0,090	0,091	0,077
Fe^{3+}	0,083	0,054	0,109	0,124	0,093	0,141	0,111	0,131	0,115	0,060	0,444	0,480	0,466	0,525	0,443	0,406	0,364	0,300	0,346
Fe^{2+}	0,621	0,640	0,590	0,573	0,599	0,541	0,602	0,629	0,566	0,647	0,643	0,688	0,648	0,440	0,492	0,504	0,476	0,548	0,483
Mn	0,036	0,037	0,037	0,036	0,032	0,031	0,031	0,036	0,031	0,031	0,008	0,008	0,008	0,001	0,001	0,003	0,004	0,003	0,005
Mg	1,692	1,600	1,507	1,516	1,535	1,568	1,560	1,652	1,765	1,675	0,432	0,417	0,467	0,187	0,194	0,211	0,238	0,243	0,291
Ca	0,627	0,741	0,845	0,833	0,774	0,802	0,820	0,647	0,632	0,614	0,004	0,002	—	—	—	—	—	—	—
f	26,1	30,1	31,7	31,5	31,0	30,3	31,4	31,5	27,8	29,7	58,7	62,3	53,5	70,2	71,7	70,5	62,6	69,4	62,4
Ca-component	21,1	24,6	28,4	28,3	26,3	27,2	27,2	21,8	21,1	20,7	—	—	—	—	—	—	—	—	—
Cr-component	26,4	25,3	23,8	20,3	18,4	16,8	14,3	8,2	6,7	4,9	57,6	52,2	40,5	14,5	12,0	10,5	7,3	7,5	6,1

Note. Amount of Fe_2O_3 has been determined after recalculation of crystallochemical formula.

(with or without olivine). In this case, the enstatite-bearing rocks
and certain garnet pyroxenites form a transitional group between the
peridotites and the eclogites, and for a further, more specific separation, it is also necessary to use the features of mineral composition.

On the basis of textural features, the eclogite xenoliths belong
to the even-grained types with an approximately equal ratio of garnet
and pyroxene, which is emphasized by the distribution of the amount
of garnet on the histograms (Rickwood et al., 1968; MacGregor and
Carter, 1970). In a small proportion of eclogite xenoliths, banded
textures are known, first discovered in the corundum eclogites of the
'Obnazhĕnnaya' pipe (Sobolev and Sobolev, 1964), being especially
characteristic of these particular rocks (Photo 4a).

The texture of the eclogites is porphyroblastic or poikiloblastic,
medium- and coarse-grained. Garnets occur both in the form of idiomorphic porphyritic phenocrysts, and as grains and intergrowths of
irregular shape. An attempt has been made to recognize two types of
eclogite on the basis of textural features (MacGregor and Carter,
1970), and although of significant interest, the features employed
by the authors for recognizing them cannot be regarded as sufficiently
discrete in view of the presence of gradual transitions from rocks
with idiomorphic garnet (first type) into those with segregations and
intergrowths of garnet of irregular shape (second type). The features
of chemical composition of the eclogites are considered in §5 along
with the remaining types of xenoliths. Here, let us stress only that
compositionally they may be regarded as the equivalents of olivine
basalts.

The chemical composition of the g a r n e t s from the bimineralic eclogites is characterized by wide variations both in the
amount of the calcium component, and in the iron index. Continuous
changes in the calcium index range from 8.8% to 33.7%, with variations in the iron index from 16.6% to 54.5% (Table 13). Still more
calcic garnets (almost up to 48%) in eclogites, containing neither
kyanite nor corundum, are known in the 'Roberts Victor' pipe
(MacGregor and Carter, 1970). Wide variations in the iron index of
the garnets of eclogites reflect significant variations in the composition of these rocks. In this respect, it is interesting to
emphasize that certain garnets studied are very similar in their iron
index to those of the peridotites, especially samples A-1 ($\underline{f}$ = 16.6%)
and M-100 ($\underline{f}$ = 25.1%). At the same time, the compositional features
of the associated clinopyroxenes, which contain respectively 2.98%
and 31.91% Na_2O, and also the textural features of the rocks indicate that they are typical eclogites, and in Sample M-100, flakes
of graphite have been found, enclosed in unaltered pyroxene, which
suggests that it is a graphite eclogite.

These wide variations in the iron and calcium indices for the
garnets from eclogites show up not only in the general assemblage
of eclogites from the kimberlites of Yakutia and South Africa, but
also in each individual kimberlite pipe. As an example, we may cite
the already-noted investigation of garnets from the eclogites of the
'Roberts Victor' pipe (MacGregor and Carter, 1970), for which the
results of numerous analyses (taken from diagrams) indicate variations
in their calcium index for the bimineralic eclogites of from 10% to
37%, and $\underline{f}$ from 20% to 55%.

Quite wide variations in the iron index have also been defined as
a result of bulk measurements of the refractive index of garnets from

TABLE 13. Chemical Composition

Oxides	'Udachnaya' Pipe	'Mir' Pipe	'Roberts Victor' Pipe					'Obnazhё- nnaya' Pipe	Lesotho	'Roberts Victor' Pipe	'Zagadoch- naya' Pipe
	Ush-1/0	A—1	K—6	K-5	B	K-4	K-3	O-270	E-16	K-2	Zg-59
SiO_2	41,96	43,16	41,34	40,40	41,27	40,31	40,92	41,37	41,50	40,98	40,47
TiO_2	0,75	0,16	0.42	0,46	0,46	0,57	0,47	0,14	0,46	0,35	0,15
Al_2O_3	22,28	22,61	22,45	21,47	22,03	21,67	21,97	22,45	21,53	23,01	22,70
Cr_2O_3	0,21	0,34	0.24	0,24	0,20	0,12	0,10	0,38	0,12	0,12	0,04
Fe_2O_3	2,64	2,09	1,09	1,68	0,83	1,22	1,10	1,10	1,68	0.70	4,66
FeO	8,77	5,76	13,01	16.57	14,94	17,66	15,02	12,20	13,81	13.23	10,37
MnO	0,27	0,23	0,54	0,45	0,45	0,58	0,45	0,11	0,27	0,43	0,33
MgO	19,20	21,39	17,54	14,56	16,11	14,02	15,33	17,18	15,33	15.50	14,88
CaO	4,04	3,63	3,39	3,64	3,89	3,91	5,01	5,01	5,30	5,63	6,34
Na_2O	0,08	0,07	—	—	—	—ᵃ	—	0,02	—	—	—
Total	100,20	99,77	100,02	100,01	99,98	100,06	100,31	99,96	100,00	99.95	99,94
Si	2.992	3.035	2.995	2.995	3,010	2.993	2,998	2,995	3,033	2,986	2.952
Ti	0,040	0.025	0,023	0,026	0,025	0,032	0,026	0,009	0,026	0,019	0.009
Al	1,878	1.876	1,917	1,880	1,893	1,896	1,892	1,924	1,852	1,977	1.953
Cr	0,012	0,017	0,014	0,014	0,011	0,007	0,006	0,026	0,009	0,007	0.005
Fe^{3+}	0,070	0,081	0,046	0,080	0,046	0,065	0 061	0,044	0,097	0,039	0,033
Fe^{3+}	0,071	0.029	0.014	0,013	—	0,003	—	0,020	—	—	0,221
Fe^{2+}	0,523	0.338	0.788	1,027	0,911	1,097	0,919	0,740	0,843	0,806	0.631
Mn	0,016	0,013	0,033	0,062	0,028	0,037	0,028	0,004	0,018	0,027	0,018
Mg	2,041	2,244	1,894	1,608	1,748	1,552	1,674	1,852	1,668	1,684	1,616
Ca	0,309	0.275	0,263	0,289	0,304	0,311	0,393	0.386	0,417	0,440	0,495
Na	0,010	0,008	—	—	—	—	—	0,001	—	—	—
Pyrope	68,8	77,3	63,3	53,7	57,9	51,7	55,6	61,6	56,6	56,9	54,1
Almandine	20,0	12,6	26,8	34,7	30,2	36,7	30,5	25,3	28,6	27,3	28,7
Spessartine	0,5	0,4	1,1	2,0	0,9	1,2	9,9	0,1	0,6	0,9	0,6
Grossular	4,6	3,4	4,6	3,6	6,8	5,1	8,3	9,0	7,5	11,5	14,1
Andradite	3,5	4,1	2 3	4,0	2,3	3,3	3,1	2,2	4,9	2,0	1,7
Ti-andradite	2,0	1,3	1,2	1,3	1,3	1,6	1,3	0,5	1,3	1,0	0,6
Uvarovite	0,6	0,9	0,7	0,7	0,6	0,4	0,3	1,3	0,5	0,4	0,2
f	24,5	16,6	31,3	41,1	35,4	42,9	36,9	30,3	37,2	33,4	35,4
Ca-component	10,7	9,7	8,8	9,6	11.0	10,4	13,0	13,0	14,2	14,9	16,6
Ca/Ca+Mg	13,5	11,2	12,2	15.2	14,8	16,7	19,0	17,2	20,0	20,7	23 4

Note. In samples O-166 and O-160₁, excess Si is respectively 0.168 and 0.180. Analyses of Samples K-1, K-2, K-3, K-A-1, M-100, A-474, A-45, A-173, based on information from A. D. Khar'kiv, A. I. Ponomarenko, and N. V. Sobole

non-olivine xenoliths in the 'Obnazhёnnaya' pipe (Fig. 8). Of 171 samples, some have been assigned to websterites, for which the increased refractive indices partially depend on the amount of Cr_2O_3. However, this amount in the garnets does not exceed 1-1.5%, and the increase in the refractive index, as a result of the presence of Cr_2O_3 in such samples, constituted no more than 0.005-0.007. The general nature of the histogram demonstrates a predominance of garnets with $\underline{f}$ = 30-40%, and the most ferriferous types are characterized by $\underline{f}$ values of from 55% to 60%, which falls well within the general range of compositions of the eclogite garnets.

The c l i n o p y r o x e n e s of the eclogites, as compared with those of the peridotites, are characterized by an increased amount of Na_2O and an increased iron index. The range of variation

Lesotho	'Mir' Pipe	'Roberts Victor' Pipe	'Mir' Pipe		'Zagadoch-naya' Pipe	'Obnazhë-nmaya' Pipe	'Mir' Pipe	'Udachnaya' Pipe	'Roberts Victor' Pipe	'Mir' Pipe	'Obnazhënnaya' Pipe		
E-4	M-100	K-1	A-474	A-45	Zg-18	O-166	A-173	U-1	37079	M-30	O-770	O-160$_1$	O-160$_2$
40,21	41,35	41,04	40,76	39,65	42,06	42,46	40,83	40,14	39,04	39,76	41,62	43,35	41,06
0,00	0,24	0,38	0,23	0,67	0,05	0,16	0,47	0,14	0,21	0,42	0,34	0,20	0,18
22,51	22,33	22,69	22,76	21,18	22,13	22,31	22,76	22,23	21,85	21,86	23,01	22,05	23,45
0,00	0,18	0,07	0,13	0,08	0,04	0,10	0,06	0,04	0,084	0,05	0,12	0,02	0,05
1,57	2,74	0,61	1,26	2,54	2,55	1,62	1,42	1,07	0,13	2,31	2,20	2,46	1,80
17,01	7,94	12,64	14,23	17,45	12,06	11,60	11,72	15,06	18,42	13,11	5,17	6,81	5,02
0,32	0,21	0,24	0,27	0,39	0,49	0,30	0,22	0,23	0,47	0,24	0,16	0,17	0,16
11,67	17,3	14,58	12,63	9,25	11,69	12,84	12,36	10,82	9,04	9,90	16,54	16,16	12,03
6,58	7,48	7,77	7,80	8,91	8,95	9,12	10,82	10,52	10,63	12,39	11,18	9,15	16,16
—	0,04	—	—	0,15	0,00	0,02	—	0,06	0,06	0,21	—	0,01	0,02
99,87	99,81	100,02	100,09	100,27	100,02	100,53	99,72	100,36	99,93	100,24	100,33	100,38	99,93
2,997	2,980	2,997	2,997	2,984	3,015	3,000	2,993	2,987	2,964	2,970	2.969	3,000	2,979
0,000	0,013	0,021	0,013	0,041	0,004	0,009	0,026	0,009	0,012	0,022	0,017	0,013	0,013
1,980	1,895	1,953	1,980	1,881	1,869	1,963	1,972	1,949	1,961	1,929	1,936	1,911	1,995
0,000	0,017	0,004	0.009	0.005	0,001	0,007	0,003	0,002	0,005	0,001	0,009	0,000	0,006
0,020	0,075	0,022	—	0,073	0,126	0,021	—	0,040	0,007	0,048	0,038	0,076	—
0,070	0,072	0,011	0,071	0,072	0,184	0,069	0,079	0,023	—	0,078	0,082	0,065	0,096
1,062	0,476	0,772	0,875	1,099	0,724	0,723	0,717	0,939	1,167	0,816	0,308	0,418	0,305
0,018	0,013	0.015	0,018	0,027	0,030	0,018	0,013	0,013	0,030	0,013	0,009	0,009	0,009
1,129	1,856	1,587	1,383	1,040	1,249	1,425	1,347	1,198	1,021	1,099	1,756	1,766	1,302
0,529	0,580	0,608	0,614	0,719	0,689	0,726	0,801	0,836	0,866	0,991	0,857	0,718	1.254
—	0,005	—	—	0,023	—	0,001	—	0,008	0,009	0,028	—	0,001	0,002
40,2	61,8	53,0	46,8	34,9	43,4	48,8	45,6	39,7	33,0	36,4	58,3	59,4	43,9
40,3	18,3	26,2	31,9	39,3	31,6	26,7	26,9	31,9	37,7	29,5	12,9	16,2	13,5
0,7	0,4	0,5	0,6	0,9	1,0	0,6	0,4	0,4	1,0	0,4	0,3	0,3	0,3
17,8	14,2	17,9	19,5	19,3	17,4	22,6	25,7	25,2	27,1	30,1	25,2	19,6	41,4
1,0	3,8	1,1	—	3,2	6,3	1,0	—	2,0	0,4	2,5	1,9	3,8	—
—	0,7	1,1	0,7	2,1	0,2	0,5	1,3	0,5	0,6	1,1	—	0,7	0,6
—	0.8	0,2	0,5	0,3	0,1	0,4	0,1	0,1	0,2	—	0,5	—	0,3
50,5	25,1	33.7	40,6	54,5	45,2	36.3	37,1	45,5	53.5	46,2	19,6	24,0	23,5
18,8	19,5	20,3	20,7	24,9	24,0	24,5	27,1	28,0	28,3	33,7	28,5	24,1	42,3
31,9	24,0	27,7	30,7	40,9	35,4	33,8	37,3	41,1	46,2	47,4	32,8	28,9	49,1

K-5, K-6, after Kushiro & Aoki, 1968; 37079, after O'Hara & Yoder, 1967; E-4, E-16, after Nixon et al., 1963. Samples

in the Na-component is quite large (from 18% to 45%) in our analyses (Table 14) with an average amount of about 32% (see §13). These limits are also almost the same for the large series of pyroxenes (about 40 samples) from bimineralic eclogites of the 'Roberts Victor' pipe (MacGregor and Carter, 1970), for which, from the diagram presented by these authors, the amount of Na-component ranges from 25% to 51%.

The iron index of the eclogite pyroxenes, like that of the garnets, varies widely (from 9.8% to 24.7%), mainly in accordance with the distribution curve (see §21).

Very many of the eclogite xenoliths contain rutile in variable amounts, sometimes approaching 1-2% down to isolated grains. Graphite is extremely rare (we have found it in Sample M-100 only), and the

TABLE 14. Chemical Composition

Oxides	'Udachnaya' Pipe	'Mir' Pipe	'Roberts Victor' Pipe					'Obnazhë nnaya' Pipe	Lesotho	'Roberts Victor' Pipe	'Zagadoch naya' Pipe
	Ush-1/0	A-1	K-6	K-5	B	K-4	K-3	O-270	E-16	K-2	zg-59
SiO_2	54,59	54,84	53.74	54,29	55.3	54.21	54,96	53,71	55.06	54.72	54,51
TiO_2	0,40	0,40	0.31	0,34	0,51	0.55	0,42	0,22	0,51	0,41	0,25
Al_2O_3	2,68	3,72	5.06	4,40	6.0	6,76	8,01	6,76	7,67	10.51	11,36
Cr_2O_3	0,20	0,28	0.34	0,11	0.15	0,15	0,06	0,18	—	0.16	0,06
Fe_2O_3	3,15	2,44	2,65	2,92	1,71	3,10	1,92	1,44	1,68	1.69	2,68
FeO	1,23	0.98	3.46	5.29	4,57	4,44	3,13	2,55	4,01	2.56	0,75
MnO	0,03	0,04	0,16	0,28	0,13	0,14	0.10	0,04	0.07	0,07	0,05
MgO	16.63	16,25	15.60	13,66	11,7	12,39	11,62	13,79	11,97	11.02	10,67
CaO	18.30	17,63	15,48	14,20	14,7	13,68	14,13	17,61	14,22	13.36	13,20
Na_2O	2,52	2,98	2,62	2,90	4,08	4,47	4,78	3,62	4,70	5,05	6,12
K_2O	0,03	0,04	0,17	0.20	0,18	0,21	0,18	0,09	0,08	0.17	0,06
H_2O^+	—	0,33	0,41	0.77	0,7	—	0,40	0,29	—	—	—
Total	99,76	99,93	100,00	99.45	99,03	100,10	99,71	100,30	99,97	99,72	99,71
Si	1,965	1 974	1,945	1,991	2.008	1,956	1,977	1.932	1.973	1.946	1,929
Al^{IV}	0,035	0.026	0,055	0.009	—	0.044	0,023	0.068	0,027	0.054	0,071
Al^{VI}	0,077	0,130	0,161	0.181	0,257	0,243	0,316	0,222	0.296	0.387	0,405
Ti	0 012	0,011	0.008	0,009	0,014	0,015	0,011	0,006	0 013	0.011	0,008
Cr	0,009	0,009	0,010	0,003	0.004	0,004	0,002	0,004	—	0,005	0,002
Fe^{+3}	0,095	0,065	0.072	0,080	0,047	0,084	0,052	0,039	0,047	0,045	0,072
Fe^{+2}	0,037	0,030	0.105	0,162	0,139	0,134	0,094	0,078	0,120	0,076	0,023
Mn	0,001	0,001	0,005	0.009	0.004	0,004	0,003	0.001	0,002	0,002	0,001
Mg	0.893	0,871	0.841	0.751	0,633	0,666	0,627	0,739	0.641	0.584	0,562
Ca	0,705	0,681	0,600	0.568	0,572	0,529	0,544	0,679	0,544	0,509	0,500
Na	0,177	0.206	0,184	0.212	0,288	0,313	0,333	0,250	0,327	0,348	0,421
K	0.001	0,001	0,007	0,009	0.008	0.010	0,008	0,004	0,004	0,008	0,003
Total of cations	4,007	4,005	3.993	3.974	3,966	4,002	3,990	4.022	3,994	3,975	3,997
f'	4,0	3,3	11,1	17,7	18,0	16,8	13,0	9,5	15,8	11,5	3,9
f	12,9	9,8	17,4	24,4	22.7	24,7	18,9	17,4	20.7	17,2	14,4
Ca/Ca+Mg	44,1	43,9	41,6	42.6	47,5	44.3	46,5	47,9	45,9	46,6	47,1
Ca/Ca+Mg+Fe	43,1	43,1	38,8	37,9	42,6	37,4	41,3	45,4	41,7	41,9	46,1

unevenness of its distribution evidently reflects the general unevenness in the amount of carbon in the deep-seated rocks.

In some eclogites, phlogopite has been identified (Kushio and Aoki, 1968), which is probably a secondary mineral. Products of alteration of pyroxene, amongst which chlorite, analcime, etc., have been identified (Berg, 1968), are also of a secondary nature.

Lesotho	'Mir' Pipe	'Roberts Victor' Pipe	'Mir' Pipe			'Zagadochnaya' Pipe	'Obnazhennaya' Pipe	'Mir' Pipe	'Udachnaya' Pipe	'Roberts Victor' Pipe	'Mir' Pipe	'Obnazhennaya' Pipe		
E-4	M-100	K-1	A-474	A-45	3r-18	O-166	A-173	У-1	37079	M-30	O-700	O-160$_1$	O-160$_2$	
54,76	52,95	54,73	54,79	56,1	55,69	52,91	54,12	55.03	52,04	53.72	48.29	51,18	48,99	
0,44	0,78	0,41	0,21	0,46	0,07	0 29	0,38	0.35	0,54	0,60	0,12	0,16	0.13	
8,33	7,41	14,10	8,92	5,87	11,35	7,38	9.07	14.47	6,93	11,25	17,85	13,21	17,78	
—	0.13	0,10	0,10	0,03	0,08	0.12	0.07	0.04	0.23	0,12	0 17	0,04	0,05	
1,29	2.23	1,17	1.80	—	3.21	2,04	1.15	2,29	2,59	3.25	1,06	1,01	0,72	
2,70	0.78	1,74	2,16	5,82	0,95	1,40	2,21	0,62	4.10	1.46	0,49	1,10	0,74	
0,04	0.04	0,04	0,04	—	0.02	0.02	0.02	0,04	0,09	0,02	0,01	0,03	0,02	
11,59	13,20	8,84	11,77	11,2	8,52	16,16	11,81	8.10	11,65	9,58	9,56	11,78	9,73	
16.35	17,30	11,62	14,58	16,6	13,54	15 86	15,19	11,58	19,49	12,94	18,68	17,58	18,23	
4.63	3,91	6,61	4,81	4,52	5,86	3,36	4,97	6,44	2 22	6,20	3,12	3.51	3,30	
0,05	—	0,24	0.08	0,10	0.12	0,04	0,07	0.07	—	0,10	0.16	0,06	0,06	
—	1,88	0,38	—	—	0,56	—	—	1.15	—	0,54	—	—	—	
100,18	100,45	99,98	99,46	100,70	99,97	99,58	99,06	100,18	99.67	99,78	99,49	99,85	99,75	
1,958	1.910	1,930	1.962	2,011	1,972	1 898	1 943	1,940	1,901	1,921	1,728	1,831	1,744	
0,042	0,090	0,070	0,038	0,000	0,028	0,102	0 057	0,060	0,099	0.079	0,272	0,169	0,256	
0,306	0.226	0.516	0 336	0,250	0,449	0,213	0,327	0.541	0.200	0,397	0,484	0.385	0,493	
0,011	0.022	0.011	0.006	0,017	0,002	0,009	0 011	0.010	0.015	0.018	0,002	0,004	0,003	
—	0,004	0.003	0.003	0,001	0,002	0 004	0.002	0.003	0.001	0.004	0,004	0,001	0,001	
0,034	0,061	0,031	0,047	0,050	0,085	0,052	0,035	0,059	0.070	0,090	0,030	0,027	0,019	
0,082	0,024	0,051	0,065	0,124	0,028	0,041	0,067	0,017	0.125	0,043	0,015	0,033	0,022	
—	0,001	0,001	0,001	0,000	0,000	—	—	0,003	—	—	0,001	0,001		
0,618	0,708	0.465	0,630	0,598	0,449	0,862	0,633	0,426	0,635	0,511	0,509	0,627	0,518	
0,627	0.667	0.439	0,559	0,637	0,515	0,610	0,585	0,438	0.764	0,496	0,718	0,673	0,695	
0.322	0,273	0.452	0,340	0.318	0,404	0,237	0,350	0,441	0.158	0,433	0,215	0,240	0,227	
0,001	—	0.011	0,004	0.004	0 005	0 001	0.003	0.004	—	0,004	0.004	0,003	0,003	
4.001	3,986	3,980	3,991	4,012	3.939	4,029	4,013	3,939	3.971	3,996	3,981	3,994	3,982	
11,7	3.3	9,9	9,4	17,2	5,9	4,5	9,6	3,8	16.4	7,8	2,9	5,0	4,1	
15,8	10.7	15,0	15,1	22,5	20,1	9,7	13,9	15,1	23,5	20,7	8,1	8,7	7,3	
47,2	48.5	48,6	47,0	51,6	53,4	41,4	48,0	50,7	54,6	49,3	58,5	51,8	57,3	
47,2	47,7	44.5	44,6	45,2	51,9	40,3	45,5	49,8	50,1	47,2	57,6	50,5	56,3	

Corundum Eclogites

The corundum eclogites belong to the rarest type of xenolith in the kimberlites. Corundum eclogite was first noted in the South African eclogites (Williams, 1932), but in the brief description only the mutual relationships of the garnet, clinopyroxene, and pink corundum were defined, without information about their composition.

52

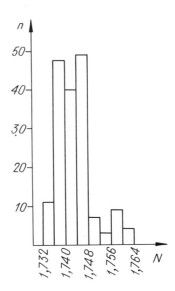

Fig. 8. Histogram showing distribution of refractive index values for garnets with low amounts of Cr_2O_3 from peridotites and eclogites from the 'Obnazhennaya' pipe.

The first detailed description of a natural corundum eclogite was given for a xenolith from the 'Obnazhĕnnaya' pipe (Sobolev, 1964b; Sobolev and Kuznetsova, 1965). This xenolith weighed about 6 kg and measured 18 X 13 X 8 cm, had a rounded-elongate shape, and contained a significant amount of violet-pink corundum (ruby) (about 15 wt %). The light-green color of the eclogite was due to the predominance of clinopyroxene. The light-orange garnet, present in minor amount, is characterized by xenomorphism, and the ruby consists of idiomorphic tabular grains, many of them showing facets with clear cross-hatching.

The clinopyroxene is often turbid, which depends on numerous ingrowths of garnet in the form of bright bands. These bands reflect the general pattern of reaction development of garnet in xenoliths from the 'Obnazhĕnnaya' pipe (replacement of breakdown textures of pyroxenes). The last phase of this process (formation of spindle-shaped bands of garnet along partings in pyroxene) has been observed in corundum eclogite 0-160, but orthopyroxene is observed only in the form of relicts.

In addition to Sample 0-160, we have found yet another eclogite in 'Obnazhĕnnaya' pipe, with corundum (Sample 0-272) of substantially smaller size, also characterized by a rounded-compressed shape (5 x 4 x 2 cm, 0.15 kg). A small xenolith (0-700) of rounded shape and about 2 cm in diameter, but in which no corundum has been found, is macroscopically similar in the color of the garnet and pyroxene and in the presence of numerous bands of garnet in the pyroxene.

The two xenoliths of corundum eclogites which we have found, represent not only the rarest type in composition and paragenesis, but also are true xenoliths of complex composition. On the polished surface of Sample 0-160 (see Photo 4a) there is in the right-hand portion a broad dark band (garnet + pyroxene), to the right of this band, non-corundum eclogite, and to the left, corundum eclogite, the zone of non-corundum eclogite comprising about 10% of the whole xenolith. In the second sample (0-272), the corundum-bearing zone comprises about 50% of the whole xenolith.

The corundum eclogite, which is a rock undersaturated in SiO_2 differs substantially from the normal eclogites in the high content of Al_2O_3 and CaO (see §5), which allies it with the grospydites. A decreased amount of these oxides characterizes the non-corundum zone of this eclogite.

The complex nature of the crystallization of the corundum eclogites 0-160 and 0-272 is also emphasized by data on the composition of their garnets. Whereas the non-corundum zone of both samples contains a garnet of uniform composition ($\underline{n}$ = 1.738; $\underline{a}_0$ = 11.559 Å), the variations in composition of the garnet in the corundum zone are quite wide, which is emphasized by changes in the value of $\underline{a}_0$ (from 11.590 to 11.647 Å) in both samples, along with an identical iron index ($\underline{n}$ = 1.737-1.738). The heterogeneity of the garnet is thus reflected only in the variable calcium index (from 28% to 42%).

In order to examine the chemical composition, we selected a garnet from a portion of Sample 0-160, most enriched in corundum, with a check for uniformity by means of X-ray analysis. The high value of $\underline{a}_0$ = 11.647 Å indicated that this garnet had the highest calcic composition. In fact, the analysis proved an amount of 42.3% Ca-component, together with $\underline{f}$ = 23.5% (see Table 13). The garnet of the non-corundum zone contains 24.1% Ca-component with $\underline{f}$ = 24.0%.

Garnets from three described samples of corundum eclogites from the South African pipes (Rickwood et al., 1968) are similar in composition to ours, differing only in the somewhat increased iron index ($\underline{n}$ = 1.741-1.748, $\underline{a}_0$ = 11.625-11.637 Å). The most interesting composition is that of a garnet from Sample CSBELL-4, with $\underline{n}$ = 1.745 and $\underline{a}_0$ = 11.720 Å, with a calcium index of about 65%. The xenolith is similar to certain grospydites, sometimes containing a substantial amount of corundum (Sobolev et al., 1968).

In comparison with the compositions of minerals from the corundum eclogite, that of the minerals from Sample 0-700 is of interest (an eclogite without corundum from this same 'Obnazhĕnnaya' pipe). Its garnet with very low $\underline{n}$ = 1.737, is also variable in composition, with $\underline{a}_0$ fluctuating from 11.547 to 11.580 Å, which corresponds to a calcium index of from 19% to 28%. It is evident that a less calcic garnet, which replaces the pyroxene, has been developed in the intergrowths, since the garnet selected away from the intergrowths appeared uniform and had 28.5% Ca-component and $\underline{f}$ = 19.6% (see Table 13), corresponding to a maximum value of $\underline{a}_0$.

The unusual paragenesis and low iron index for the garnets suggests a special composition for the clinopyroxene of the corundum eclogite. In fact, the clinopyroxene is characterized by an exceptionally large Al_2O_3 content in general and Al^{IV} in particular, which shows up markedly amongst the pyroxenes of the normal eclogites (see Table 14). The low content of Na_2O suggests that its composition is intermediate between that of the sodium-rich types of the grospydites and the pyroxenes synthesized in the system $MgSiO_3$-$CaSiO_3$-Al_2O_3 in equilibrium with corundum and pyrope (Boyd, 1970). The pyroxene of the non-corundum zone contains significantly less Al^{IV}, but still much more than the normal eclogites, even those associated with magnesian garnets (see Table 14, Sample A-1). It is evidently possible to regard this feature as an indicator of special eclogites, undersaturated in SiO_2. In spite of the absence of corundum, Sample 0-700 may be assigned to this type, because the composition of its pyroxene is almost identical with that of corundum eclogite 0-160.

Thus, the paragenetic features of the corundum eclogite and also the composition of its minerals and those of the clearly associated

54

SiO$_2$-undersaturated eclogites, suggest that they are related cate-
gories, which are characterized by a complicated crystallization
history. An analogous complex crystallization is also evident for
Sample 372, briefly described by Williams (1932). The banded segregations of garnet in the pyroxene of this sample, frequently regarded
as dissociation textures of pyroxene, undoubtedly also belong to
later reaction materials.

§4. THE GROSPYDITES AND KYANITE ECLOGITES
OF THE 'ZAGADOCHNAYA' PIPE

Amongst the numerous varieties of xenoliths of garnet-bearing
rocks, discovered in the kimberlite pipes of Yakutia, those of the
garnet-clinopyroxene-kyanite rocks occupy a very special place. They
were first discovered in the 'Zagadochnaya' pipe (Bobrievich et al.,
1960), and as a result of the completely unusual composition of the
associated minerals (grossular garnet, clinopyroxene of diopside-
jadeite composition, and kyanite), they were named grospydites,
derived from the names of these minerals (grossular, pyroxene, and
disthene [kyanite]). During the examination of these xenoliths, it
has been shown, on the basis of several samples, that the garnet has
a constant composition, corresponding to the invariant point on the
paragenetic diagram for the eclogite facies.

In view of the exceptional rarity of these rocks and the restric-
tion on material, available for such study, we made an additional
attempt in 1965 to collect these unique rocks from the 'Zagadochnaya'
pipe. In the old workings at the pipe, a substantial number of xenoliths were collected, the macroscopic features of which (high density,
presence of kyanite grains, and the yellow color of the garnets) provisionally assigned them to the grospydites. Moreover, a large quantity of concentrate of the heavy fraction from the kimberlite was collected, in which small fragments of deep-seated xenoliths and intergrowths of garnet and kyanite were found. A study of this collection
enabled us to recognize completely new, unexpected features in the
composition of the minerals which formed the xenoliths, especially
the presence of a continuous series in the garnets of pyrope-grossular
composition, with addition of almandine (Sobolev, Zyuzin and Kuznetsova, 1966; Sobolev et al., 1968), and to demonstrate that the earlier-
constructed diagram applies to an exceptional case of a definite value
of sodium concentration.

Thus, besides the grospydites with variable garnet composition, we
also recognized kyanite eclogites, which include parageneses with garnet, containing less than 50% of calcic component.

The grospydites and associated kyanite eclogites are characterized
by the rounded, smoothed surface of the xenoliths typical of many
deep-seated inclusions in kimberlites (see Photo 1). The high density, light color, the presence of numerous light-colored grains of
kyanite and a light orange-yellow garnet amongst the larger xenomorphic segregations of light-green clinopyroxene, distinguishes the
xenoliths of these rocks from the various peridotites and eclogites
from the Yakutian kimberlite pipes.

The dimensions of the grospydite and kyanite-eclogite xenoliths
vary widely (from 0.5 to 15 cm along the long axis) (see Photo 1).
The most common are samples up to 5 cm in size. In the samples
examined, wide variations have been observed in the amounts of garnet, pyroxene, and kyanite. The amount of garnet in the various

samples varies from about 20% to 50%, pyroxene from 10% to 30%, and kyanite from 1 vol % to 25 vol %. In all the samples, alteration products of pyroxene and various secondary minerals are present in variable amounts.

In addition to the grospydites and kyanite eclogites, our collection includes a xenolith of a non-garnet rock, consisting of pyroxene and kyanite, which represents an association previously unknown amongst the deep-seated rocks. This rock has been termed a kyanite pyroxenite.

Although many of the samples are characterized by a massive fabric, there are cases of clear alternation of bands, which consist of garnet-pyroxene and garnet-kyanite associations (Photo 4b). The texture is medium- and coarse-grained, poikiloblastic (Photo 5), often with idiomorphic segregations of fine-grained kyanite amongst the large xenomorphic pyroxene grains.

Along with the principal rock-forming minerals (garnet, pyroxene, and kyanite), the grospydites contain a platy corundum (sapphire) varying in amount from isolated grains to considerable quantities (about 0.5 vol %). The secondary minerals include structureless hydrated products, which have not been studied in detail.

The garnet is characterized by equant grains, vermicular aggregates, and large masses of irregular shape, in a number of cases containing numerous kyanite inclusions. In some samples, there are diablastic intergrowths with kyanite (Photo 6), the latter probably occasionally being replaced by garnet at the boundary with the pyroxene.

The clinopyroxene is characterized by xenomorphic grains, sometimes quite large (up to 5-10 mm), most commonly intensely altered by the development of secondary products. Sometimes, completely fresh idiomorphic kyanite grains have been observed in an almost totally altered pyroxene groundmass, in which only relicts of fresh pyroxene are seen. In some cases, a rim, consisting of brown secondary products, is developed around the grains of relatively fresh pyroxene (see Photo 5). Kyanite is present in the form of idiomorphic grains enclosed both in garnet and pyroxene. Kyanite has been observed in the form of diablastic intergrowths with garnet, and it is also possible to regard the textural features as the result of the process of metamorphic crystallization.

Corundum is present in the form of isolated grains, evenly distributed through the sample, being found in intergrowths with garnet, pyroxene, and kyanite. In such cases, it gives the impression of being a mineral in equilibrium with them. This may apply, for example, to Samples Zg-52 and Zg-24, which have comparatively increased amounts of corundum. However, in most cases, the position of corundum with respect to the other minerals is less clear, and it is often confined to the the peripheral zones of the xenoliths, sometimes even forming discordant veins.

Rutile is present in some samples in the form of rare grains. There are large segregations of rutile (up to 0.5 cm across) in one of the kyanite-eclogite xenoliths.

In the heavy-fraction concentrate from the kimberlite, there are numerous small fragments of grospydites and kyanite eclogites, and also a multitude of grains of yellow-orange garnets. When the grains are crushed, inclusions of kyanite are often revealed, which indicates that such garnets belong to the paragenesis with kyanite. In the concentrate, many intergrowths of garnet with kyanite and corundum have been found. In some of the intergrowths, a blue cor-

56

undum makes up 5 vol % to 10 vol %. In the small fragments of xenoliths and intergrowths, individual rutile grains are sometimes
observed.

Among the secondary minerals, the most common are zoisite and a
basic plagioclase. They are present in significant amounts in the
altered samples, but in the fresh samples they are developed in the
peripheral zones. Sometimes quartz, along with plagioclase, appears
as alteration products of the kyanites. Corundum is also a possible
alteration product of kyanite.

In the general scheme of mineral formation, the above secondary
minerals probably appear after the entry of the xenoliths into the
kimberlitic magma. The last, possibly exogenous products, are the
earthy, structureless products of alteration of pyroxene.

Chemical analyses of the rocks studied, which contain an identical
mineral paragenesis (garnet + clinopyroxene + kyanite ($\pm$ corundum)),
are discussed in §5. The amount of calcic component will be used as
a basis for separating the rocks into kyanite eclogites (less than
50%) and grospydites (more than 50%). The former are represented by
five analyses, and the latter, by ten.

Comparison of the data (see §5) reveals a significant difference
between the recognized rock types both on the basis of amount of
Al_2O_3, and on the amount of MgO and CaO. The clearest difference is
manifested in the value of the Ca/Ca + Mg + Fe ratio, which decreases
for the kyanite eclogites (32%) and increases for the grospydites
(49%), along with equal values for the iron index. The high calcination losses emphasize the presence of secondary minerals, clearly
water-bearing, because calcite has not been recorded in the thin-
sections.

The large amount of Al_2O_3, MgO, and CaO, together with the
decreased amount of SiO_2 is a specific common feature of the rocks
examined. They may be regarded as equivalent in chemical composition to certain anorthosites or gabbro-anorthosites with increased
MgO content at the expense of spinel. Of all the known rocks,
assigned to the eclogite group, only the undersaturated corundum
eclogites are most similar in composition.

Garnet

The garnets of the rocks studied are colored in various orange tints:
from yellow-orange to reddish-orange in the most ferriferous types.
Determination of the refractive index (Fig. 9) has shown the narrow
limits of variation in composition on the basis of iron index, with
variations in n from 1.738 to 1.762, with the largest n values
recorded in isolated cases. With similarity in paragenesis and positive correlation in color intensity of the garnets with the n values,
an impression is created that the differences between the garnets are
insignificant in all the samples. However, mass determinations of
unit-cell parameters (a_0) have at once revealed marked differences
in their composition (Sobolev, Zyuzin and Kuznetsova, 1966; Sobolev
et al., 1968), which has demonstrated that the earlier discovered
composition of a grossular-rich garnet in grospydite (Bobrievich et al.,
1960), is only one case in a continuous series of pyrope-grossular garnets with addition of the almandine component. The histogram (see
Fig. 9) demonstrates the wide fluctuations in values of a_0 from 11.613
to 11.812 Å. The distribution of a_0 values for 77 samples of garnets

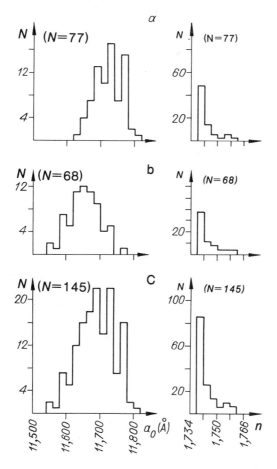

Fig. 9. Histogram of values of a_0 and n for garnets from grospydite and kyanite-eclogite xenoliths from the 'Zagadochnaya' pipe (a), garnets from intergrowths with kyanite from the same pipe (b), and both groups together (c). N is the number of samples.

from xenoliths of grospydites and associated kyanite eclogites approximates to a normal curve with a broad maximum. The most frequent compositions are those with a_0 = 11.681-11.780 Å, but garnets with extreme a_0 values have been much less frequently recorded. The most calcium-rich is the garnet from Sample Zg-119 with a_0 = 11.834 Å, approximating pure grossular in composition. The refractive indices, as noted above, vary within a very small range and for most of the garnets from the xenoliths (more than 70% of the samples) they are practically identical, ranging from 1.738 to 1.742 (see Fig. 9).

The results of measuring a_0 and n, enable us to determine the composition of the garnets, based on the well-known composition-properties diagram in the system pyrope-almandine-grossular (Gnevushev et al., 1956; Giller, 1962). An estimate of the composition of the garnets from the data presented in the histogram (see Fig. 9), demonstrates a variation in the amount of grossular of from 30% to 81%, and in the iron index, of from 30% to 60%.

In order to obtain the most complete definition of possible fluctuations in composition, we have in addition studied the physical
properties of a significant number of garnets from the kimberlite
concentrate, in which inclusions of kyanite and corundum have been
found. This feature enables us to assign them either to the grospydite paragenesis or to the hypothetical garnet-kyanite paragenesis,
not found, however, in the larger-sized rock fragments. It is clear
that these garnet grains and the intergrowths of garnet with kyanite
and corundum, may be regarded as the result of complete disintegration of the grospydite and kyanite-eclogite xenoliths during the
process of their incorporation in the kimberlite.

The construction of such histograms as that for the garnets in the
xenoliths, demonstrates the similarity in physical properties of these
garnets, especially in respect to $\underline{n}$. The values of the unit-cell
parameter for the garnets from the concentrate tend towards a decrease,
and the general nature of the histogram seemingly defines a shift in
the compositions towards a decrease in the amount of calcium. In the
xenolith collection, only the garnet from kyanite eclogite Zg-60 is
characterized by a decreased value for $\underline{a}_0$ = 11.613 Å. In the garnets
from the concentrate, almost 25% of the total number of grains
examined are defined by a lower value for $\underline{a}_0$, corresponding to less
than 30% of calcic component. The maximum amounts of calcium arc
analogous to those of the garnets in the grosypdites.

Such displacement of the histograms, which show the distribution
of calcium in the garnets, in our opinion, may indicate for the xenoliths and for individual grains with kyanite inclusions, a predominant disintegration of the kyanite-eclogite xenoliths as compared with
those of the grospydites, possibly dependent on the lower content of
garnet and kyanite in the sample, these minerals being more stable with
respect to secondary processes as compared with pyroxene. The overall
assemblage of results, involving 145 samples, is quite representative,
and evidently reflects practically all the possible variations in the
composition of garnet in paragenesis with kyanite in the complex of
rocks from the 'Zagadochnaya' pipe. Composite histograms are also
shown in Figure 9.

Samples were selected for chemical analysis on the basis of data
from the determination of the physical properties. The possibility
of an extremely precise determination of the composition of these garnets based on physical properties enabled us to select a representative series of garnets (21 samples) for further investigation, amongst
which were garnets taken from the concentrate (Sample Zg-103) with
inclusions of kyanite with the lowest content of calcium and a large
series of samples of kyanite eclogites and grospydites down to a few
samples with exceptionally large amounts of grossular.

The results of chemical analyses and X-ray microanalyses of the garnets, shown in Table 15, emphasize the features, established on the
basis of physical properties (wide fluctuations in the amount of calcium and a comparatively narrow range in the iron index). At the same
time, the low content of TiO_2, Cr_2O_3, and MnO has been established.
The values of the measured and calculated constants, $\underline{a}_0$ and $\underline{n}$, coincide completely satisfactorily, giving an additional control to the
quality of the analyses. As might be expected, arising from data on
the physical properties, the garnet of Sample Zg-119, which is characterized by the largest content of grossular (94%), is one of the
closest to the end member of this series of garnets ever found. However, the conditions of formation and the paragenesis of this grossu-

lar, like other garnets of the grospydites, are distinguished from
those of similar composition in the marbles and skarns.

The smallest amount of calcic component in the analyzed garnets
has been recorded in Sample Zg-103. Apparently, the range of compositions from practically pure grossular to pyrope-almandine garnet,
containing 21% calcic component, characterizes practically all the
variety of garnet compositions from the 'Zagadochnaya' pipe, associated with kyanite. This is well illustrated by the diagram
(Fig. 10), on which are plotted the compositional points for the
analyzed garnets and individual compositions determined on the basis
of physical properties, being distinguished either by the iron index, or by the content of calcium.

Thus, the garnets from the grospydites and the associated kyanite
eclogites occupy a well-defined field on the pyrope-almandine-grossular diagram, which is characterized by very wide fluctuations in the
content of calcium (from 21% to 94%) and in the iron index (from 30%
to 60-65%). Those compositions, containing more than 40% of calcic
component (more than that in the corundum eclogite), are essentially
the most common in the series of rocks described.

We noted above that amongst the xenoliths of the 'Zagadochnaya'
pipe, there is a marked predominance of true grospydites with garnet,
containing more than 50% of grossular. For illustration, we shall present data of approximate quantitative calculations based on the results
of measuring a_0 (see Fig. 9). The chemical analyses of the garnets
have shown that those of the true grospydites are characterized by
values of a_0 in excess of 11.680 Å (Samples Zg-49 and Zg-63). Thus,
the following relationships have been derived from determinations of
a_0 for 145 garnets, associated with kyanite in the samples and individual grains (see Fig. 9): they comprise 85% in the xenoliths of
true grospydites; amongst the grains and intergrowths of garnet with
kyanite from the concentrate, there is a marked increase in the percentage of true eclogite compositions, and here the garnets of disintegrated grospydites comprise only 28%.

It is evident that the field of garnet compositions in paragenesis
with kyanite, recognized on the basis of extremely representative data
(more than 160 analyses and determinations of physical properties),
embraces almost the entire possible range of garnet compositions of
the paragenesis indicated from the xenoliths of the 'Zagadochnaya'
pipe. However, taking the paragenesis of garnet with kyanite as a
basis, necessary for assigning the garnets examined to the grospydites, kyanite eclogites, or kyanite-garnet rocks, it must be emphasized that the garnets of all the compositions indicated (see Fig. 10)
are also possible in paragenesis without kyanite. The rare occurrence of xenoliths of such type is explained by the predominant disintegration of the non-kyanite rocks as a result of intense alteration of pyroxene.

C h r o m i u m - b e a r i n g   g a r n e t s. In addition to
the garnets, colored in various shades of orange, which are non-chromium varieties of the pyrope-almandine-grossular series, very rare
finds of chromium-bearing garnets of similar composition have been
made in the xenoliths from the 'Zagadochnaya' pipe and in the heavy-
fraction concentrate, and these are distinguished by a marked alexandrite effect. The assignment of these garnets to the grospydites
and kyanite eclogites is in no doubt, since they have been found in
paragenesis with pyroxene and kyanite, which, however, contain an
increased amount of chromium. The color of the garnets varies from

TABLE 15. Chemical Composition of Garnets from Grospydites

Oxides	Zg103*	Zg-60	Zg-15	Zg-52	Zg-24	Zg-48	Zg-49	Zg-63	Zg-3	Zg-43
SiO_2	41,3	41,58	40,43	40,76	41,95	39,68	40,04	39,19	41,60	39,57
TiO_2	0,06	0,15	0,08	0,30	0,19	0,38	0,34	0,44	0,10	0,08
Al_2O_3	22,7	21,87	22,32	22,76	22,20	22,88	22,06	22,85	22,28	22,94
Cr_2O_3	0,02	0,02	0,02	Сл.	0,02	0,04	0,06	0,03	0,08	0,12
Fe_2O_3	—	4,17	1,44	1,71	1,93	1,38	2,22	2,59	2,77	1,44
FeO	10,4	12,55	11,13	7,86	12,18	11,00	8,58	10,35	5,80	6,37
MnO	0,18	0,20	0,23	0,27	0,26	0,14	0,56	0,16	0,22	0,08
MgO	15,6	8,49	9,62	10,57	6,55	7,18	7,52	5,17	7,01	6,96
CaO	8,20	10,58	14,7	15,85	14,88	17,50	18,50	19,40	19,92	21,97
Total	98,52	99,71	99,98	100,08	100,16	100,18	99,88	100,18	99,53	99,53
Si	3,023	3,000	2,999	2,984	3,000	2,959	2,984	2,946	3,000	2,951
Ti	0,004	0,009	0,004	0,018	0,011	0,022	0,018	0,023	0,006	0,004
Al	1,963	1,979	1,957	1,972	2,017	2,014	1,944	2,030	1,982	2,015
Cr	0,001	0,002	0,001	0,000	0,001	0,004	0,003	0,003	0,005	0,009
Fe^{3+}	0,033	0,010	0,038	0,010	0,000	—	0,035	—	0,007	—
Fe^{3+}	—	0,230	0,042	0,087	0,112	0,081	0,090	0,144	0,150	0,081
Fe^{2+}	0,605	0,806	0,692	0,484	0,787	0,685	0,537	0,650	0,365	0,399
Mn	0,013	0,018	0,013	0,018	0,017	0,009	0,036	0,009	0,014	0,004
Mg	1,703	0,972	1,068	1,154	0,752	0,801	0,837	0,577	0,787	0,775
Ca	0,642	0,871	1,175	1,246	1,226	1,397	1,478	1,561	1,607	1,755
Pyrope	57,6	33,6	35,7	38,6	26,0	26,9	28,3	19,6	26,9	25,8
Almandine	20,4	35,7	24,5	19,1	31,1	25,8	20,6	27,0	17,6	15,9
Spessartine	0,4	0,6	0,4	0,6	0,6	0,3	1,2	0,3	0,5	0,1
Grossular	19,6	20,0	37,2	40,3	41,8	45,7	47,0	51,8	54,1	57,6
Andradite	1,7	0,5	1,9	0,5	—	0,0	1,8	—	0,4	—
Ti-Andradite	0,2	0,5	0,2	0,9	0,5	1,1	0,9	1,2	0,3	0,2
Uvarovite	0,1	0,1	0,1	0,0	0,0	0,2	0,2	0,1	0,2	0,4
f	27,3	51,8	42,0	33,5	54,5	48,9	43,0	57,9	39,9	38,1
Ca-component	21,6	30,1	39,3	41,7	42,3	47,0	49,9	53,1	55,0	58,2
n measured	—	1,758	1,761	1,742	1,759	1,751	1,745	1,757	1,741	1,741
n calculated	—	1,763	1,754	1,747	1,761	1,754	1,753	1,760	1,747	1,745
a_0 measured	—	11,613	11,631	11,653	11,647	11,672	11,683	11,683	11,696	11,709
a_0 calculated	—	11,604	11,635	11,640	11,659	11,661	11,676	11,685	11,689	11,699

Note. For Samples Zg-60, Zg-24, Zg-3, Zg-37, Zg-6, and Zg-33, the excess of Si is: 0.194, 0.230, 0.132, 0.150, 0.198

pale-violet to green, with concurrent increase in n as compared with the garnets of the normal grospydites. The amount of Cr_2O_3 in these garnets varies from 1.5% to 7.2%, and the calcic component, from 27% to 62%, in 17 analyzed garnets from xenolith fragments and from the concentrate (Tables 16 and 17).

In order to stress the rarity of the chromium-bearing garnets, we noted that during the examination of more than 2 kg of garnet con-

and Associated Kyanite Eclogites from the 'Zagadochnaya' Pipe

Zg-37	1	Zg-6	Zg-8	Zg-34	Zg-38	Zg-28	Zg-33*	Zg-1	Zg-128*	Zg-119*	Uz-132*	Al-2*
41,45	40,70	39,68	41,96	40,09	39,95	41,42	39,7	39,74	39,9	39,9	40,4	40,8
0,22	0,34	0,35	0,19	0,10	0,62	0,11	0,08	0,44	,08	0,19	0,29	0,29
21,71	20,41	23,16	22,72	22,26	21,78	21,59	21,2	22,09	21,0	21,5	21,9	21,3
0,03	Сл.	0,02	0,04	0,03	0,03	0,02	0,04	Сл.	0,02	0,00	0,06	0,03
2,93	0,80	1,01	0,93	1,96	2,86	2,56	—	0,73	—	—	—	—
5,52	8,02	7,53	5,02	4,20	4,54	2,22	4,19	4,57	2,77	2,42	11,9	13,4
0,13	0,15	0,11	0,14	0,12	0,15	0,06	0,06	0,04	0,03	0,00	0,13	0,18
5,95	4,58	4,09	4,66	4,32	2,94	3,29	2,29	1,59	1,33	0,81	7,67	6,35
21,73	23,57	24,36	24,00	26,87	26,95	28,26	30,7	31,03	33,3	34,2	16,6	17,4
99,67	100,32	100,31	99,66	99,95	99,53	99,53	98,26	100,23	98,43	99,02	98,95	99,75
3,000	3,000	2,964	3,000	2,995	3,001	3,000	3,037	2,987	3,048	3,030	3,041	3,073
0,013	0,020	0,022	0,011	0,004	0,036	0,006	0,005	0,025	0,005	0,014	0,018	0,016
1,945	1,889	2,043	2,046	1,964	1,929	1,950	1,911	1,961	1,891	1,925	1,946	1,892
0,002	—	0,003	0,003	0,002	0,001	0,001	0,002	—	0,001	0,000	0,004	0,002
0,040	0,047	—	—	0,030	0,034	0,043	0,082	0,014	0,103	0,061	0,032	0,090
0,127	—	0,054	0,053	0,080	0,128	0,104	—	0,028	—	—	—	—
0,350	0,524	0,471	0,321	0,260	0,284	0,142	0,184	0,287	0,075	0,090	0,719	0,752
0,008	0,010	0,004	0,009	0,006	0,009	0,004	0,005	0,003	0,001	0,000	0,009	0,014
0,674	0,537	0,457	0,530	0,479	0,329	0,375	0,262	0,178	0,151	0,091	0,860	0,715
1,771	1,984	1,945	1,964	2,152	2,167	2,318	2,513	2,504	2,727	2,783	1,339	1,403
23,0	17,6	15,6	18,4	16,1	11,3	12,7	8,8	5,9	5,1	3,1	29,4	24,8
16,3	17,2	17,9	13,0	11,4	14,1	8,4	6,2	10,5	2,5	3,0	24,6	26,2
0,3	0,3	0,1	0,3	0,2	0,3	0,1	0,2	0,1	0,1	0,1	0,3	0,5
57,7	61,5	65,2	67,7	70,5	70,7	76,2	80,2	81,5	86,8	90,1	43,0	43,1
2,0	2,4	—	—	1,5	1,7	2,2	4,2	0,7	5,2	3,1	1,6	4,5
0,6	1,0	1,1	0,5	0,2	1,8	0,3	0,3	1,3	0,3	0,7	0,9	0,8
0,1	—	0,1	0,1	0,1	0,1	0,1	0,1	—	—	—	0,2	0,1
41,4	49,4	53,4	41,4	41,5	55,5	39,6	41,3	63,9	33,2	50,0	46,6	54,1
60,4	64,9	66,4	68,3	72,3	74,3	78,8	84,8	83,5	92,3	93,9	45,7	48,5
1,746	—	1,745	1,741	1,742	1,743	1,740	1,740	—	1,737	1,738	—	—
1,749	—	1,746	1,743	1,745	1,747	1,743	1,743	—	1,744	1,742	—	—
11,724	—	11,737	11,737	11,751	11,767	11,777	11,790	—	—	11,831	—	—
11,712	—	11,731	11,736	11,754	11,754	11,778	11,788	—	—	11,838	—	—

and 0.169 respectively. Analysis 1, after Bobrievich *et al.*, 1960. Sample Al-2, from information by B. S. Nai.

centrate from the 'Zagadochnaya' pipe, only 20 grains of chromium-
bearing, calcium-rich garnets were found, which represents about
0.02 wt %. The compositional points for the chromium-bearing garnets, associated with kyanite in the xenolith fragments and intergrowths, have been plotted on a diagram along with the points for
garnets that do not contain chromium, from this same paragenesis
(see Fig. 10). All the points are located within the recognized
compositional field.

62

TABLE 16. Chemical Composition of Chromium-Bearing Garnets from 'Zagadochnaya' Pipe

Oxides	Z-54	Z-41	Z-1	Z-53	Z-47	Z-51	Z-52
SiO_2	39,3	39,9	40,8	39,6	41,1	40,4	39,0
TiO_2	0,26	0,17	0,12	0,13	0,21	0,11	0,19
Al_2O_3	19,8	20,0	19,5	17,7	17,7	17,1	16,2
Cr_2O_3	2,30	3,50	3,86	5,22	6,01	6,69	7,37
FeO	9,83	9,82	9,80	8,22	8,31	9,58	8,25
MnO	0,10	0,12	0,11	0,08	0,40	0,06	0,05
MgO	7,47	11,6	8,88	11,0	14,1	10,3	5,62
CaO	19,6	14,7	17,6	15,5	12,3	15,4	22,2
Total	98,66	99,8	100,67	97,45	100,13	99,64	98,88
Si	3,001	3,977	3,037	3,030	3.035	3,045	3,019
Ti	0,015	0,009	0,004	0,005	0.013	0,005	0,014
Al	1,780	1,757	1,709	1,600	1,544	1,523	1,479
Cr	0,138	0,206	0,233	0,313	0,346	0,399	0,456
Fe^{3+}	0,067	0,028	0,054	0,082	0.097	0,073	0,051
Fe^{2+}	0,562	0,582	0,554	0,442	0,418	0,534	0,484
Mn	0,006	0,008	0,004	0,005	0,027	0,005	0,005
Mg	0,850	1,291	0,988	1,255	1.553	1,156	0,646
Ca	1,606	1,174	1,404	1,269	0,972	1,246	1,842
Pyrope	28,1	42,2	33,5	42,1	52,3	39,2	21,6
Almandine	18,6	19,1	18,8	15,1	14,1	18,3	16,3
Spessartine	0,2	0,3	0,1	0,2	0,9	0,2	0,2
Grossular	42,1	26,2	33,0	22,5	9,8	18,3	35,8
Andradite	3,4	1,4	2,7	4,1	4,9	3,7	2,6
Ti-andradite	0,7	0,5	0,2	0,3	0,7	0,3	0,7
Uvarovite	6,9	10,3	11,7	15,7	17,3	20,0	22,8
f	42,5	32,1	38,1	29,5	24,9	34,4	45,3
Mg-component	28,1	42,2	33,5	42,1	52,3	39,2	21,6
Ca-component	53,1	38,4	47,6	42,6	32,7	42,3	61,9
Cr-component	6,9	10,3	11,7	15,7	17,3	20,0	22,8

Clinopyroxenes

The pyroxenes are characterized by a light-green or grass-green color. In individual samples, as a result of its extremely light color, the pyroxene is difficult to distinguish from kyanite. The distinguishing feature in such cases is the perfect cleavage of the kyanite. A detailed study of the chemical composition has been carried out on the pyroxenes, which are associated with the most varied garnet compositions in the grospydites and kyanite eclogites, distinguished both by calcium content, and by iron index (Table 18).

The data obtained enable us to distinguish certain features in the composition of the pyroxenes from the grospydites, which separate them from those of the deep-seated eclogites in general and from the eclog-

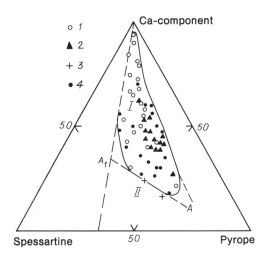

Fig. 10. Garnet compositions from grospydites and associated kyanite eclogites of the 'Zagadochnaya' pipe. 1) analyzed garnets; 2) chromium-bearing analyzed garnets; 3) garnets from eclogites without kyanite; 4) garnet compositions, plotted according to data on n and a_0: I) field of garnets, associated with kyanite; II) assumed field of garnet without kyanite; A_1-A) line separating fields.

ites of the metamorphic complexes. In first place amongst these features is the very high content of Al_2O_3 in the pyroxenes of the true grospydites. In these pyroxenes, there is a clearly-defined excess of Al over Na, with a very low content of Al^{IV}. In some of the pyroxenes, associated with the most calcic garnets, the excess reaches 0.25, which comprises about 25% of the total Al_2O_3. Such an unusually large amount of Al_2O_3, not combined in the normal components (minals) of the pyroxenes (Tschermak's component and jadeite), suggests the possibility of mechanical contamination by kyanite, difficult to distinguish in some samples from pyroxene, in the analyzed material. However, a check by X-ray methods on the amount of kyanite has given a negative result, whereas in artificially prepared mechanical mixtures of pyroxene and kyanite, the latter could be identified with more than 5% present. The identification of a similar large amount of Al_2O_3 in the pyroxene, with the aid of the X-ray microprobe finally gave the clue to the structural nature of the excess Al_2O_3 in the pyroxenes. Such an amount, interpreted as a solid solution of kyanite, has been noted by O'Hara and Yoder (1967). This amount has been assumed by them for a pyroxene, obtained during the crystallization of a kyanite eclogite at $\underline{P}$ = 30 kb and $\underline{T}$ = 1800°C, when all the kyanite present in the rock, disappeared into the final products. Although the large content of Al_2O_3, including excess Al^{VI}, has been identified in all the samples examined, it is extremely variable for the pyroxenes, associated with garents with varying calcic and iron indices. In particular, the lowest amount of Al^{VI} may be regarded as that in the pyroxenes of the true kyanite eclogites as compared with those of the grospydites.

On the basis of the paragenetic features of the pyroxenes with garnets of variable calcic index, we have recognized three types of pyroxene compositions (see §13). The first type consists of pyrox-

TABLE 17. Results of Partial Chemical Analysis of Chromium-Bearing Garnets from 'Zagadochnaya' Pipe (Sobolev et al., 1968)

Oxides	Z-9	Z-5	Z-43	Z-3	Z-16	Z-17	Z-45	Z-2	Z-44	Z-13
Cr_2O_3	4,9	5,9	4,7	4,5	4,6	3,9	1,4	4,4	4,0	1,5
FeO	9,5	9,7	9,5	8,4	11,1	10,1	8,2	7,9	10,1	7,5
MgO	14,6	11,2	10,3	11,7	9,6	8,1	9,4	11,2	8,4	7,4
CaO	10,3	13,4	13,9	15,0	14,4	15,1	15,4	18,1	17,8	18,6
Pyrope	53	43	40	43	37	33	37	40	31	30
Almandine	20	20	21	17	23	23	19	15	21	16
Grossular	13	19	24	27	27	31	39	32	36	49
Uvarovite	14	18	15	13	13	13	5	13	12	5
f	27	33	34	29	29	41	33	27	40	37
Ca-component	27	37	39	40	40	44	44	45	48	54
n measured	1,754	1,773	1,769	1,755	1,762	1,766	1,742	1,762	1,759	1,742
n calculated	1,761	1,769	1,767	1,759	1,766	1,767	1,742	1,758	1,764	1,741
a_0 measured	11,612	11,676	11,657	11,668	11,660	11,697	11,653	11,681	11,691	11,705
a_0 calculated	11,604	11,650	11,653	11,652	11,655	11,671	11,655	11,670	11,683	11,693
Associated kyanite										
Ng	1,782	—	1,773	—	1,762	1,760	1,733	1,760	1,741	
Np	1,765	—	1,757	—	1,747	1,745	1,719	1,745	1,725	
Cr_2OSiO_4, mol %	17,5	—	16	—	13	12	2,5	12	5	

enes of true kyanite eclogites, for which, as a result of the large variations in composition of the associated garnets based on f values of from 32% to 55% and a calcic index of from 30% to 50%, extremely broad limits on the amount of Al^{VI}, and also Na, have been recorded. The second type may be taken to include the pyroxenes of the grospydites, the garnet of which occupies an intermediate position with respect to the calcic index between those of the grospydites proper and those of intermediate composition, that is, they contain from 50% to 75% of the calcium component. And finally, the third type of composition consists of pyroxenes, associated with garnets, most rich in calcium. The pyroxenes of the second and the true third type are characterized by a clearly-defined increased Ca/Ca + Mg + Fe ratio and by the presence of excess Al^{VI}, which distinguishes them from the pyroxene of the first type. The earlier recognized types of pyroxene compositions (more precisely, one combined) from the grospydites and kyanite eclogites (Sobolev et al., 1968) have been over averaged, although they also define the features indicated. In the present stage of study of these most interesting xenoliths, it is convenient to recognize the above three types of pyroxenes.

The identification of a significant amount of excess Al^{VI} and Ca in the pyroxenes makes it necessary to introduce additional components into the pyroxene composition. One of these components ($CaAl_{2/3}Si_2O_6$) has been assumed during the statistical treatment of pyroxene analyses

(Dobretsov et al., 1970). However, when recalculating the portion of Al^{VI}, allowing for this component, there still remains excess Al^{VI}, not associated with Ca. Therefore, it is possible to assume the entry of an isomorphous amount of Al^{VI} in the form of an hypothetical component, with the composition $Al_{2/3}SiO_3$, in some degree similar to the 'solution of kyanite' in pyroxene (O'Hara and Yoder, 1967). Another variant of the calculation may be used, especially for pyroxenes associated with garnets, similar to grossular, and to be precise, regarding the excess of Ca and Al as a special 'anorthite' component (Mao, 1971). In spite of the variety of variations, the principal feature of these pyroxenes is the presence of isomorphism on the basis of the $3Mg^{2+} \rightleftarrows 2Al^{3+}$ scheme.

In the kyanite eclogites and grospydites of the 'Zagadochnaya' pipe, there is a clearly-defined association between the amount of calcium in the garnets and their iron index and the amount of sodium in the associated pyroxenes (Sobolev, Zyuzin and Kuznetsova, 1966; Sobolev et al., 1968), which enables us to assign the entire series of variable compositions of the kyanite-bearing xenoliths of the 'Zagadochnaya' pipe to rocks, formed under similar P-T conditions. This opinion is expressed in greater detail in §6.

C h r o m i u m - b e a r i n g   p y r o x e n e along with the chromium-bearing garnets and kyanites, has been recorded in peculiar, extremely rare fragments of kyanite eclogites and grospydites, in which the amount of Cr_2O_3 is more than one order of magnitude greater than in the normal xenoliths of this type. The characteristic difference of the chromium-bearing pyroxene is the emerald-green color. The chemical compositions of four pyroxenes with variable content of Cr_2O_3 from 0.92% up to 2.89% are presented in Table 18. The unusual composition of these pyroxenes is emphasized by the increased amount of Na_2O from 4.79% to 5.98%. Such pyroxenes have been termed chrome-ophacites (Sobolev et al., 1968).

Kyanite

The kyanite in all the xenoliths studied is distinguished by idiomorphism and resistance to alteration as compared with the garnets and especially with the pyroxene. The colorless tabular grains of kyanite with well-developed cleavage serve as a diagnostic feature of these xenoliths. The optical properties of the colorless kyanites ($\alpha = 1.712$; $\gamma = 1.728$; $\gamma-\alpha = 0.016$) are not distinguishable from those of kyanites from various metamorphic rocks.

Special attention is drawn to the kyanites of the chromium-bearing grospydites and kyanite eclogites, described in greater detail in §16. Here we shall dwell only briefly on the general features of their composition, expressed in the markedly increased content of Cr_2O_3, ranging almost unbrokenly up to 16.8% in a series of samples, with concurrent increased refractive indices (α up to 1.778, and γ up to 1.794). Examination of these unique kyanites (Sobolev, Zyuzin and Kuznetsova, 1966; Sobolev et al., 1968) has enabled us to clarify the pattern in the successive increase in Cr_2O_3, which has made it possible to estimate the amount of Cr_2O_3 in kyanite, without carrying out an analysis. In the chromium-bearing layer of a grospydite (Sample Zg-13), study of the kyanite has shown the markedly uneven distribution of chromium, the amount of which varies even within a single grain of kyanite. In this sample, individual grains are found with a clear change from colorless to light-blue tints, indicating

TABLE 18. Chemical Composition of Clinopyroxenes from Grospydites and

Oxides	Chromium-bearing grospydites and kyanite eclogites				Kyanite eclogites			
	Zg-13	Z-54*	Z-42*	Z-41*	Zg-60	Zg-52	Zg-24	Zg-49
SiO_2	53,32	54,9	55,4	55,2	55,50	51,99	55,25	54,37
TiO_2	0,23	0,37	0,13	0,15	0,18	0,28	0,23	0,13
Al_2O_3	12,40	11,1	10,6	9,41	15,63	15,75	16,17	14,47
Cr_2O_3	0,92	1,65	2,48	2,89	0,04	0,03	0,04	0,03
Fe_2O_3	1,48	—	—	—	1,94	1,55	1,64	2,38
FeO	1,40	1,70	2,04	1,68	0,96	0,58	1,36	0,95
MnO	0,00	0,00	0,00	0,00	0,02	Сл.	0,04	0,02
MgO	10,40	8,66	9,13	10,5	7,17	10,72	7,23	8,95
CaO	14,67	14,7	14,2	15,4	10,52	14,75	12,34	13,35
Na_2O	4,79	5,36	5,98	5,57	7,61	4,22	6,36	5,21
K_2O	0,16	0,00	0,00	0,00	0,08	0,22	0,07	0,13
Total	99,77	99,44	99,96	100,80	99,65	100,09	100,73	99,99
Si	1,902	1,970	1,969	1,952	1,944	1,828	1,920	1,910
Al^{IV}	0,098	0,030	0,031	0,048	0,056	0,172	0,080	0,090
Al^{VI}	0,418	0,440	0,413	0,343	0,589	0,482	0,582	0,509
Ti	0,006	0,010	0,003	0,004	0,005	0,008	0,006	0,003
Cr	0,026	0,047	0,068	0,081	0,001	0,001	0,001	0,001
Fe^{3+}	0,039	—	—	—	0,050	0,042	0,043	0,063
Fe^{2+}	0,041	0,051	0,060	0,051	0,028	0,017	0,039	0,028
Mn	0,000	0,000	0,000	0,000	0,001	0,000	0,001	0,001
Mg	0,547	0,463	0,485	0,552	0,374	0,561	0,374	0,468
Ca	0,555	0,565	0,540	0,584	0,395	0,555	0,459	0,502
Na	0,328	0,372	0,414	0,382	0,516	0,287	0,428	0,355
K	0,007	—	—	—	0,004	0,008	0,003	0,006
Total of cations	3,969	3,948	3,982	3,997	3,963	3,961	3,936	3,936
f_1	7,0	9,9	11,0	8,5	7,0	2,9	9,4	5,6
f	12,8	9,9	11,0	8,5	17,3	9,5	18,0	16,3
Ca/Ca+Mg	50,4	55,0	52,7	51,4	51,4	49,7	55,1	51,8
Ca/Ca+Mg+Fe	48,6	52,4	49,8	49,8	49,6	49,0	52,6	50,3

marked enrichment in Cr_2O_3 only in one portion of the kyanite grain, whereas the colorless portion is almost completely devoid of chromium. In this sample, a kyanite grain has been analyzed and has a chromium content varying from 5.55% to 8.41% (Table 19). Occurrences of such kyanite are quite isolated and so far apply only to the 'Zagadochnaya' pipe.

C o r u n d u m of blue color and tabular habit, has often been

Grospydites

Zg-3	Zg 43	Zg-37	Zg 6	Zg 8	Zg-34	Zg-38	Zg-28*	Zg-33	Zg-1*	Zg 119*
54,82	54,03	54,70	54,06	55,16	54,56	55,31	54,9	55,18	57,9	58,3
0,12	0,04	0,18	0,42	0,10	0,10	0,38	0,00	0,05	0,18	0,18
14,19	12,91	14,41	17,64	15,22	16,50	19,26	17,1	19,07	22,5	22,7
0,09	0,08	0,02	0,03	0,03	0,04	0,02	0,00	0,08	0,00	0,00
1,63	1,53	1,53	1,56	1,01	1,56	0,82	—	0,66	—	—
0,64	0,75	0,61	0,83	1,00	0,40	1,10	1,14	0,76	1,20	0,53
0,03	0,05	Сл.	Сл.	0,02	0,01	Сл.	0,00		0,01	0,01
9,08	10,12	8,67	6,87	8,22	7,26	5,57	5,79	6,10	1,69	1,56
14,94	16,58	14,17	11,98	14,55	12,94	10,11	12,2	10,98	7,00	6,10
4,68	4,62	5,31	6,20	5,12	6,00	7,30	7,36	7,00	9,73	9,85
0,07	0,12	0,05	0,12	0,16	0,07	0,11	—	0,06	—	—
100,29	100,83	99,65	99,71	100,59	99,44	99,98	98,49	99,94	100,21	99,23
1,918	1,900	1,923	1,890	1,921	1,912	1,922	1,938	1,911	1,967	1,987
0,082	0,100	0,077	0,110	0,079	0,088	0,078	0,062	0,089	0,033	0,013
0,503	0,433	0,520	0,617	0,546	0,594	0,711	0,650	0,689	0,869	0,901
0,001	0,001	0,005	0,011	0,003	0,002	0,010	0,000	0,001	0,006	0,006
0,002	0,002	0,001	0,001	0,001	0,001	0,001	0,000	0,002	0,000	0,000
0,043	0,038	0,040	0,042	0,026	0,042	0,021	—	0,017	—	—
0,019	0,023	0,018	0,023	0,029	0,013	0,031	0,034	0,023	0,035	0,014
0,001	0,001	0,000	0,000	0,001	0,001	0,000	0,000	0,000	0,000	0,000
0,473	0,530	0,454	0,357	0,426	0,379	0,286	0,305	0,314	0,086	0,080
0,560	0,625	0,534	0,449	0,543	0,486	0,376	0,462	0,408	0,255	0,223
0,317	0,313	0,362	0,420	0,345	0,408	0,492	0,505	0,468	0,645	0,651
0,003	0,006	0,002	0,005	0,007	0,004	0,005	—	0,003	—	—
3,924	3,972	3,935	3,925	3,926	3,930	3,933	3,956	3,925	3,896	3,875
3,9	4,2	3,8	6,1	6,4	3,3	9,8	10,0	6,8	31,3	14,9
11,6	10,3	11,3	15,4	11,4	12,7	15,4	10,0	11,3	31,3	14,9
54,2	54,1	54,0	55,7	56,0	56,2	56,8	60,2	56,5	74,8	73,6
53,2	53,1	53,1	54,2	54,4	55,4	54,3	53,7	54,8	67,8	70,3

recorded in the grospydites in the form of isolated grains, sometimes
up to 1-2% in amount, and evenly distributed through the rock. In
many cases, the secondary nature of the corundum has been clearly
identified; it concentrates towards the peripheral altered zones of
the xenoliths, and occasionally forms discordant veins.

Extremely unusual compositions, representing a continuous series
of solid solutions of Al_2O_3-Cr_2O_3, have been identified in fragments

TABLE 19. Chemical Composition of Kyanites from the 'Zagadochnaya' Pipe

Oxides	Zg-8	Z-54	Zl-1	Z-13$_2$	Z-13$_1$	Zl-2	Z-53	Z-52	Zl-9	Z-51
SiO_2	37,5	36,7	36,0	36,6	36,5	36,3	35,7	35,3	35,6	35,2
TiO_2	0,01	0,07	0,02	0,04	0,04	0,07	0,08	0,05	0,07	0,09
Al_2O_3	60,5	58,0	55,8	55,0	54,3	52,4	50,0	49,8	47,9	46,5
Cr_2O_3	0,04	3,53	4,96	5,55	8,41	9,67	13,1	13,5	15,2	16,8
Fe_2O_3	0,18	0,30	0,42	0,42	0,42	0,42	0,48	0,47	0,54	0,52
MnO	0,00	0,00	0,00	0,00	0,00	0,00	0,00	0,00	0,00	0,00
MgO	0,03	0,00	0,04	0,05	0,06	0,05	0,06	0,05	0,07	0,06
CaO	0,00	0,01	0,00	0,00	0,00	0,00	0,00	0,00	0,00	0,00
Total	98,26	98,61	97,24	97,66	97,30	98,91	99,37	99,17	99,38	99,12
Si	1,029	1,017	1,016	1,030	1,015	1,023	1,014	1,006	1,018	1,015
Al^{VI}	1,956	1,894	1,855	1,824	1,782	1,741	1,673	1,673	1,617	1,580
Ti		0,001	0,000	0,001	0,001	0,002	0,002	0,001	0,002	0,002
Cr	0,001	0,077	0,112	0,123	0,185	0,215	0,294	0,305	0,344	0,385
Fe^{3+}	0,003	0,006	0,010	0,010	0,010	0,010	0,010	0,010	0,010	0,010
Mg	0,002	0,000	0,002	0,002	0,003	0,002	0,003	0,003	0,003	0,003
Cr_2OSiO_4	—	3,9	5,7	6,3	9,3	10,9	14,8	15,3	17,4	19,5
Al_2OSiO_4	99,7	95,8	93,7	93,1	90,0	88,4	84,5	84,1	82,0	79,9
Ng	—	—	—	—	—	—	—	1,780	1,782	1,794
Np	—	—	—	—	—	—	—	1,763	1,765	1,778

of chromium-bearing grospydites and kyanite eclogites. They may be regarded both as primary (intergrowths with kyanite and inclusions in it), and secondary formations, developed in the alteration products of the chromium-bearing kyanite.

Summarizing the data on the compositional features of the minerals of the grospydites and kyanite eclogites of the 'Zagadochnaya' pipe, we emphasize once again the importance of their exceptionally wide fluctuations, especially that of the garnets. The compositional points for the garnets fill the previously empty gap in the pyrope-almandine-grossular system (see Fig. 10).

Xenoliths of such rocks have been found in the 'Zarnitsa' pipe (Bobrievich et al., 1960). Detailed investigations of deep-seated material have revealed isolated xenoliths also in the 'Udachnaya' and 'Aikhal' pipes. The composition of the garnet from a xenolith in the 'Udachnaya' pipe (Sample Uz-132) is given in Table 15. On the basis of the calcium component (45%), it may be assigned to the kyanite eclo-gites. Values for the calcic index close to that indicated are also characteristic on the basis of a preliminary study of the garnets from altered kyanite eclogites of the 'Aikhal' pipe.

Thus, occurrences of xenoliths of kyanite eclogites in the 'Udach-naya' and 'Aikhal' pipes have significantly extended the geographical distribution of the kyanite-bearing parageneses of the eclogites in the Yakutian pipes, though it is true, only within the one Daldyn-Alakit diamond-bearing region. In the other regions, even in pipes examined in such detail as the 'Obnazhённaya' and 'Mir', eclogite-grospydite associations have not so far been identified.

§5. CHEMICAL COMPOSITION OF DEEP-SEATED ECLOGITES

The xenoliths of the deep-seated garnet-bearing rocks are characterized by an exceptional variation in chemical composition. Amongst
the xenoliths, there are both typical lherzolites with a very low content of Al_2O_3, and also various eclogites ranging to corundum and
kyanite types and grospydites. The variety of compositions of the
xenoliths is illustrated on the graphs constructed on the basis of
ratios of principal oxides, and also on those, one of the coordinates
of which is the average iron index of the rocks, and the other, the
amount respectively of MgO, Al_2O_3, or CaO (wt %) (Figs. 11, 12). Comparison of the available information on the South African (Williams,
1932; Carswell and Dawson, 1970; Kushiro and Aoki, 1968; and O'Hara's
data) and Yakutian kimberlites (Bobrievich and Sobolev, 1962), and including our new data (see Table 20), demonstrates that a sharp boundary
between the ultramafic and basic compositions is absent, although the
typical lherzolites are clearly separated from the eclogites. There
is also a well-defined area of transitional rocks, mainly websterites.
The association between the lherzolites and the pyrope websterites and
pyroxenites may be illustrated by the xenoliths showing complex structures and compositions in which typical lherzolites and pyroxenites
coexist in different sectors of the sample. Such samples were first
identified in the Yakutian kimberlites. They comprise a banded peridotite (Bobrievich et al., 1959) and also xenoliths of complex structure and composition from the 'Obnazhёnnaya' pipe (Milashev et al.,
1963; Sobolev and Sobolev, 1964). In spite of the repeatedly observed
clearly banded nature of the zonation, the xenoliths from the 'Obnazhёnnaya' pipe have sometimes, without sufficient reason, been interpreted as segregations of minerals of concentrically-zoned construction (Milashev et al., 1963).

Quantitative-mineralogical observations have shown that peridotite
xenoliths, enriched in both garnet and olivine, are extremely rare.
This conclusion is based on data, presented by Milashev et al. (1963),
and also information from a study of xenoliths from the South African
pipes (Rickwood et al., 1968). Attention is drawn to the view of
the latter authors about the limiting amount of garnet in the peridotites, which does not exceed 19 vol %. The minimum amount of garnet
in the eclogites is about 28%, although there are a number of exceptions here (MacGregor and Carter, 1970).

Thus, although it is possible to select samples with a continuous
change in composition from peridotites to eclogites, intermediate compositions, containing from 4% to 8% Al_2O_3 along with a low iron index,
and also 4% to 8% CaO, are very rare. The pyrope websterites (enstatite eclogites), in which olivine is almost or completely absent, are
seemingly a connecting link between the peridotitic and eclogitic compositions, although on features of paragenesis and composition of the
rock-forming minerals, they are usually similar to the eclogites.
This is seen particularly from the figures for the average composition
of pyroxenes in these rocks (see §13).

Among the peridotites, the lherzolites are characterized by the
most constant composition. However, in this exceptionally uniform
group of rocks, containing very little Al_2O_3, it has been possible to
recognize different types of composition amongst the peridotites from
the 'Udachnaya' pipe, which contain chromium-rich garnets. The compositions of these rocks (Table 20), after recalculation to anhydrous
forms, differ significantly on average from those of the garnet lher-

70

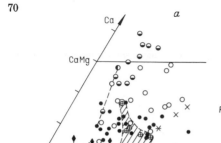
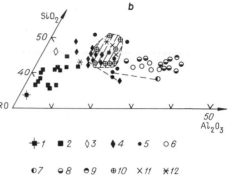

Fig. 11. Compositions of pyrope-bearing xenoliths from kimberlite pipes based on ratios of Mg, Ca, and Fe (a) and RO, SiO_2, and Al_2O_3 (b). 1) dunites; 2) lherzolites; 3) enstatites; 4) websterites; 5) eclogites; 6) kyanite eclogites; 7) corundum eclogite; 8) grospydites; 9) kyanite pyroxenite; 10) diamond-bearing eclogites; 11) average compositions of olivine tholeiites (after Kutolin, 1968); 12) average compositions of olivine basalts (after Nockolds, 1954); compositional field of diamond-bearing eclogites is cross-hatched; compositional points for individual zones of complex xenoliths are joined by dashed lines.

zolites from the various pipes of Yakutia and South Africa (Table 21), on the basis of MgO and SiO_2 content. The pseudoporphyritic peridotites of the 'Udachnaya' type, are consequently richer in olivine and contain less Al_2O_3 (and hence, garnet also) as compared with the peridotites of other pipes. A noteworthy indicator of this is the composition of their garnets, which are extremely rich in Cr_2O_3.

Although the amount of Cr_2O_3 is not distinctive in all the peridotites, the Cr/Cr + Al ratio in the peridotites of the 'Udachnaya' pipe is much greater, which clearly emphasizes the regular connection between the amount of Al_2O_3 in the rock and the garnet composition (Fiala, 1965; Sobolev and Sobolev, 1967).

The eclogites, beginning with the enstatite-bearing varieties and the pyroxenites with pyrope garnet, form a clear unbroken series with the kyanite and corundum eclogites and grospydites. The compositions of the normal bimineralic eclogites, which are analogous to certain olivine basalts (Forbes, 1965), agree especially well with the average compositions of the olivine basalts (Nockolds, 1954). However, their iron index is somewhat lower than the average iron index for the oceanic basalts, according to Kutolin (1968), although many individual analyses from this author's set, overlap in composition with some eclogites from the pipes.

The kyanite eclogites, the corundum eclogite, and the grospydites fall in the compositional field of enrichment in CaO and Al_2O_3 (Bobrievich et al., 1959) and restricted iron-index values (Sobolev et al., 1968; Godovikov, 1968). Starting with the features of chemical composition, the analogues of these rocks may be the high-alumina basalts, and for the grospydites, the peculiar gabbro-anorthosites or anorthosites enriched in magnesium.

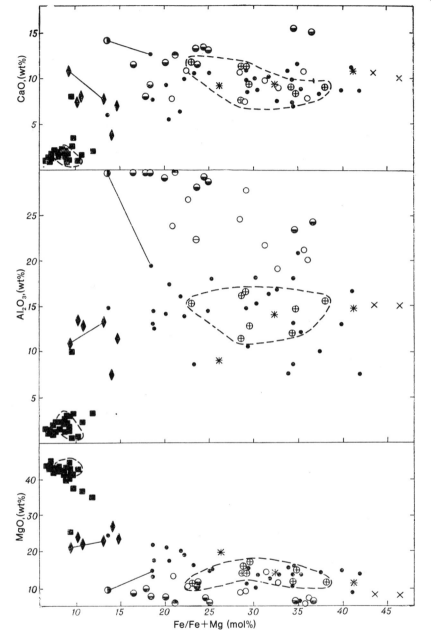

Fig. 12. Association between iron index (average) of deep-seated xenoliths and amount of MgO, Al_2O_3, and CaO (symbols as for Figure 11).

The diamond-bearing eclogites are not distinctive in compositional features from the remaining xenoliths from the kimberlite pipes (with the exception of an increased amount of Na_2O), although it is important to note that this variety of eclogites with diamonds, both in iron index and in content of Al_2O_3 and CaO, is of special importance, emphasizing the presence of individual segregations or layers of more ferriferous composition in the substrate of the mantle below the stability boundary for the diamond (Bobrievich et al., 1959b; Sobolev and Kuznetsova, 1966).

The problem of distribution of Na_2O in deep-seated rocks is of great interest. Unfortunately, chemical analyses of the eclogites cannot provide representative information for comparison of the amounts of Na_2O, since the pyroxenes of absolutely all the known samples have been altered to a significant degree. This may be judged from analyses of diamond-bearing eclogites with an average amount of Na_2O of 2.26% (see Table 21). On the basis of the average composition of the pyroxenes of the diamond-bearing eclogites or even that of the completely analyzed samples alone, and the approximate ratio of garnet and pyroxene, about 3.5% Na_2O has been recorded. As seen from the comparison, this figure is half as much again as that established from the rock analyses. In the same way, it can be shown that the amount of Na_2O for the normal eclogites is about 1-1/2 to 2 times greater than that determined from the rock analyses. Therefore, we consider a much more effective comparison of the composition of the eclogites to be that starting with the composition of the analyzed minerals and their ratio in the rock, which reflects the primary compositional features of the eclogites (O'Hara and Yoder, 1967).

Still more complex is the problem of the amount of K_2O in the eclogites, the increase in which, as a rule, is associated with secondary alterations (Berg, 1968). If we start from the identified amount of K_2O in the pyroxenes, then a stable amount of K_2O is identified only in the diamond-bearing eclogites (of the order of 0.06-0.10).

Thus, the garnet peridotites and eclogites of the kimberlite pipes are equivalent in composition in most cases to the shallower peridotites and olivine basalts. The wide fluctuations in the composition of these rocks indicate the complexity of the composition of the substrate, from which they have been extracted.

§6. CONDITIONS OF FORMATION OF ROCKS OF THE FACIES

The deep-seated pyrope-bearing rocks of the kimberlite pipes comprise a wide series of compositions, ultramafic and basic. All of them may be almost completely characterized by the system $CaO-MgO-Al_2O_3-SiO_2$. The presence of a magnesian garnet enables us to plot a sharp boundary in the conditions of formation between these rocks and their equivalents in chemical composition, the plagioclase-bearing and spinel-bearing peridotites and olivine basalts.

The presence of such phases as a pyrope garnet, associated with a sodium-bearing pyroxene, indicates the clearly-defined role of pressure during their crystallization.

Let us trace the principal phases of transformation in the ultramafic and basic rocks with change in pressure. In the ultramafic rocks, the first important reaction, demonstrated experimentally, is the replacement of the anorthite-olivine paragenesis by that of spinel and diopside (Kushiro and Yoder, 1965). The pressure, corresponding to this reaction, is more than 3 (which corresponds to the reaction cordierite + forsterite = enstatite + spinel), but less than 10 kb. Ultramafic rocks of both types are well known.

The next phase, which defines one of the boundaries of this facies
on the basis of pressure, is that of garnetization and eclogitization.
It must be stressed that in contrast to eclogitization of rocks with
excess of SiO_2, the reaction of formation of garnet proceeds here not
with separation but with absorption of SiO_2. In the absence of in-
flux of SiO_2, it must be compensated by the transition of portion of
the enstatite into olivine, and in the purely olivine rocks, spinel
will be preserved. Therefore, the decrease in SiO_2 potential here
will be inhibited by the formation of garnet and garnetization will
require a pressure greater than in the more acid rocks.

In the rocks without spinel, the eclogitization reaction proceeds
with separation of SiO_2 and consequently at lower pressures, of the
order of 17 kb (Yoder and Tilley, 1962). The increase in SiO_2 potential here will restrict the area of stability of garnet. In these
rocks, the existence of eclogite-like parageneses with plagioclase is
possible.

The position of the line of reaction of garnetization of the spinel
peridotites is quite similar according to the experimental results of
various authors, and some insignificant discrepancies have been due
to differences in original compositions.

Experimental data in the system SiO_2-$MgSiO_3$-Al_2O_3 at $\underline{T}$ = 1200°C
and $\underline{P}$ = 30 kb have shown the close similarity between the compositions of the phases obtained (analyzed with the aid of the microprobe) and those of analogous minerals in natural lherzolites (Boyd,
1970). This similarity is particularly clear with respect to the
calcic index of the garnet, in equilibrium with two pyroxenes. These
investigations have convincingly stressed the role of the bipyroxene
geothermometer (see §22).

Although all the experimental data have also emphasized that the
natural ultramafic compositions have been well described by the system CaO-MgO-SiO_2-Al_2O_3, it is necessary to allow for the possible
effect of additional components on the position of the equilibrium
line. An abridged variant of such a calculation has been given by
O'Hara (1967) in the form of a petrogenetic grid for determining the
conditions of pyroxene crystallization. The effect of the addition
of Cr_2O_3 has also been estimated (MacGregor, 1970; Sobolev, 1970),
and it has been conclusively shown that the position of the boundary
of the garnetization reaction for the ultramafic rocks has been displaced towards the side of higher pressures up to 10 kb, with an
increase in the Cr_2O_3/R_2O_3 ratio in the system from 4% up to 52%
(MacGregor, 1970). Hence the coexistence of pyrope and spinel peridotites, containing a chrome-spinel, is possible through a wide
range of pressures in the mantle. It must be stressed that an increase
in the role of Cr_2O_3 in natural deep-seated associations is not connected with an increase in its amount in the system, but depends on a
decrease in Al_2O_3, which is exemplified by the relationship between
the composition of the garnets and the rocks in the pyrope peridotites
of the Czech Massif and the kimberlites of Yakutia (Fiala, 1965; Sobolev and Sobolev, 1967; Sobolev, 1970).

The effect of the amount of CaO in the system on the position of
the boundary of the garnetization reaction has also been demonstrated
experimentally (MacGregor, 1965$\underline{a}$, 1970). Increase in pressure up to
that required to achieve the reaction, takes place only in the case of
the wehrlitic paragenesis (without enstatite). In this paragenesis
(diopside-forsterite), the field of garnet compositions significantly
expands (increase in the amount of Ca-component up to 45%) with in-

TABLE 20. Chemical Composition of Deep-Seated Xenoliths from Kimberlite Pipes

| Oxides | 'Udachnaya' Pipe Lherzolites | | | | | | 'Obnazhënnaya' Pipe Webstertes | | | | | Eclogites | | Corundum eclogite | 'Newland' Pipe | Diamond-bearing eclogites | | 'Mir' Pipe | |
	Ud-3	Ud-5	Ud-18	Ud-101	Ud-4	Ud-92	O-466cв	O-466г	O-465	O-467	O-172	O-166	O-160a	O-1606	84661	FP-1508	M-180	M-180a	M-52
SiO_2	43,36	43,46	42,68	44,34	42,32	43,16	46,64	48,57	47,87	49,14	47,30	44,51	45,09	40,26	43,84	43,26	46,64	49,06	46,27
TiO_2	0,04	0,18	0,29	0,18	0,19	0,13	0,41	0,50	0,40	0,24	0,30	0,15	0,06	0,18	0,76	0,88	0,10	0,80	0,75
Al_2O_3	2,33	3,04	0,73	1,77	1,19	1,27	13,08	10,73	10,10	12,70	11,15	15,43	19,20	29,43	14,54	12,54	15,21	11,90	11,13
Cr_2O_3	0,37	0,89	0,30	0,58	0,18	0,21	0,23	1,00	1,10	—	0,37	—	0,02	0,04	0,07	0,08	0,04	0,07	0,10
Fe_2O_3	1,55	1,54	6,93	1,81	2,01	5,70	1,28	0,63	1,20	2,14	1,73	1,27	0,80	0,23	3,72	3,37	3,87	3,65	4,44
FeO	5,62	6,00	2,74	6,31	6,05	2,97	4,90	3,32	3,61	2,66	5,53	7,86	4,99	2,50	10,77	9,05	8,73	7,17	5,49
MnO	0,08	0,12	0,12	0,11	0,10	0,11	0,17	0,11	0,14	0,06	0,20	0,15	0,11	0,06	0,27	0,25	0,19	0,28	0,13
MgO	43,48	42,18	43,74	42,56	45,88	45,12	22,57	20,86	25,31	21,72	23,30	14,68	13,94	9,48	14,65	16,67	11,25	11,49	13,66
CaO	1,47	2,11	0,91	1,09	1,29	0,99	7,71	10,72	7,95	7,94	6,93	11,96	12,54	14,02	8,19	9,12	8,70	8,92	11,02
Na_2O	—	0,31	—	—	—	0,17	0,94	1,89	0,81	0,95	1,20	1,46	1,71	1,63	1,17	1,33	1,92	3,87	2,67
K_2O	—	0,14	—	—	—	0,16	0,18	0,12	0,08	0,15	0,14	0,08	0,07	0,18	0,96	0,74	1,56	1,00	0,28
H_2O^+	—	—	—	—	—	—	—	—	—	—	—	—	—	—	0,75	2,75	2,30	1,55	3,21
П. п. п.	—	—	—	—	—	—	1,31	1,42	1,32	1,45	1,89	1,90	1,37	1,55	0,64	0,36	—	0,33	1,05
total	98,30	99,97	98,44	98,75	99,21	99,99	99,42	99,87	99,89	99,15	100,04	99,45	99,90	99,61	100,33	100,40	100,51	100,09	100,20

Oxides	'Mir' Pipe — Diamond-bearing eclogites			'Zagadachnaya' Pipe — Kyanite eclogites					'Zagadachnaya' Pipe — Grospydites									Kyanite pyroxemite
	M-33	M-5	M-45	Zg-15	Zg-24	Zg-48	Zg-52	Zg-17	Zg-37	Zg-28	Zg-8	Zg-11	Zg-33	Zg-7	Zg-56	Zg-3	1	2	Zg-35
SiO_2	44,19	49,43	45,28	42,31	41,86	43,46	42,91	42,46	44,32	44,48	45,31	42,83	41,99	42,00	42,75	44,45	44,96	44,18	44,48
TiO_2	0,49	0,33	0,38	Сл.	0,40	0,35	Сл.	0,57	Сл.	Сл.	Сл.	Сл.	Сл.	Сл.	Сл.	Сл.	0,35	0,32	Сл.
Al_2O_3	16,27	15,00	16,15	22,89	21,19	24,11	26,27	27,08	27,79	28,69	28,70	28,85	29,10	31,04	31,40	21,93	23,20	24,07	29,70
Cr_2O_3	0,25	0,07	0,05	0,02	0,02	0,02	0,02	0,01	0,02	0,01	0,01	0,01	0,01	0,01	0,01	0,05	0,01	0,01	0,01
Fe_2O_3	3,30	2,06	2,98	1,72	1,92	1,99	1,59	1,48	0,95	1,03	0,95	1,09	0,87	1,40	0,95	1,42	3,14	4,84	1,41
FeO	7,06	4,20	8,93	4,81	7,77	4,82	4,13	5,36	3,66	2,23	2,55	3,07	2,22	2,49	2,19	4,18	3,30	1,76	1,29
MnO	0,31	0,07	0,21	0,16	0,29	0,17	0,12	0,15	0,11	0,08	0,08	0,10	0,10	0,07	0,10	0,15	0,05	0,03	0,06
MgO	13,96	10,93	15,97	13,24	11,38	9,05	10,53	8,87	8,06	7,05	5,71	6,94	8,33	9,54	7,12	9,97	6,42	6,11	5,15
CaO	10,93	11,54	7,54	7,54	9,67	10,55	10,72	7,30	11,46	11,61	13,17	13,36	11,22	7,98	8,94	13,14	15,47	15,06	12,29
Na_2O	1,61	3,86	1,63	1,81	1,88	1,87	1,47	1,93	1,65	1,52	2,37	1,38	0,79	1,22	1,66	1,81	1,71	1,22	1,44
K_2O	0,32	0,41	0,16	0,66	0,36	0,53	0,47	0,54	0,69	0,79	0,47	0,27	1,22	0,30	1,26	0,45	0,26	0,85	1,04
H_2O^+	1,61	1,78	0,95	—	—	—	—	—	—	—	—	—	—	—	—	—	—	—	—
P_2O_5	—	—	—	0,15	0,15	0,11	0,06	0,16	0,07	0,09	0,09	0,07	0,06	0,12	0,05	0,09	0,11	0,14	0,05
ign. loss	—	—	—	4,23	3,03	2,65	1,87	3,52	1,88	2,33	0,86	2,03	3,62	3,77	3,75	2,09	1,04	1,42	3,11
Total	100,30	99,68	100,23	99,54	99,92	99,68	100,16	99,43	100,66	99,91	100,27	100,00	99,53	99,94	100,18	99,73	100,02	100,01	100,03

Note. Analyses 1 and 2, after Bobrievich, Smirnov & Sobolev (1960). Analyses recalculated to dry weight.

TABLE 21. Average Amounts (above line) and Average Standard Deviations (below line) for Rock-Forming Oxides in Lherzolites and Eclogites of Kimberlite Pipes

Rock types	N	SiO₂	TiO₂	Al₂O₃	Fe₂O₃	Cr₂O₃	FeO	MnO	MgO	CaO	Na₂O	K₂O
Pyrope lherzolites of 'Udachnaya' pipe	6	$\frac{43,22}{0,48}$	$\frac{0,17}{0,07}$	$\frac{1,72}{0,78}$	—	$\frac{0,42}{0,25}$	$\frac{7,89*}{0,54}$	$\frac{0,11}{0,02}$	$\frac{43,83}{1,20}$	$\frac{1,31}{0,41}$	0,24(2)	0,15(2)
Pyrope lherzolites	15	$\frac{46,13}{0,45}$	$\frac{0,12}{0,10}$	$\frac{1,89}{0,75}$	—	$\frac{0,35}{0,09}$	$\frac{6,78}{0,77}$	$\frac{0,11}{0,02}$	$\frac{41,80}{1,83}$	$\frac{1,61}{0,68}$	$\frac{0,17}{0,07}$	$\frac{0,14}{0,09}$
Eclogites	21	$\frac{46,13}{2,12}$	$\frac{0,60}{0,29}$	$\frac{14,34}{2,80}$	$\frac{4,49}{2,11}$	$\frac{0,06}{0,07}$	$\frac{7,15}{1,40}$	$\frac{0,20}{0,15}$	$\frac{13,54}{2,03}$	$\frac{9,60}{1,97}$	$\frac{1,28}{0,86}$	0,49
Diamond-bearing eclogites	8	$\frac{46,00}{2,20}$	$\frac{0,56}{0,26}$	$\frac{14,08}{1,90}$	$\frac{3,42}{0,67}$	$\frac{0,09}{0,06}$	$\frac{7,68}{1,67}$	$\frac{0,21}{0,08}$	$\frac{13,57}{2,27}$	$\frac{9,50}{1,50}$	$\frac{2,26}{1,01}$	$\frac{0,68}{0,44}$
Kyanite eclogites of 'Zagadochnaya' pipe	5	$\frac{42,61}{1,27}$	—	$\frac{24,31}{2,38}$	$\frac{1,74}{0,22}$	—	$\frac{5,38}{1,25}$	—	$\frac{10,61}{1,66}$	$\frac{9,16}{1,20}$	$\frac{1,79}{0,19}$	$\frac{0,51}{0,11}$
Grospydites of 'Zagadochnaya' pipe	10	$\frac{43,73}{1,74}$	—	$\frac{27,48}{2,96}$	$\frac{1,67}{1,22}$	—	$\frac{2,77}{0,70}$	—	$\frac{7,53}{1,12}$	$\frac{12,14}{2,25}$	$\frac{1,53}{0,41}$	$\frac{0,66}{0,35}$

Note. All iron in form of FeO.

crease in pressure up to 45 kb and temperatures up to 1300°C. In other words, the definitely established wehrlitic parageneses (even without diamond) may be assigned to the field of greatest pressures in the present facies of pyrope peridotites and eclogites with graphite.

The structural-textural features enable us to distinguish clearly two types of pyrope peridotites, in spite of the similarity in paragenesis: those with a complex history of crystallization (reaction of garnetization in the solid phase), and porphyritic types, crystallized directly from the melt in the stability field of pyrope (see §1). In the rocks of the second type, often very rich in olivine, there are reaction rims around the garnet. The composition of these rims, based on examples from Austria and the Czech Massif, had already been studied in detail earlier on (Mrha, 1900), and it was shown that these rims are formed as a result of the reaction, pyrope + olivine = spinel + pyroxene.

During the examination of a fresh rim around pyropes from the peridotites of Krzemze in the Czech Massif (Sobolev and Lodochnikova, 1962), along with chrome-spinel, enstatite, and diopside in the rim, we found a magnesian amphibole and identified the constant presence of a selvedge of enstatite around the kelyphitized garnet. It has been established that the composition of the rim corresponds to the sum of the garnet and olivine, and it is assumed that the above reaction has been achieved as a result of melting at the boundary between the garnet and olivine grains during a decrease in pressure. These data, and also the results of later investigations (Fiala, 1965, 1966; Mikhailov and Rovsha, 1965) have proved the reversible nature of the garnetization reaction. Subsequently, analogous patterns have also been recorded on the basis of an investigation of the kelyphitic rims of garnets from kimberlite pipes (Ilupin et al., 1969).

In spite of the numerous similarities in the pyrope peridotites known in place to the xenoliths of similar rocks from the kimberlites, there are a number of differences that indicate the lower-temperature nature of the former. Distinguishing features of the minerals of the peridotites from the metamorphic assemblage are as follows: the presence of zonation in the garnets based on the iron index with increase in the peripheral zones (Fediukova's data); the rarity of dissociation textures in the pyroxenes, together with an increase in the value of the Ca/Ca + Mg ratio in the clinopyroxenes; the nature of the distribution of Ca, Mg, and Al in the coexisting pyroxenes (O'Hara, 1967); and the complete absence of such rocks as garnet harzburgites (O'Hara, 1970).

It must be emphasized that in the xenoliths of pyrope peridotites from the kimberlites, dissociation textures of pyroxenes are absent only in the rocks with a low Ca/Ca + Mg ratio in the clinopyroxene.

Thus, one of the principal arguments of the supporters for a shallow-depth, crustal origin for all the xenoliths of pyrope peridotites, based on the complete similarity of the pyrope peridotites of the pipes to those of the metamorphic complexes (Davidson, 1967; Trofimov, 1966), cannot be regarded as conclusive.

The compositions of the natural eclogites from the kimberlite pipes vary quite widely in the amounts of FeO and Na_2O, which makes it quite difficult to make an experimental model for them, comparatively with the peridotites. However, the basalt-eclogite transition has been studied experimentally in natural samples of basalts and eclogites of

78

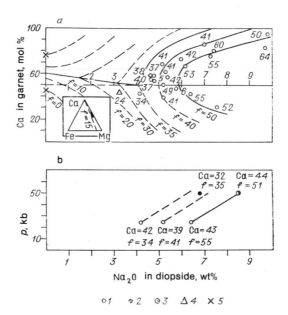

Fig. 13. Connection between calcic index of garnets and amount of Na$_2$O in associated pyroxenes from grospydites and kyanite eclogites from the 'Zagadochnaya' pipe (1), from a diamond-bearing kyanite eclogite from the 'Roberts Victor' pipe (2), from a kyanite eclogite from the 'Roberts Victor' pipe (3), and from a corundum eclogite from the 'Obnazhënnaya' pipe (4), and experimental points of F. R. Boyd (5) (T = 1200°C and P = 30 kb for a). Ca-calcic index of garnet (for b).

complex composition. Whereas the first experiments (Yoder and Tilley, 1962) only revealed the transition field, later investigations of the most varied compositions (Green and Ringwood, 1967) demonstrated the dependence of the value of the transition interval on the composition, mainly on the amount of SiO$_2$. The transition interval increases with increase in the amount of aluminum and alkalies.

The position of the garnetization-reaction line for the ultramafic rocks, and also the disappearance of plagioclase for the basic rocks, define the lower boundary of the facies based on pressure. The upper boundary is defined by the graphite-diamond transition (Leipunsky, 1939). Thus, the range of pressures in the facies (e.g. for $\underline{T}$ = 1200°C) is extremely broad (see Fig. 50), from 20 to about 50 kb.

There is significant interest in defining the position of the grospydites and associated kyanite eclogites within the framework of the present facies. The discovery of a continuous series of pyrope-grossular garnets with addition of almandine for the first time under natural conditions in one paragenesis, has suggested several reasons for this feature (Sobolev et al., 1968). The presence of pyrope-grossular garnets in these rocks may be associated, first, with changes in the conditions of crystallization of the rocks, primarily total pressure; second, with a change in the chemical potential of SiO$_2$ (O'Hara and Mercy, 1966); and finally, third, with a change in the potentials of other components.

The first assumption cannot be accepted, because xenoliths of all these rocks have been found in a single kimberlite pipe and they are

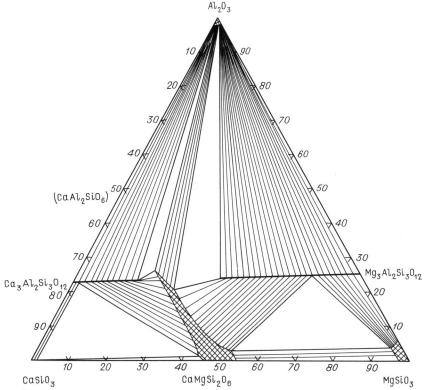

Fig. 14. Portion of the system $CaSiO_3$-$MgSiO_3$-Al_2O_3, wt%, after
Boyd (1970), with experimentally-determined compositions of garnets
and pyroxenes in bimineralic and trimineralic parageneses (T = 1200°C,
P = 30 kb).

characterized by completely similar structural-textural features.
The presence of corundum in equilibrium with other minerals in a number of samples, excludes the possibility of a connection between the
change in composition of the garnets and a change in the potential
of SiO_2 alone.

Consequently, the most likely reason for the broad series of pyrope
garnets is the increase in the chemical potential of Na_2O, which leads
to the dissociation of garnet of intermediate composition (50% grossular), with the formation of a clinopyroxene-kyanite paragenesis.
Analyses demonstrate the limits of variation in the amount of Na_2O in
the pyroxenes of these rocks from 4.5-5.0% in association with garnets
of intermediate composition (with $\underline{f}$ = 40%) up to 7-7.5% in a parage-
sis with garnets containing 30% and 70-80% grossular. In the paragenesis with garnet approaching pure grossular, the amount of Na_2O in
the pyroxene increases to 9.0%.

Thus, for garnets of identical iron index, a clear association has
been established between their calcic index and the amount of Na_2O in
the associated pyroxenes (Fig. 13). In the garnets, which contain
less than 50% calcic component, the calcic index decreases with increase in the amount of Na_2O in the pyroxene, and in the garnets of

the true grospydites (more than 50% Ca-component), it consequently
increases (Sobolev, Zyuzin and Kuznetsova, 1966; Sobolev et al., 1968).
The diagram constructed (Fig. 13), and also data on the composition of
the garnets (see Fig. 10), demonstrate that the continuity of the series of garnets is defined even for compositions with $\underline{f}$ = 30%. With
increase in the iron index of the garnets, an even greater amount of
Na_2O is necessary in the pyroxenes in order to dissociate the garnet
of intermediate composition (see Fig. 13).

Banno (1967), who has supported the correctness of the above pattern, has in his calculations shifted the corresponding point of the
curve into the field where the calcic index for the garnets is about
60%. Boyd (1970) has shown that in the non-ferriferous garnets, for
which a break in composition at $\underline{P}$ = 30 kb and $\underline{T}$ = 1200°C has been
established, the indicated boundary, determined both on the basis of
asymmetry of the break (46% and 75% Ca-component), and from the calcic index of the pyroxene, lies in the area of 60-65% calcic index
(see Fig. 13). However, with increase in the amount of Na_2O in the
pyroxenes, this value falls, approaching 50%, and does not pass above
55% for almost the entire series of compositions. Consequently, for
the actual rocks (with $\underline{f}_{gar}$ = 30%, and Na_2O_{di} = 4-5%), our earlier-
defined boundary based on Ca/Ca + Mg + Fe = 50%, cannot be retained.

The natural pyrope-grossular garnets have been regarded previously
as compositions with a break between 40% and 60% grossular (Bobrievich et al., 1960), which reflected the exceptional case of a definite value for the chemical potential of Na. In a non-sodic system,
analogous compositions have been reproduced experimentally (Boyd,
1970). The experimentally-obtained break is explained by insufficient pressure as compared with that (32.6 kb, corrected) in which
the continuous series was defined (Chinner et al., 1960). In the
interpretation of the compositions of natural garnets, the constant
presence of iron with the lowest iron index in the garnets of intermediate compositions ($\underline{f}$ about 30%), must be emphasized. Such an
amount must substantially lower the pressure necessary for the stability of the intermediate garnets, and thus, the estimate of the
pressures for the grospydites and associated kyanite eclogites of
the 'Zagadochnaya' pipe, which we gave earlier (between 20 and 30 kb)
(Sobolev, Zyurin and Kuznetsova, 1966) remains correct. The presence
of quartz in the form of inclusions in kyanite in some samples also
limits the field of stability of the grospydites on the basis of pressure (to the left of the quartz-coesite equilibrium line; see §24).

There is significant interest in the possibility of estimating the
pressure of formation of these rocks on the basis of the composition
of the chromium-bearing kyanite, in equilibrium with minerals of the
Al_2O_3-Cr_2O_3 solid solution series, based on the results of experimental
investigations (Seifert and Langer, 1970). It has been established
that in this association, the limiting amount of Cr-component in kyanite depends on pressure, and comprises 10% at $\underline{P}$ = 10 kb, and 20-30% at
$\underline{P}$ = 30 kb. Detailed investigations of the concentrate from the 'Zagadochnaya' pipe (Sobolev, Kuznetsova and Zyuzin, 1966), have led to the
establishment of a series of compositions of chromium-bearing kyanites
in paragenesis with corundum (Al_2O_3-Cr_2O_3) with the addition of up to
20% Cr-component (see §16). Taking the available data as sufficiently
representative, we may conclude that, in all probability, kyanites
would be found with a Cr-component close to the limiting value for the
present conditions, which also corresponds to the limits of possible
pressures recognized by us. It is most likely that the possible fig-

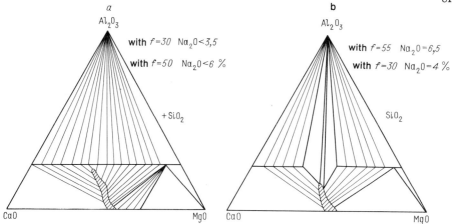

Fig. 15. Examples of paragenetic diagrams (a, b) for the system
CaO-MgO-Al$_2$O$_3$-SiO$_2$ for parageneses with corundum at P = 30 kb and
T = 1200°C, and set values for the iron index of garnet and Na$_2$O content in pyroxene.

ure for the pressure is close to 25 kb, which may be supported by the
discovery of a Cr-kyanite with an appropriate amount of Cr-component
during further investigations.

Since the effect of the amount of Na$_2$O in the pyroxene on the calcic index of the garnets at constant pressure has been conclusively
established, it must also be taken into account in constructing a
paragenetic diagram for the eclogite facies. The earlier-constructed
diagram (Bobrievich, Smirnov and Sobolev, 1960) must be regarded as
an exceptional case, corresponding to a definite amount of Na$_2$O in
the pyroxenes when defining the iron index of the garnets. An actual
variant for the system without sodium has been given by Boyd (1970)
(Fig. 14).

Figure 15a shows that the continuous series of garnets may be individual variations in the value of the iron index of the garnets and
the amount of Na$_2$O in the clinopyroxenes. The data in Figure 13 form
a factual basis. In order to tie in with the experimental results
(Boyd, 1970), we will take $\underline{T}$ = 1200°C and $\underline{P}$ = 30 kb. It must, however, be emphasized that a decreased value for the pressure (about
25 kb) is more likely for the grospydites of the 'Zagadochnaya' pipe,
since the continuous series of garnets has been identified here,
beginning with $\underline{f}$ = 30%.

Figure 15a shows that the continuous series of garnets may be
stable (at the ascribed $\underline{P}$ and $\underline{T}$) both for the iron-poor ($\underline{f}$ = 30%),
and the more iron-rich garnets ($\underline{f}$ = 50%), with simultaneous increase
in the concentration of sodium. A still greater concentration of
sodium at these same values for the iron index of the pyrope-grossular garnets leads to a break in their compositions, expressed in
Figure 15b. The value of $\underline{f}$ = 55%, as indicated by information on
the composition of the garnets of the grospydites and kyanite eclogites, is close to the limit for the 'Zagadochnaya' pipe. On this diagram, there is a trend towards a decrease in the calcic index of the
garnet in the peridotites along with an increase in the amount of Na$_2$O
in the pyroxenes.

We emphasize once again that both variants presented are accurate for the defined $\underline{P}$ and $\underline{T}$, and their difference has been controlled only by the compositional features (change in $\underline{f}$ and concentration of Na_2O). In this case, the presence of corundum is obligatory in the corresponding eclogites and grospydites.

A clear relationship has been established between the increase in Na_2O in the pyroxenes of the kyanite eclogites and the increase in pressure (Sobolev, 1970), which is well supported by the data from a diamond-bearing kyanite eclogite from the 'Roberts Victor' pipe, and in general, from the kyanite eclogites of this pipe. Such a relationship is clearly manifested in the association with garnets of identical calcic index and iron index in the 'Zagadochnaya' and 'Roberts Victor' pipes (see Fig. 13). The presence of diamond has enabled us to accept a pressure value for the kyanite eclogite from the 'Roberts Victor' pipe of about 50 kb in contrast to the value of approximately 25 kb for the xenoliths of the 'Zagadochnaya' pipe.

The pressure range of 20-30 kb, which we assumed for the grospydite subfacies of the 'Zagadochnaya' pipe (Sobolev, Zyuzin and Kuznetsova, 1966; Sobolev et al., 1968), has also been confirmed by the results of the first experiments with compositions, analogous to the grospydites (Green, 1967).

Thus, within the framework of the facies of pyrope peridotites and eclogites, we may recognize a grospydite subfacies, the field of which is limited by the pressure of the dissociation reaction of anorthite into grossular + kyanite + quartz (Hariya and Kennedy, 1966; Boettcher, 1970), and also by the quartz-coesite transition (Holm et al., 1967). In the scheme for the facies of the upper mantle (see §24), the field of the grospydite subfacies borders the field of the subfacies of garnetized peridotites.

The established features of the composition of the porphyritic peridotites of the 'Udachnaya' pipe, for which a pressure of more than 30 kb at $\underline{T}$ = $\underline{ca}$ 1200°-1300°C may be assumed on the basis of constant presence of traces of K_2O in the pyroxenes, the increased role of Cr_2O_3 in the garnets, and the subcalcic nature of the pyroxenes, enables us to assign them on the basis of pressure to the subfacies of coesite eclogites, which is transitional to the facies of the diamond-bearing rocks. The known rocks, assigned to this subfacies, are limited to ultramafic compositions, including the pyrope wehrlites, the very appearance of which (with Ca-rich garnets) is associated mainly with increase in pressure (MacGregor, 1965).

CHAPTER IV

THE FACIES OF DIAMOND-BEARING ECLOGITES AND PERIDOTITES

(the diamond-pyrope facies)

The data of metamorphic petrology convincingly indicate the presence of a maximum upper limit to pressures reached during metamorphism in the crust at 20 kb (Sobolev, 1955, 1960a, b; 1964a; Dobretsov et al., 1970). The principal indicator of this is the complete absence of coesite in natural metamorphic formations, in spite of the extremely widespread distribution of silica in the crust.

Besides the metamorphic rocks of the crust, fragments of deep-seated rocks, caught up from the upper mantle during the displacement of the kimberlites, are available for direct study. A study of the mineral associations in these rocks has permitted, with varying degrees of precision, establishment of possible limits of variation in pressure during their formation. However, the direct quartz-coesite transition has also not been identified here in diamond free assemblages, though this is associated, not with the impossibility of formation of coesite under upper mantle conditions, but with the compositional features of the associations themselves, in which free silica is practically absent. At the same time, during examination of deep-seated parageneses, the possibility exists of an objective recognition of an assemblage of minerals, which were formed under conditions of still greater pressures than have been established for coesite. The uniquely accessible group of materials for such investigations are minerals, directly associated with natural diamonds. The correct recognition of such a group of minerals enables us to understand features of their composition and paragenesis for pressures exceeding 40 kb, at which diamond is formed in its stability field. Such a viewpoint on the conditions of formation of natural diamonds, repeatedly emphasized by features of their synthesis and results of a study of the different properties of natural and synthetic diamonds, has been accepted at present by the great majority of investigators.

We may recognize three types of possible association between diamonds and other minerals: 1) minerals of xenoliths of diamond-bearing rocks (peridotites and eclogites); 2) crystalline inclusions in diamonds; and 3) intergrowths of diamonds and other minerals.

An important task in the investigation of these types of association is the identification of the syngenesis of diamond with the associated minerals and exclusion of various secondary minerals from consideration. The greatest difficulties in this respect arise during investigation of the third type of association (intergrowths), although the second type also requires a careful analysis of features that allow recognition of primary and secondary inclusions. Such features will be considered below when defining each type of association.

§7. XENOLITHS OF DIAMOND-BEARING PERIDOTITES

Xenoliths of pyrope peridotites with diamonds are amongst the rarest of fragments of diamond-bearing rocks.

The first sample of a xenolith of diamond-bearing rock found in a kimberlite pipe (Bonney, 1899), represents an eclogite, and not the most widely distributed type of xenolith (pyrope peridotite). Information on xenoliths of pyrope peridotites with diamonds is limited only to general references to the South African deposits (Williams, 1932), without the description, it seems, of a single actual occurrence of such a rock. It had even been stated that such rocks cannot exist.

The first actual discoveries of xenoliths of pyrope-bearing ultramafic rocks with diamonds, were made in Yakutia by B. S. Nai in the 'Aikhal' kimberlite pipe (Sobolev, Nai et al., 1969). In addition to two fragments of completely serpentinized rocks with pyrope and diamond (Samples A-2a and A-4a), three small fragments were found in this pipe, in which diamonds are located within separate segregations of serpentine (Samples A-1a, A-3a, and A-5a). These segregations may be regarded either as peridotite fragments, in which garnet has not been involved, or as pseudomorphs after olivine grains with inclusions of diamond.

Sample A-2a is a rounded segregation of yellow-green serpentine at the contact with kimberlite. In the serpentine, there are idiomorphic grains of a crimson garnet, which suggests that this sample is a fragment of a garnet peridotite. The garnet grains are 1-2 mm in size, without a kelyphitic rim, except for a single grain in which the rim occurs where the grain is in contact with kimberlite. Diamond has been found in the sample deeply immersed in serpentine, its visible dimensions being 0.7 x 0.3 mm. The visible dimensions of the flat octahedral facets are 0.3-0.5 mm. The diamond crystal is surrounded by a thin friable carbonate selvedge.

Sample A-4a is a fragment of a serpentinized pyrope peridotite about 0.5 cm^3 in size with a clear negative impression of a triangular facet of a flat-facetted diamond octahedron. The facet is about 0.4 mm in size. The serpentine is dark-green in color, and the garnet has a significantly more intense color than in Sample A-2a, approximating to violet, and is characterized by a clear alexandrite effect; no kelyphitic rim is present.

In Sample A-1a, a diamond, 1.4 x 1.0 mm in size, half projects from serpentinized olivine. The serpentine segragation of rounded shape, about 1.5 cm in diameter, occurs in a kimberlite fragment.

Sample A-3a is a fragment of serpentine with an octahedral chromite crystal, measuring a little more than 1 mm across. In the center of the chromite crystal is a diamond, measuring 0.7-0.8 mm. It is apparently an intergrowth of several small crystals.

Sample A-5a is a fragment of serpentinized olivine with a negative impression of a diamond measuring 3.0 x 1.5 mm, which is apparently an intergrowth of several crystals.

During further investigation of the collection of xenoliths of ultramafic rocks, collected by B. S. Nai from the 'Aikhal' pipe, we found a small diamond, about 2 mm in size, in crushed material produced from a small fragment of an ultramafic xenolith, consisting of serpentine, garnet, and chromite (Sample A-90a). None of the samples (A-2a, A-4a, and A-90a) has revealed any enstatite or chromediopside.

We paid particular attention to clarifying the composition of the garnets in Samples A-2a, A-4a, and A-90a, and the chromite in Sample A-3a. The results of garnet analyses, carried out with the aid of the microprobe, are given in Table 22. In addition, we also measured the n values for the garnets of Samples A-2a and A-4a, which were 1.757 and 1.784 respectively.

As is seen from the analytical results for the garnets from the diamond-bearing peridotites, the garnet from Sample A-2a has quite a

TABLE 22. Chemical Composition of Garnets from Peridotites
of the 'Aikhal' Pipe

Oxides	A-90a	A-2a	A-4a	A-1
	Garnets			
SiO_2	41,5	43,0	41,5	42,5
TiO_2	0,07	0,04	0,19	0,04
Al_2O_3	17,4	17,6	12,5	19,8
Cr_2O_3	8,00	8,20	14,1	4,91
FeO	7,22	6,82	7,28	6,63
MnO	0,24	0,32	0,40	0,28
MgO	19,9	22,9	18,8	22,1
CaO	5,19	1,93	6,24	3,36
Na_2O	0,02	—	—	—
Total	99,54	100,8	101,0	99,62
Si	3,021	3,051	3,047	3,033
Ti	0,004	0,004	0,013	0,004
Al	1,495	1,473	1,085	1,664
Cr	0,460	0,460	0,820	0,275
Fe^{3+}	0,041	0,066	0,095	0,061
Fe^{2+}	0,396	0,339	0,355	0,274
Mn	0,013	0,017	0,026	0,017
Mg	2,160	2,420	2,055	2,351
Ca	0,407	0,149	0,490	0,257
Na	0,002	—	—	—
Pyrope	60,9	61,3	40,5	71,0
Almandine	13,3	11,6	12,1	11,3
Spessartine	0,5	0,6	0,9	0,6
Grossular	—	—	—	—
Andradite	2,1	3,3	4,8	3,1
Ti-andradite	0,2	0,2	0,7	0,2
Uvarovite	11,4	1,6	11,2	5,4
Knorringite	11,6	21,4	29,8	8,4
f	16,8	13,9	18,0	12,4
Mg-component	72,5	82,7	70,3	79,4
Ca-component	14,7	5,1	16,7	8.7
Cr-component	23,0	23,0	41,0	13,8

Note. A-90a, A-2a, and A-4a are diamond-bearing peridotites

peculiar composition, characterized by a large amount of Cr_2O_3 and a low content of CaO, and it is completely identical to the garnets known as inclusions in diamonds. This similarity is also supplemented by the low iron index of the garnet (14.3%). The amount of CaO in the garnet from Sample A-90a is significantly larger, with almost an identical amount of Cr_2O_3, although in this garnet there is also a significant amount of knorringite (about 13%). A still larger amount of CaO is typical of the garnet from Sample A-4a, but here the amount of Cr_2O_3 is markedly increased. In any case, along with two garnets from diamonds, with about 16% Cr_2O_3 and a garnet from a kimberlite concentrate from South Africa with 17.5% Cr_2O_3 (Nixon and Hornung, 1969), the garnet from Sample A-4a is amongst the most chromium-rich from ultramafic parageneses. The total amount of Cr-component in it exceeds 40%, and knorringite amounts to 29.1%.

In spite of the increased CaO content in two of the garnets studied (A-90a and A-4a), as compared with the pyropes enclosed in diamonds, their compositions especially of the latter are distinguished from those of the chrome-rich pyropes from peridotite xenoliths, which contain a bipyroxene paragenesis (see §1) by the markedly lower content of calcic component, and their compositional points must consequently occupy a position on the paragenetic diagram for the ultramafic rocks (Bobrievich et al., 1960) below the typical peridotite point. This agrees completely with the paragenetic features of the garnets examined, in which clinopyroxene is absent.

There is considerable interest in the results of a preliminary study of the chromites. Later on, in the account of the composition of the syngenetic inclusions in the diamonds (see §9), we shall stress the clear typomorphism of the composition of the chromites, enclosed in diamonds, which are the most chromium-rich types as compared with all the chromites of terrestrial origin. This feature is also clearly defined for the chromites of Samples A-90a and A-4a, in which about 6.5% Al_2O_3 and more than 60% Cr_2O_3 have been identified. Both these chromites are associated with diamonds in peridotite xenoliths.

Thus, it may be suggested that chromite, analogous in composition to that enclosed in diamond, may serve as an indicator of the possible presence of diamond in xenoliths of pyrope-bearing rocks of ultramafic composition, and it may be used for more extensive prospecting for xenoliths of diamond-bearing peridotites, in which diamond may simply not have been caught up by chance because of the small dimensions of the xenoliths.

During 1975-1976 years new discoveries of diamond-bearing pyrope peridotites including lherzolites have been described (Dawson, Smith, 1975; Eggler, McCallum, 1975; Pokhilenko et al., 1976). A sample from Wyoming (Eggler, McCallum, 1975) containing Ca-poor chrome pyrope and chromite in an altered serpentinized matrix, relates in our opinion to harzburgitic paragenesis, and seems to be similar to the 'Aikhal' xenoliths. A xenolith from the 'Udachnaya' pipe (Pokhilenko et al., 1976) is of special interest and contains the five mineral paragenesis: garnet + olivine + enstatite + clinopyroxene + ilmenite. Pyrope contains moderate Ca with CaO = 5.4%, Cr_2O_3 = 6% and high Ti with TiO_2 = 1.9%. Ilmenite is enriched in Cr with 5.5% Cr_2O_3.

§8. XENOLITHS OF DIAMOND-BEARING ECLOGITES

Amongst the various eclogite xenoliths, discovered at different times during the study of kimberlite pipes in South Africa and Yakutia, those containing diamonds occupy a special place. A detailed study of such eclogites, along with an investigation of the syngenetic inclusions in the diamonds, may provide exceptionally useful information about the conditions of mineral formation at the highest pressures under natural conditions.

Individual consideration of such eclogites is even more important, since the presence of diamond allows us to estimate the lower limit of pressure during their formation and convincingly demonstrates their peculiar deep-seated, mantle origin (Sobolev and Kuznetsova, 1966; Sobolev, 1970).

Over a lengthy period of time, from 1899 to 1959, a single description of actual eclogite xenoliths containing diamonds concerned two samples found in 1897 in the small 'Newland' pipe in West Griqualand, 70 km from the Kimberley group of pipes (Bonney, 1899, 1900). It was noted that one of the xenoliths in the sample, measuring approximately 10 x 7.5 x 5 cm and containing on the surface 10 visible diamond crystals, is a portion (possibly a quarter or a third) of a larger xenolith, which evidently had an ellipsoidal shape.

The second sample, in which only two diamond crystals were recorded, is also a fragment, having a more elongate shape as compared with the first sample. The outward similarity of both samples has been noted in regard to the general porphyritic nature of the structure and the color of the garnet and pyroxene (Bonney, 1899). The first of the samples is preserved in the British Museum (Natural History), London (Coll. No. 84661); the second is in the Freiberg Mining Academy (Coll. No. 001508). In the original description of these samples, unfortunately, there is no information about the minerals comprising the eclogites.

Although no single description of a new discovery of such eclogites is known from the South African deposits up until recently, individual references to diamond-bearing eclogites in various pipes have been made by various authors. An analysis of the literature on the South African kimberlites (Rickwood et al., 1969) has shown that individual references to discoveries of such eclogites are available for the 'Crown', 'Jagersfontein', 'Roberts Victor', and 'Premier' pipes, and for the 'Newland' pipe, to a further discovery of a diamond-bearing eclogite, containing graphite (Wagner, 1909). However, except for the above-noted first description of diamond-bearing xenoliths, no further account of such samples is known until quite recently.

The first detailed description of an eclogite with diamond, together with an analysis of the garnet, has been given for a sample from the 'Mir' pipe, found immediately after the discovery of this kimberlite occurrence (Bobrievich et al., 1959b), and the special position of such eclogites among the deep-seated inclusions of kimberlite pipes has been indicated from the ferruginous composition of the garnet.

Intense prospecting for diamond-bearing eclogites in the workings of the 'Mir' pipe has led to discoveries of new samples, mainly of small size, not exceeding 20-30 mm in length. Fifteen such samples have been examined in varying degrees of detail, depending on the amount of available material (Sobolev, Botkunov et al., 1966; Botkunov and Kuznetsova, 1969; Sobolev, Botkunov, Lavrent'ev and Pospelova, 1971). At the same time, an additional sample, found in 1959 in the 'Mir' pipe, has been studied, and the composition of the minerals of xenoliths from two samples from the 'Newland' pipe has

been investigated. A study of the composition of the garnet and
pyroxene in these samples, material of which was kindly loaned by
Dr. G. Claringbull (London) and Professor H. Rosler (Freiberg),
showed the complete identity of their compositions (Sobolev and
Kuznetsova, 1966), later confirmed with respect to the garnet by
repeated examination of its physical properties by Rickwood and
Mathias (1970). The discovery of a large eclogite fragment with
diamonds, measuring 15 x 10 x 6 cm in the 'Mir' pipe, has been described by Mikheenko et al., 1970.

In the case of the South African kimberlites, individual references to eclogites with diamonds are contained, as noted above, in
the reports on the geology of diamonds (Corstorphine, 1908; Wagner,
1909; Williams, 1932; Holmes and Paneth, 1936). After a lengthy
hiatus, new information appeared in 1967 about new discoveries of
diamond-bearing eclogites in various South African pipes. In the
first of these (Davidson, 1967), devoted to a review of three
occurrences of diamond-bearing eclogites, known previously from the
'Newland' and 'Mir' pipes, the discovery of a kyanite eclogite with
diamonds in the South African kimberlites was noted.

Subsequently, a considerable number of xenoliths of diamond-
bearing eclogites have been found in the South African kimberlites,
often of large dimensions, and for the first time, diamond-bearing
kyanite eclogites and a corundum eclogite were identified (Dawson,
1968; Gurney et al., 1969; Rickwood and Mathias, 1970). The greatest
number of discoveries were made at the 'Roberts Victor' pipe, the
working of which after a lengthy recess, has only recently been
renewed. In this pipe, in addition to several discoveries of bimineralic, garnet-pyroxene, diamond-bearing eclogites, graphite is present
in one of them, and a kyanite eclogite with diamonds has also been found
(Switzer and Melson, 1969). The diamond occurrence in the eclogite
from this pipe has also been referred to by MacGregor (MacGregor and
Carter, 1970), who pointed out that eclogite xenoliths in general
markedly predominate in number over peridotite types (over 90%) in
the 'Roberts Victor' pipe.

Thus, the total number of accurately confirmed xenoliths of
diamond-bearing eclogites has risen by almost an order of magnitude
as compared with the 1967 data (Davidson, 1967). For the 'Mir' pipe
alone, the latest discoveries make a total of 18, and here a xenolith
of diamond-bearing eclogite has been found, which contains corundum
and graphite (Sobolev, Botkunov and Lavrent'ev, in press). In the
South African pipes, more than 10 occurrences in toto are now known
(Rickwood and Mathias, 1970). A series of diamond-bearing kyanite
eclogites and even grospydites has been described in the 'Udachnaya'
pipe, Yakutia (Pokhilenko et al., 1976).

The diamond-bearing eclogites are not distinctive in texture from
most of the eclogite xenoliths in the pipes. They are coarse-grained
rocks, often with porphyritic aspect, with an idiomorphic garnet which,
judging by some of the larger samples, makes up more than 50% of the
rock. In individual samples, the pyroxene occupies a markedly minor
position (Mikheenko et al., 1970), which is expressed in the chemical
composition of the sample.

A study of the distribution of diamonds in eclogites, with respect
to the garnet and pyroxene, is of importance in explaining the features
of the conditions of formation and growth of the diamond crystals,
since the diamond-bearing xenoliths are unique formations, in which
the diamond has not undergone displacement after its crystallization.

From this viewpoint, there is particular interest in a sample found
in 1959 in the 'Mir' pipe (Bobrievich et al., 1959). This xenolith,
which was originally quite large, has been crushed during the process
of extraction from the kimberlite into a large number of fragments.
At present, its major portion is preserved in the form of numerous
fragments of different size in the collection (No. 180) of the
Amakinsk Expedition of the Yukutian Territorial Geological Survey
(YaTGU) at Nyurba in the Yakutian ASSR. Each of the fragments contains one or more diamonds, including a polycrystalline intergrowth,
consisting of approximately 80 individuals (Bartoshinsky, 1960).
This sample is a good example of the arbitrary arrangement of the
diamonds in any part of a xenolith, and during the crushing of small
eclogite fragments, with the object of selecting minerals for
analysis, many new diamonds were revealed in them. Such discoveries,
in particular, were made by ourselves.

It should be noted that some authors (Vasil'ev et al., 1968;
Mikheenko et al., 1970) have used individual described cases of the
occurrences of diamonds on the outer part of eclogite xenoliths as
a basis for assuming the later formation of the diamonds in eclogites,
with respect to that of garnet and pyroxene. It has been stated that
' . . . in all the eclogite xenoliths known from descriptions, the
diamonds occur only on their spheroidal surface', and that ' . . . there
is not a single crystal that is a confirmed endogenic inclusion in
eclogite' (Mikheenko et al., 1970). In this respect, we emphasize
that, although the position of the diamonds was not indicated in
detail for the diamond-bearing eclogite from the 'Mir' pipe (Bobrievich et al., 1959), it is possible even from the illustrations that
accompany this description, to be convinced that the diamonds do not
occur on the outer part of the sample alone. It is sufficient to
refer to Figures 1 and 2 (Bobrievich et al., 1959, p. 616), in which
the diamond in the eclogite and the negative impression of this
same diamond are shown, that is, a readily direct proof of the
position of an actual diamond crystal within the sample.

In some large xenoliths, which have not been subjected to crushing,
diamonds have in fact been found only on the outer surface, for
example, ten diamonds in a sample from the 'Newland' pipe (Bonney,
1899), but this does not necessarily indicate the absence of diamonds
within the sample. Here, we must allow for the extremely irregular
distribution of diamonds in xenoliths, using these same samples from
the 'Newland' pipe as an example; these samples, with almost identical dimensions and identical mineral composition respectively contain
10 and 2 visible diamonds. Small xenoliths from the 'Mir' pipe,
10-30 mm in size, resembling pebbles, also contain different numbers
of visible diamonds: from 1 (in most cases) up to 6 (Sobolev,
Botkunov and Lavrent'ev, 1969). In one of the xenoliths (Sample
M-59), the cross-section of the diamond on the surface reaches 8 mm,
which indicates that with appropriate reconstruction of the crystal,
its possible weight should be 4-5 carats.

A comparatively large diamond crystal (about 6 mm across) has
been found by N. V. Sobolev and A. I. Botkunov inside a xenolith of
diamond-bearing corundum eclogite from the 'Mir' pipe (Sample M-75).
The negative impression of the octahedral facet and one of the
apices of this crystal are shown in Photo 7, which clearly indicates
its position below the surface of the xenolith.

Preliminary investigations of new discoveries of diamond-bearing
eclogites in the kimberlites of South Africa (Gurney et al., 1969)

has enabled the identification of a multitude of small diamonds in a large xenolith, about 10 cm in diameter, extracted from the 'Roberts Victor' pipe, and also a number of features concerning the distribution of diamonds in eclogites. In this respect, an extract from the work indicated is of interest: ' . . . the authors of this paper have seen several samples of diamondiferous eclogite. In one, the diamond was present as an inclusion in a garnet. In a second, the diamond was near the center of a nodule approximately 10 inches long and 6 inches wide. Recently, a nodule was broken open at the Roberts Victor Mine and a 5 3/4 carat perfect octahedral diamond was disclosed' (Gurney et al., 1969, p. 354).

Such variation in the content of diamonds in the known eclogite xenoliths, both in the size of the crystals (from 0.01 up to 5.75 carats) and in quantity, is undoubtedly associated with the uneven distribution of carbon in the diamond-forming environment. It is perfectly natural to conclude that eclogites occur, which do not contain diamonds, but which have been formed in the field of diamond stability. Such xenoliths may be found especially in those pipes where diamond-bearing eclogites have already been discovered. Therefore, a detailed study of the composition of the garnet and pyroxene in all the known occurrences of diamond-bearing eclogites is necessary in order to recognize the stable features of their differences from eclogites fromed at relatively lower pressures.

The compositions of the diamond-bearing eclogites, as shown by us earlier (Sobolev, 1968a, 1970), are practically indistinguishable from those of the eclogite xenoliths which do not contain diamonds. On the composite diagram, which reflects the features of the chemistry of the deep-seated xenoliths (see Fig. 10), their compositional points cannot be separated from those of the other eclogites. The same opinion is obtained from a comparison of the average compositions (see Table 21), which are insignificantly distinguished only by the amount of Na_2O. These figures have been confirmed, insofar as in selecting eclogite xenoliths without diamonds, we excluded compositions assigned to garnet-pyroxene associations, which contain the most magnesian garnets, and to those most widely distributed in the pipes, the garnet-pyroxenite type. The preliminary data demonstrate the extremely narrow limits of variation in the amount of the principal components, and in general, they correspond to the picritic basalts.

In spite of the fact that the information about the composition of the diamond-bearing eclogites has been obtained from the analysis of a very limited amount of material, and in some cases, different small portions of the same sample have been analyzed, the average figures are characterized by insignificant standard deviations in the amounts of the principal components. This indicates the comparatively constant composition of the diamond-bearing eclogites studied and the representative nature of the material used.

This conclusion does not, of course, apply to the compositions of the kyanite and corundum diamond-bearing eclogites, which by analogy with those that do not contain diamonds, must be characterized by an increased amount of Al_2O_3 and CaO. In particular, for the corundum eclogite with diamonds (Sample M-75), the amount of Al_2O_3 in the garnet and pyroxene, close to 20%, must be regarded as the lowest. Of special interest in this sample is the increased amount of Na_2O, judging by the composition of the pyroxene, which contains 10.7% Na_2O, and in every case, not less than 5%.

The principal rock-forming minerals of the diamond-bearing
eclogites, like those of other types, are a garnet of pyrope-
almandine composition, with variable amount of calcium, and a
jadeitic pyroxene. Kyanite- and corundum-bearing varieties of diamond-
bearing eclogites have been identified. Rutile is present as a constant additive. In some samples, graphite has been identified along
with the diamond.

Garnet

The garnet of the diamond-bearing eclogites is colored yellow-orange
to orange, according to the amount of almandine component present.
The results of determining the compositions of garnets from 17 xenoliths, mainly from the 'Mir' pipe (Table 23) indicate the wide
variations in their iron index and amount of calcic component present.
Comparison of the garnet compositions from bimineralic eclogites
from separate samples of kyanite (Switzer and Melson, 1969) and corundum eclogites from the 'Mir' pipe, demonstrates that, in the eclogites
without kyanite and corundum, the variations in CaO content embrace
practically the entire range of garnet compositions from the eclogites
(Sobolev, 1964a), from 9 to 38% calcic component.

Data on the physical properties of garnets from the diamond-bearing
eclogites of South Africa (Rickwood and Mathias, 1970) emphasize the
similarity between their compositions and those from the corresponding
eclogites from the 'Mir' pipe. The garnet from the diamond-bearing
eclogites from the 'Frank Smyth' pipe (South Africa) contains the
maximum amount of calcic component at approximately 45%.

As is known, the garnets of eclogites from kimberlite pipes and
from metamorphic assemblages are often of similar composition. The
overall difference in the compositions of the eclogitic garnets in
general amounts to different values for the iron index and the amount
of calcic component, determined mainly by the composition of the
medium in which the garnets have been formed (Sobolev, 1964a; Dobretsov et al., 1970), since all eclogites are formed at high pressures.
Thus, only the compositional features may explain the presence of
ferriferous garnets in the diamond-bearing eclogites (f = 55-63%),
whereas in the formal classification of all garnets from eclogites
(Coleman et al., 1965; Smulikowski, 1964), their compositions not
only pass beyond the limits of the field of garnets from deep-seated
eclogites, but are also located throughout the entire field of garnets from eclogites of the metamorphic complexes. Here, we must
once again emphasize that garnet, rich in almandine, may be stable
at high pressures, and the presence of ferriferous garnets in diamond-
bearing eclogites and diamonds, with f up to 69%, not only does not
refute the possibility of their crystallization at high pressures,
but on the contrary, supports the hypothesis of considerable differences in the compositions of the substrate of the upper mantle
(Sobolev and Kuznetsova, 1966).

In spite of the emphasized similarity between the compositions of
the garnets from the diamond-bearing eclogites and those from the
eclogites of the kimberlite pipes and even those of the metamorphic
complexes, it has been possible recently to detect a stable feature,
on the basis of which all these garnets may be distinguished. This
is the distribution of disseminated amounts of sodium in the garnets
of the eclogite associations. The method of such analysis with the
aid of the microprobe and the detailed results of an analysis of more

TABLE 23. Chemical Compositions

Oxides	M-45	F-1508	A-45/2*	M-76*	TM-8*	M-180	M-52
SiO_2	41,26	40,96	39,2	39,9	41,3	39,14	40,04
TiO_2	0,24	0,40	0,44	0,68	0,26	0,17	0,53
Al_2O_3	22,52	21,66	21,1	21,5	22,3	21,61	21,16
Cr_2O_3	0,04	0,08	0,06	0,05	0,01	Сл.	0,12
Fe_2O_3	3,15	3,90	—	—	—	2,39	3,85
FeO	12,73	16,70	22,7	19,1	18,9	20,80	17,31
MnO	0,29	0,39	0,40	0,47	0,25	0,33	0,40
MgO	16,82	11,60	8,32	11,0	12,2	9,29	10,12
CaO	3,11	4,91	5,81	5,98	6,08	6,30	7,02
Na_2O	0,13	0,11	0,2	0,2	0,13	0,18	0,15
Total	100,29	100,71	98,23	98,88	101,43	100,21	100,30
Si	2,988	3,030	3,043	3,024	3,033	2,973	2,994
Ti	0,013	0,022	0,026	0,039	0,013	0,009	0,031
Al	1,923	1,893	1,932	1,922	1,934	1,936	1,870
Cr	0,002	0,004	0,003	0,002	—	—	0,006
Fe^{3+}	0,062	0,081	0,039	0,037	0,053	0,055	0,093
Fe^{3+}	0,112	0,132	—	—	—	0,082	0,123
Fe^{2+}	0,770	1,030	1,436	1,174	1,108	1,324	1,083
Mn	0,017	0,027	0,028	0,032	0,018	0,023	0,027
Mg	1,814	1,280	0,961	1,243	1,338	1,055	1,128
Ca	0,239	0,387	0,481	0,487	0,481	0,511	0,562
Na	0,018	0,016	0,028	0,027	0,019	0,026	0,022
Pyrope	61,0	44,6	32,8	42,0	45,1	34,9	38,3
Almandine	29,7	40,6	48,9	39,6	37,4	46,5	41,0
Spessartine	0,6	0,9	1,0	1,1	0,6	0,8	0,9
Grossular	4,8	8,5	13,8	13,3	13,6	14,5	13,2
Andradite	3,1	4,1	2,0	1,9	2,6	2,8	4,7
Ti-andradite	0,7	1,1	1,3	2,0	0,7	0,5	1,0
Uvarovite	0,1	0,2	0,2	0,1	—	—	0,3
f	34,2	49,3	60,6	49,3	46,5	58,0	53,5
Ca-component	8,7	13,9	17,3	17,3	16,9	17,8	19,8
Ca/Ca + Mg	11,6	23,2	33,4	28,2	26,4	32,6	33,3
Excess Si	—	—	—	—	—	—	—

of Garnets from Diamond-Bearing Eclogites

M-46*	M-53*	M-50	Mb-1*	M-49*	M-51*	M-33	M-59*	M-821*	M-54
39,7	39,2	39,86	40,4	40,2	40,1	40,17	40,4	40,2	41,72
0,43	0,40	0,42	0,24	0,28	0,41	0,45	0,53	0,35	0,38
21,5	21,00	21,73	22,1	21,8	21,8	21,67	21,7	21,9	21,96
0,07	0,04	0,15	0,08	0,07	0,04	0,05	0,04	0,07	0,05
—	—	2,19	—	—	—	2,50	—	—	2,17
18,7	21,0	18,22	17,3	16,8	13,9	11,70	14,5	13,9	11,25
0,39	0,36	0,35	0,27	0,31	0,20	0,35	0,24	0,16	0,20
9,79	7,22	8,48	9,74	10,3	10,1	12,26	9,35	8,71	8,65
8,53	8,83	8,98	9,56	9,82	11,3	11,55	12,9	13,3	13,70
0,17	0,13	0,21	0,10	0,11	0,15	0,17	0,11	0,13	0,15
99,28	98,18	100,59	99,79	99,69	98,00	100,88	99,77	98,72	100,23
3,014	3,059	3,001	3,032	3,018	3,034	2,951	3,025	3,036	3,000
0,025	0,023	0,023	0,014	0,018	0,023	0,027	0,032	0,023	0,023
1,924	1,933	1,928	1,958	1,931	1,947	1,882	1,917	1,952	1,962
0,004	0,002	0,009	0,007	0,006	0,002	0,003	0,001	0,003	0,003
0,047	0,042	0,040	0,023	0,045	0,028	0,088	0,050	0,022	0,012
—	—	0,087	—	—	—	0,053	—	—	0,115
1,139	1,328	1,145	1,064	1,011	0,850	0,720	0,859	0,854	0,713
0,027	0,023	0,023	0,018	0,018	0,014	0,022	0,014	0,009	0,014
1,108	0,840	0,950	1,092	1,150	1,137	1,343	1,044	0,980	0,972
0,693	0,736	0,724	0,767	0,789	0,919	0,910	1,035	1,076	1,108
0,027	0,021	0,031	0,014	0,018	0,024	0,024	0,018	0,018	0,022
37,0	28,5	32,1	37,0	38,5	38,6	43,7	35,1	33,4	33,0
38,0	45,0	41,6	36,0	33,9	28,9	25,2	28,9	29,1	28,1
0,9	0,8	0,8	0,6	0,6	0,5	0,7	0,5	0,3	0,5
20,2	22,3	21,8	24,1	23,5	29,3	24,4	31,3	34,7	36,4
2,3	2,1	2,0	1,2	2,3	1,4	4,4	2,5	1,1	0,6
1,3	1,2	1,2	0,7	0,9	1,2	1,4	1,6	1,2	1,2
0,2	0,1	0,5	0,4	0,3	0,1	0,2	0,1	0,2	0,2
51,7	62,0	57,2	49,9	47,9	43,5	39,1	46,5	44,9	46,4
24,0	25,7	25,5	26,4	27,0	32,0	30,4	35,5	37,2	38,4
	46,7	43,2	41,3	40,7	44,7	40,4	49,8	52,3	53,3
	—	—	—	—	—	—	—	—	0,152*

than 100 different garnets have been presented in Chapter I. At this point, we shall only stress that, on the basis of Na_2O content, a significant difference has been established for the garnets associated with diamonds, on the one hand (0.10-0.22% Na_2O), and those from the eclogites of the metamorphic complexes, on the other (0.01-0.05% Na_2O) (Sobolev and Lavrent'ev, 1971). The garnets from the eclogite xenoliths that do not contain diamonds, may fall into the diamond-bearing group on the basis of their sodium content. Thus, in some cases, the assignment of individual eclogite xenoliths to the diamond-bearing (or more precisely diamond-pyrope) facies may also be established without diamonds being present.

Clinopyroxenes

The c l i n o p y r o x e n e s from the diamond-bearing eclogites are distinguished by several compositional features from those of other eclogites (Table 24). These features, particularly the increased amount of sodium and excess Mg + Fe^{2+}, as compared with those of the pyroxenes from normal eclogites, that is, an isomorphous amount of clinoenstatite-ferrosilite solid solution, have served as a basis for recognizing them as a separate paragenetic type (Sobolev, 1968b). Subsequent investigations of the pyroxenes from diamond-bearing eclogites have stressed the presence of these features.

We earlier recorded that a certain underestimate of SiO_2 occurs in individual analyses of clinopyroxenes from these xenoliths (Sobolev, Zyuzin and Kuznetsova, 1966), leading to under-compensation for sodium during the formation of the jadeite or aegirine components. We associated such an underestimate with the exceptionally small amount of material available for chemical analysis, and subsequently, such analyses were checked with the aid of the microprobe. Comparison of the results of the chemical and X-ray analyses emphasized their overall similarity with the exception of the difference in the amount of SiO_2.

A tendency towards a general, almost complete absence of Al^{IV} has also been emphasized, which agrees well with the theoretical views on the increased role of Al^{VI} in silicates as the pressure increases (Sobolev, 1947, 1955, 1965).

The absence of Al^{IV} in the clinopyroxenes studied eases the task of clarifying the role of Fe^{3+} in their composition, since most analyses demonstrate the high content of total iron (recalculated as FeO). The compositional points for the pyroxenes based on the amount of Al^{VI} and Na have been plotted on a diagram (Fig. 16), and their position is quite graphically expressed by the diagonal line, that is, the complete compensation of Na and Al^{VI} with the formation of the jadeite component. In half of the analyses, some deficiency of Al^{VI} is discernible, that is, an addition of Fe_2O_3 is possible. However, in some analyses a certain excess of Al is even observed. For almost all the compositions, the points of which are located below the diagonal, there is typically a practically identical $Fe^{3+}/(Fe^{3+} + Fe^{2+})$ ratio (about 30%). The chemical analytical data in some cases clearly give overestimated figures for Fe_2O_3, evidently as a result of partial oxidation of iron during preparation of material for analysis.

The new information also completely emphasizes the increased amount of Na_2O in the clinopyroxenes of the diamond-bearing eclogites within the range of 4.5 to 9.0%.

TABLE 24. Chemical Compositions of Clinopyroxenes from Diamond-Bearing Eclogites

Oxides	M-45	F-1508	A-45/2	TM-8	M-180	M-52	M-46	M-53	M-50	Mb-1	M-49	M-51	M-33	M-59	M-821	M-54
SiO_2	54,6	54,5	55,3	58,4	54,9	55,6	55,5	57,3	55,7	56,5	54,1	55,1	55,6	56,0	56,7	55,6
TiO_2	0,56	0,84	0,52	0,36	0,61	0,47	0,56	0,39	0,43	0,26	0,25	0,41	0,40	0,42	0,35	0,34
Al_2O_3	6,81	7,34	9,07	13,7	9,31	8,60	9,39	12,0	11,7	12,6	12,2	11,4	10,0	12,5	16,2	12,5
Cr_2O_3	0,05	0,14	0,07	0,00	0,08	0,05	0,06	—	0,03	0,06	0,04	0,04	0,05	0,01	0,05	0,03
FeO	4,66	6,65	6,15	4,08	7,18	5,64	4,48	4,72	4,27	2,97	2,81	4,41	3,03	2,98	1,72	2,28
MnO	0,06	0,14	0,07	0,05	0,05	0,08	0,04	0,04	0,00	0,00	0,00	0,04	0,00	0,06	0,06	0,00
MgO	12,7	10,6	8,93	6,93	8,08	9,78	8,96	7,66	7,65	7,90	7,45	7,58	9,50	8,16	5,79	8,17
CaO	13,8	13,6	12,7	9,20	11,6	13,5	12,7	11,5	11,5	11,3	12,0	11,2	13,4	12,2	9,86	12,4
Na_2O	4,76	4,76	7,29	8,99	7,00	5,84	6,82	7,62	7,54	7,12	6,85	7,47	6,00	7,43	8,99	7,03
K_2O	0,11	0,15	0,07	0,10	0,06	0,08	0,08	0,06	0,08	0,09	0,09	0,07	0,05	0,05	0,15	0,15
Total	98,10	98,71	100,24	101,81	98,85	99,66	98,59	101,33	98,94	98,81	95,81	97,72	98,3	99,78	99,89	98,50
Si	1,992	1,992	1,989	2,016	2,002	2,001	2,007	2,005	1,996	2,002	1,986	1,999	2,004	1,979	1,976	1,986
Al IV	0,008	0,008	0,011	—	—	—	—	—	0,004	—	0,014	0,001	—	0,021	0,024	0,014
Al VI	0,286	0,308	0,374	0,556	0,399	0,364	0,400	0,496	0,491	0,528	0,516	0,487	0,425	0,501	0,644	0,514
Ti	0,015	0,023	0,014	0,008	0,017	0,013	0,015	0,011	0,011	0,009	0,009	0,011	0,011	0,011	0,009	0,009
Cr	0,001	0,004	0,002	0,000	0,001	0,002	0,002	0,000	0,001	0,001	0,001	0,011	0,001	0,000	0,001	0,001
Fe^{3+}	0,044	0,015	0,069	0,039	0,080	0,028	0,065	0,010	0,035	—	—	0,030	—	—	0,050	0,069
Fe^{2+}	0,098	0,188	0,117	0,079	0,139	0,141	0,072	0,129	0,092	0,089	0,086	0,103	0,091	0,089	0,002	0,069
Mn	0,002	0,004	0,002	0,002	0,002	0,000	0,001	0,000	0,000	0,000	0,000	0,000	0,000	0,000	0,002	0,000
Mg	0,699	0,578	0,480	0,357	0,439	0,526	0,482	0,397	0,409	0,417	0,408	0,410	0,511	0,429	0,301	0,434
Ca	0,539	0,534	0,489	0,340	0,453	0,521	0,491	0,431	0,441	0,430	0,472	0,434	0,518	0,463	0,368	0,474
Na	0,337	0,336	0,510	0,611	0,496	0,407	0,478	0,517	0,525	0,490	0,490	0,523	0,420	0,510	0,607	0,485
K	0,004	0,007	0,007	0,004	0,004	0,004	0,004	0,004	0,007	0,004	0,004	0,004	0,002	0,004	0,009	0,008
Total of cations	4,016	3,997	4,064	4,002	4,032	4,009	4,017	4,001	4,012	3,970	3,986	4,003	3,983	4,007	3,991	4,063
i'	12,4	24,5	19,6	18,1	24,0	21,1	22,1	24,5	18,4	17,6	17,4	20,1	15,1	17,2	14,2	13,8
i	17,1	26,0	27,9	24,8	33,2	24,3	—	25,9	23,5	17,6	17,4	24,6	15,1	17,2	14,2	13,8
$Ca/Ca+Mg$	43,9	48,1	50,5	48,8	50,8	49,8	50,5	52,1	51,9	50,8	53,6	51,4	50,3	51,9	55,0	52,2
$Ca/Ca+Mg+Fe$	39,3	41,1	45,0	41,7	43,9	42,8	—	45,0	46,8	45,9	48,9	44,2	46,3	47,2	51,2	48,5
K_0	31	7	37	33	37	16	—	7	28	—	—	23	—	—	—	—

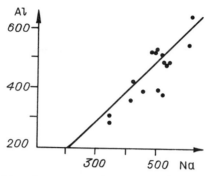

Fig. 16. Connection between amounts of Na and Al (based on 6000 atoms of oxygen in clinopyroxenes from diamond-bearing eclogites.

Thus, at present, we have available the data from 16 analyses of clinopyroxenes from bimineralic diamond-bearing eclogites, found, with the exception of the sample from the 'Newland' pipe, in the 'Mir' pipe, Yakutia (see Table 24). The information based on the possible variation in pyroxene compositions may be regarded as sufficiently representative, because the garnets associated with the pyroxenes, based on variation in composition, characterize almost all the possible variants of garnets from eclogites (see §12). Some of the results, after repeated analysis on the microprobe according to the above-noted methods, have been replaced by new ones. This applies to Samples M-33, M-52, and M-46. In these cases, when even a close coincidence in results has been obtained, we have accepted the X-ray data for the purposes of generalization.

The average composition of the clinopyroxenes from diamond-bearing eclogites (see §13) supports the earlier-recognized typomorphic features of their composition (Sobolev, 1968b), still further stressing the common feature of increased sodium content. Comparison with the average figures for pyroxene from eclogite inclusions in kimberlites (Dobretsov et al., 1970, §56) demonstrates the clear differences in the amount of sodium, total iron index, and the value of the Ca/(Ca + Mg + Fe) ratio.

It has been shown above that the role of Fe^{3+} in the clinopyroxenes of the diamond-bearing eclogites cannot be regarded as significant. In this respect, we emphasize that the average value of the iron index for the pyroxenes from diamond-bearing eclogites (see §13), in combination with the extremely large amount of sodium, markedly separates them amongst the pyroxenes of all types of eclogites, with the exception of the low-temperature types only (Dobretsov et al., 1970), amongst which they are distinguished on the basis of increased amount of Al^{VI} and the complete absence of Al^{IV}.

Along with the data published, we have the results of a study of minerals from 15 xenoliths of diamond-bearing eclogites from the South African kimberlites (Reid et al., 1973). Eleven of the 15 garnets examined, contain 20 to 40% calcic component and only 4 samples contain less than 20%. In 10 clinopyroxenes, Na_2O varies from less than 2 up to 7%, and Al_2O_3 from 2 to 18%. Whereas the composition of the garnets is generally similar, there is some decrease in the amount of sodium in the South African diamond-bearing eclogites.

A comparison of the composition of the pyroxenes in the bimin-

eralic diamond-bearing eclogites with those of the corresponding
corundum and kyanite eclogites (see §13) is of interest. Although
the compositions of the garnets of the types indicated overlap, the
pyroxenes are clearly distinguished on certain compositional features.
In first place, the pyroxenes of the kyanite and corundum diamond-
bearing eclogites show a significant decrease in the amount of iron,
which brings them close to the pyroxenes of the kyanite eclogites of
the 'Zagadochnaya' pipe (Sobolev et al., 1968) and the corundum
eclogite from the 'Obnazhĕnnaya' pipe (Sobolev and Kuznetsova, 1965).
The amount of jadeite component in the pyroxenes of the diamond-
bearing kyanite and corundum eclogites is greatly increased in
comparison with that in the pyroxenes of the corresponding non-
diamond-bearing eclogites from the pipes noted. This is especially
characteristic of the corundum eclogites from the 'Mir' pipe
(Sobolev, Botkunov and Lavrent'ev, in press), the pyroxene of which
contains 70% jadeite component along with the complete absence of
Al^{IV}.

Other Minerals

R u t i l e is constantly present in the diamond-bearing eclogites, including the corundum-bearing sample M-75. Of particular
interest is the above-noted discovery of rutile inclusions in a
number of diamonds from a diamond-bearing eclogite in the 'Roberts
Victor' pipe (Gurney et al., 1969), and rutile has been recorded
both in the diamonds and outside them.
D i a m o n d is the principal typomorphic mineral, the presence
of which may be regarded as the principal classificatory feature of
these unique eclogites. We have already discussed the distribution
of diamonds in eclogites and have demonstrated the arbitrary nature
of their arrangement within (or on the outside of) the xenoliths.
Even during the study of the first sample of a diamond-bearing
eclogite from the 'Mir' pipe (Sample M-180), it was shown that the
numerous diamonds in it have similar morphological features: they
consist of plane-facetted octahedra with trigonal growth layers
(Bartoshinsky, 1960). Examination of the diamonds from this eclogite even suggested a possible quantitative relationship between
the diamonds of this type and all the diamonds of the 'Mir' pipe,
and the possible proportion of 'eclogitic' diamonds was taken to be
extremely low (about 1.5% of the total extracted) (Bartoshinsky, 1960).
As a result of new discoveries of diamond-bearing eclogites in the
'Mir' pipe, along with emphasis on the octahedral nature of the plane-
facetted diamonds, it was concluded that there was substantially
greater variations in the fine morphologic features of these crystals
(Bartoshinsky et al., 1973), which has increased the probable
percentage of 'eclogitic' diamonds in the 'Mir' pipe. However, the
number of such diamonds in this pipe can scarcely be large, since
during the investigation of the inclusions in the diamonds, the
eclogitic association of minerals (orange garnet--light-green
clinopyroxene) clearly occupies a minor place as compared with that
of the ultramafic association (Sobolev et al., 1971a, b).
Information from a study of the detailed composition of diamond from
eclogites, especially the amount of nitrogen (E. V. Sobolev et al.,
1967), suggests that it is similar to that of other diamonds from the
kimberlites.
The diamonds from the eclogites are characterized by the most varied

dimensions. The smallest evidently include crystals (not exceeding 0.2 mm in size, and a possible weight of the order of 0.005 carats), which we found when examining Sample M-180 from the 'Mir' pipe. In this sample, there is a wide variation in the size of the diamond crystals from 0.01 up to 0.5 carats (Bartoshinsky, 1960).

The largest known diamond from the eclogites is probably that from the 'Roberts Victor' pipe, with a weight of 5.75 carats (Gurney et al., 1969), and also, after appropriate reconstruction, the diamond from Sample M-59 from the 'Mir' pipe. Two relatively large crystals, weighing about 1.5 carats each, have been identified in the corundum eclogite M-75.

In addition to single crystals in the eclogites, there are isolated occurrences of polycrystalline diamond aggregates. Such an aggregate, comprising about 80 individuals, has been found in eclogite M-180 (Bartoshinsky, 1960), which is of importance in the general inter- pretation of the conditions of formation of diamond aggregates, as compared with that of single crystals, and demonstrates the possibil- ity of equilibrium between these items, found in a single eclogite sample.

G r a p h i t e has been found in some diamond-bearing eclogites, and sometimes in large amounts. The well-known (and the first) description of graphite and diamond is that from Sample M-180 from the 'Mir' pipe (Bobrievich et al., 1960). Although features indicat- ing the later nature of the graphite with respect to the diamond (especially pseudomorphs), were lacking in this sample, the relation- ships of these minerals, even when directly associated, are vague. The carbon isotope composition of the graphite is analogous to that of the diamond from this eclogite (Kropotova and Fedorenko, 1970). In some samples, for example M-75, single-crystal flaky segregations of graphite are present within the altered pyroxene.

K y a n i t e. The only precise description of a kyanite eclogite with diamonds so far is that for a single sample, 111060 (Switzer and Melson, 1969) without a detailed account of the kyanite. References are available to a very few occurrences of kyanite- bearing eclogites with diamonds (Rickwood and Mathias, 1970).

C o r u n d u m. At present, it is possible to provide a brief description of corundum only from a sample from the 'Mir' pipe. In contrast to the corundum from the non-diamond eclogite of the 'Obnazhěnnaya' pipe, which is colored pink by the addition of Cr_2O_3, that from Sample M-75 consists of light-blue tabular crystals, which may be assigned to sapphire.

Thus, the main rock-forming minerals of the diamond-bearing eclogites (garnets and clinopyroxenes) are characterized by clear typomorphic features. These acquire special significance when con- sidering the conjugate features of the composition of the garnets and pyroxenes. The distribution of Fe and Mg and also Ca and Mg in the corresponding garnets and pyroxenes of the diamond-bearing eclogites, as compared with information on other eclogite parageneses, is discussed in §21 and §22.

The Kyanite Eclogites of the 'Roberts Victor' Pipe, South Africa

Until recently, data on the composition of the minerals from xenoliths of kyanite eclogites and grospydites have been obtained from the rocks of the 'Zagadochnaya' pipe alone (Bobrievich et al., 1960; Sobolev et al., 1966). Although occurrences of kyanite eclogites

(Williams, 1932) were known early on in some of the South African
pipes, especially the 'Roberts Victor' and 'Jagersfontein' pipes,
the first data on the composition of their garnets were obtained later
(O'Hara and Mercy, 1966; Sobolev et al., 1968).

On the basis of literature data and analogy with the composition of
the rocks from the 'Zagadochnaya' pipe, these same South African pipes
were regarded as a possible site for typical grospydites (Bobrievich
and Sobolev, 1962; Sobolev et al., 1968). In fact, with more detailed
investigations, the 'Roberts Victor' pipe has yielded a grospydite
with garnet, containing about 65% grossular, and also a whole series
of kyanite eclogites with garnets, containing from 20 up to 48% of
calcic component (12 garnets) (MacGregor and Carter, 1970).

The compositional features of the garnets from the kyanite eclogites of the 'Roberts Victor' pipe from six simples, including one
diamond-bearing type, are shown in Table 25. The variations in the
calcic index of these garnets and in their iron index correspond to
the limits of variation in the compositions of analogous garnets in
the kyanite eclogites from the 'Zagadochnaya' pipe for a calcic
component of 32.6 to 48.1%, and for an iron index of 34.7 to 58.5%.

The compositions of the pyroxenes of these xenoliths have been
much less studied than those of the garnets, owing to their intense
alteration as a result of secondary processes. The presence of rare
relicts of fresh pyroxene has made it possible to determine its
composition in three samples, including a diamond-bearing eclogite
(Switzer and Melson, 1969) and an eclogite kindly loaned to us by
Professor H. Rösler, Director of the Mineralogical Museum of the
Freiberg Mining Academy (Sample RV-15).

An important feature of the composition of these pyroxenes, which
distinguishes them from those of analogous rocks in the 'Zagadochnaya'
pipe, is the increased content of sodium, and consequently, the jadeite component, in association with garnets, completely similar in
calcic index and iron index to those of the 'Zagadochnaya' pipe.
This feature is also emphasized by the results of examining a series
of samples of kyanite eclogites (7 samples) (MacGregor and Carter,
1970). The quantitative data of these authors for the composition
of the pyroxenes, taken from their diagram, demonstrate that on
average, the pyroxenes of the 7 samples contain approximately 50%
jadeite. This agrees completely with the data obtained on the basis
of averaging the composition of the 3 pyroxenes shown in Table 25.

The data presented support the hypothesis (Sobolev, 1970) concerning
the increase in the role of sodium in the pyroxenes of the kyanite
diamond-bearing eclogites and agree with the general theoretical view
of the increase in the role of Al^{VI} in silicates as pressure increases
(Sobolev, 1947, 1965; Thompson, 1947).

The kyanite-eclogite and gropydite xenoliths of the 'Roberts
Victor' pipe are also distinguished from the corresponding rocks of
Yakutia by certain features of their mineral composition. In first
place, a diamond has been found in a kyanite eclogite (Sample 111060)
from this pipe (Switzer and Melson, 1969). A special characteristic
of the depth of these xenoliths is also the increased amount of Na_2O
in the garnets (Sobolev and Lavrent'ev, 1971), as compared with those
of the analogous rocks of Yakutia (see §12).

A peculiar feature of the kyanite eclogites of the 'Roberts Victor'
pipe is also emphasized by the identification of the phenomenon of
partial melting of these rocks with the appearance of glass and a
series of newly-formed phases (Switzer and Melson, 1969). It has been

TABLE 25. Chemical Compositions of Garnets and Pyroxenes from a Diamond-Bearing Eclogite from the 'Mir' Pipe (M-75) and from Kyanite Eclogites from the 'Roberts Victor' Pipe

Oxides	111060*	87375	RV-15	RV-1	37077	110752	M-75*	111060*	RV-15*	110752*	M-75*
	Garnets							Clinopyroxenes			
SiO_2	41,6	39,47	40,23	39,87	40,27	40,16	40,9	52,1	56,3	53,3	57,6
TiO_2	0,3	0,39	0,35	0,41	0,51	0,35	0,55	0,3	0,29	0,3	0,48
Al_2O_3	24,0	22,27	22,34	21,94	22,32	22,70	22,2	17,5	17,3	17,9	19,3
Cr_2O_3	—	—	0,03	0,05	0,06	—	—	—	0,04	—	—
Fe_2O_3	3,3	2,77	2,44	2,45	0,98	2,17	—	—	—	—	—
FeO	8,48	13,76	11,43	11,06	11,08	9,36	11,2	1,5	1,95	1,8	1,38
MnO	0,1	0,59	0,21	0,16	0,14	0,19	0,10	—	0,04	—	—
MgO	12,0	6,55	7,95	7,23	7,40	7,17	11,5	7,2	5,30	6,4	4,27
CaO	12,3	14,39	15,95	16,59	17,24	18,12	12,9	11,9	9,26	12,1	6,78
Na_2O	0,10	0,16	0,16	0,10	—	0,12	0,17	6,8	8,51	6,4	10,7
K_2O	—	—	—	—	—	—	—	0,3	0,05	0,2	—
Total	102,20	100,35	100,39	99,86	100,00	100,34	99,52	97,6	99,04	98,40	100,57
Si	2,968	2,968	2,995	2,987	3,004	2,977	3,021	1,869	1,971	1,891	1,973
Δ Al ^{IV}	—	—	—	—	—	—	—	0,131	0,029	0,109	0,027
Ti	0,017	0,023	0,020	0,023	0,029	0,022	0,031	0,009	0,008	0,009	0,012
Al ^{VI}	2,016	1,978	1,958	1,934	1,963	1,987	1,934	0,611	0,686	0,641	0,751
Cr	—	—	0,001	0,003	0,003	—	—	—	0,001	—	—
Fe^{3+}	—	—	0,021	0,040	0,005	—	0,035	—	—	—	—
Fe^{3+}	0,180	0,163	0,122	0,104	0,050	0,187	—	—	—	—	—
Fe^{2+}	0,506	0,867	0,711	0,693	0,691	0,579	0,657	0,045	0,057	0,053	0,039
Mn	0,004	0,036	0,013	0,009	0,009	0,013	0,004	—	0,002	—	—
Mg	1,278	0,732	0,805	0,810	0,823	0,793	0,264	0,386	0,275	0,339	0,218
Ca	0,939	1,161	1,274	1,340	1,378	1,439	1,020	0,457	0,347	0,460	0,249
Na	0,014	0,023	0,023	0,014	—	0,018	0,024	0,474	0,576	0,439	0,712
K	—	—	—	—	—	—	—	0,013	0,001	0,009	—
Total of cations	7,922	7,951	7,943	7,957	7,955	8,015	7,990	3,995	3,953	3,950	3,981
Pyrope	43,8	24,5	27,3	27,3	27,9	26,2	42,6	—	—	—	—
Almandine	23,5	34,6	28,3	26,8	25,1	25,3	22,1	—	—	—	—
Spessartine	0,1	1,2	0,4	0,3	0,3	0,4	0,1	—	—	—	—
Grossular	31,8	38,6	41,9	42,2	44,8	47,0	31,8	—	—	—	—
Andradite	—	—	1,1	2,0	0,3	—	1,8	—	—	—	—
Ti-andradite	0,8	1,1	1,0	1,2	1,4	1,1	1,6	—	—	—	—
Uvarovite	—	—	—	0,2	0,2	—	—	—	—	—	—
f	34,9	58,5	51,5	50,8	47,5	34,7	35,4	10,4	17,2	13,5	15,2
Ca-component Ca/Ca + Mg for pyroxenes	32,6	39,7	44,0	45,6	46,7	48,1	35,2	51,5	51,1	54,0	49,2

Note. Samples 111060, 85375, and 110752 after G. Switzer. Sample 37077, after O'Hara & Mercy (1966). For Sample 37077, SiO_2 has been determined by difference from 100%. Sample 111060 is diamond-bearing.

assumed that the most probable reason for such melting is the rapid lowering of pressure at constant temperature, that is, through the rapid displacement of the xenoliths from great depths.

The differences between the mineral composition of the kyanite eclogites of the 'Roberts Victor' pipe and those of the eclogite and grospydites of the 'Zagadochnaya' pipe, permit us confidently to assign the former to a significantly more deep-seated association, and the studied xenoliths from the 'Roberts Victor' pipe, to the facies of diamond-bearing rocks.

The special depth of some part of these rocks suggests the possibility of finding diamond-bearing grospydites in the 'Roberts Victor' pipe, and also inclusions of grossular garnets, and possibly, kyanite in the diamonds from this pipe.

§9. CRYSTALLINE INCLUSIONS IN DIAMONDS

An investigation of the mineral-captives, enclosed by diamond during growth, may provide valuable information not only about the composition of the medium in which natural diamonds were formed, but also the general features of mineral-formation at $\underline{PT}$ conditions appropriate to the stability field of the diamond. Mineral inclusions have been visually recorded comparatively early on in diamonds (Bauer and Spencer, 1904; Sutton, 1921; Williams, 1932), but their precise unequivocal diagnosis became possible only with the use of X-ray methods. The first result was the identification of the orientation pattern of the diamond and of an olivine enclosed in it (Mitchell and Giardini, 1953). At almost the same time, extensive investigations began into diagnosing solid inclusions in the diamonds from Yakutia and the Urals (Futergendler, 1956, 1958, 1960; Orlov, 1959; Gnevushev and Nikolaeva, 1958, 1961; Futergendler and Frank-Kamenetsky, 1961; Yefimova, 1961), as a result of which the principal inclusions in diamonds could be separated into syngenetic and epigenetic types.

The data obtained were later supplemented by investigations of Harris (1968), and up till recently, the syngenetic inclusions have included olivine, chrome-spinel, garnet, enstatite, clinopyroxenes, coesite, magnetite, ilmenite, rutile, and diamond itself.

The most reliable basis for assigning minerals to the syngenetic group at present is the similarity to the minerals of the diamond-bearing eclogites (peridotites (pyrope, olivine, and chromite) and eclogites (garnet, omphacite, rutile, and finally, diamond itself)).

On the basis of X-ray data, the epigenetic inclusions include 11 mineral species: calcite, goethite, graphite, hematite, kaolinite, muscovite, pentlandite, pyrrhotite, quartz, sellaite (MgF_2), and xenotime (Harris, 1968).

We have been occupied with an investigation of the exclusively syngenetic inclusions, and data on epigenetic inclusions have been collected only in passing.

The features of contemporaneity of growth and paragenetic association between mineral inclusions and diamond itself are as follows: 1) morphologic relations between the inclusions and diamond; 2) regular connection between orientation of many mineral inclusions and that of the diamond host; 3) absence of fragments of crystals amongst the inclusions in diamonds, in spite of their extremely widespread distribution in the kimberlites; 4) almost complete absence of secondary alterations of minerals enclosed in diamond, with ex-

ceptionally widespread development of such alterations in the minerals
of the kimberlites; 5) identical composition of different grains of
a particular mineral in one diamond with extremely widespread variations in the composition of inclusions from different diamond
crystals; 6) presence of regular paragenetic relationships between
inclusions of different minerals in one diamond; and 7) regular
differences between minerals and parageneses enclosed in diamond,
from those of lower-pressure facies.

1) As correctly shown in the works of Grigor'ev (1961), Shafranovsky (1960, 1961), and other investigators, important direct
indicators of the contemporaneity of the growth of the crystals
are the induction surfaces of combined growth. A study of such
surfaces in inclusion in diamonds is subject to great difficulties,
requires special methods of observation, and is at present in a
preliminary stage.

Up till now, no one has described typical induction surfaces of
the combined growth of an inclusion and diamond. It has also not
been possible so far to find accurate data during our observations,
even during a goniometric examination of some of the inclusions
(Sobolev, Bartoshinsky et al., 1970).

As indirect indicators of combined growth of inclusion and diamond,
it is evidently possible to consider the so-far little investigated
cases of distortion of the shape of the crystals of inclusions and
even the clear bending of the facets. One of the first such
distortions described was the case of an unusual elongation of
olivine: along [101], and not along [001] (Hartman, 1954).

Unusual elongated forms have also been found in the garnets. Thus,
Orlov (1959) has figured a garnet crystal from a diamond, markedly
elongated along one of the axes L_3. Elongated grains of pyrope have
also been found in the diamonds that we have studied (Photo 8).

We must especially note an inclusion of pyrope of tabular habit
in one of the Yakutian diamonds (Photo 9). In the photograph, the
inclusion is shown in two dimensions, and it is clearly seen that the
length of the oval platelet is approximately 20 times greater than
its thickness. It is possible that the numerous facets are induction-
al in type. It is not surprising that this very inclusion displays
the rarest type of orientation of garnet and diamond, namely, the
unidimensional coincidence of the L_4 axis of garnet and diamond
(Zyuzin, 1969). Another example of the tabular habit of the pyropes,
enclosed in diamond, has been identified in one of the samples (AV-16)
from the 'Mir' pipe. In Photo 10a, it is seen that the tabular
individuals of garnet have been oriented in the diamond in two
different directions (two individuals are shown in the photograph
in the form of narrow bands). The orientation of the garnets in this
sample has not been checked by X-ray methods, but it is likely that
the inclusions of garnet here have an epitaxic nature. It is
possible that the garnet grew in the cleavage direction of the
octahedron. Another almost analogous case of arrangement of tabular
inclusions of garnet parallel to the octahedral faces is shown in
this same photograph.

The rarest crystallographic shapes ever identified in garnet include
octahedra (Goldschmidt, 1897). In the pyrope—almandine garnets, such
a shape was first recorded in an inclusion of pyrope rich in calcium
and chromium from a diamond in the 'Udachnaya' pipe (Sample 57/9)
(Sobolev, Bartoshinsky et al., 1970). In addition, the garnet
crystal is elongated along one of the L_2 axes.

For both inclusions of alivine and garnet, rounded facets have been recorded in addition to the smooth and plane types. In individual cases, a tendency has been noted towards some distortion of shape of the garnet crystals (see Photo 8). Truncated apices and ribs are typical of many crystal-inclusions, which may be associated with features of their combined growth with diamond.

We shall dwell on the characteristics of an unusual morphologic type of inclusion, the aspect of which is the shape of the negative crystal of diamond. This type was first identified in the diamonds of the 'Mir' pipe (Sobolev, Botkunov et al., 1972). Under macroscopic observation, the facets of many crystalline inclusions are seen through the octahedral facets of the diamond in the form of triangles, strictly parallel to each diamond facet, that is, they are the facets of the same octahedron within the diamond. They are often complicated by the narrow facets of the trisoctahedron and less frequently, the tetrahexahedron, and are overlain by lineation, typical of the surface of the diamond itself. These forms have been identified for the chrome-bearing pyropes, orange pyrope-almandine garnets, chrome-diopside, olivine, and chromite, that is, for the whole assemblage of minerals accepted, on the basis of a number of features, as syngenetic inclusions in diamonds. This same type probably also includes the earlier-studied octahedron of garnet in Sample 57/9 (Sobolev, Bartoshinsky et al., 1970).

This type of inclusion often comprises both single octahedra and parallel intergrowths of several octahedra (Photo 10b, 11a, 11b), sometimes intensely compressed parallel to the facets of those pyramids of diamond growth to which they are restricted.

In some of the diamonds, there are different minerals enclosed, characterized by forms of negative crystals. Thus, in Sample AV-11, such forms have been recorded both for orange garnet and for light-green omphacite (eclogitic paragenesis of inclusions). In Sample AV-14, the coexisting pyrope, chrome-diopside, and enstatite also consist of inclusions of octahedral shape. A similar picture has also been observed in Sample AV-10 (chrome-pyrope and chrome-diopside). At the same time, in Samples AV-34, AV-14, and AV-10, the olivine inclusions (sometimes numerous, and up to 9 in number in AV-14) consist of idiomorphic crystals.

Especially interesting are the cases of discovery of complex (polymineralic) crystalline inclusions in diamonds, which have an octahedral shape, that is, the shape of negative diamond crystals. Such a shape has been identified for intergrowths of chromium-bearing garnet and enstatite in Sample AV-93 (Photo 12), pyrope and chrome-diopside (AV-34), and chrome-diopside and enstatite (AV-14). In the last sample, a further intergrowth in the form of an octahedron has been recorded, which consists of garnet and chrome-diopside.

Whereas the enclosed mineral individuals and their intergrowths in the above samples have an octahedral shape, there are a large number of cases where octahedral shapes occur in combination with numerous, probably specific facets of the enclosed mineral species. This applies especially to a series of compressed inclusions of olivine in Sample AV-37 (Photo 13). The best-developed facets of these inclusions are parallel to those of the diamond octahedron. In addition, some of the facets are apparently those of olivine itself. Similar features have also been observed in some garnet inclusions.

An interesting group of inclusions has been identified in a diamond, taken from a diamond-bearing eclogite in Sample M-46 from the 'Mir' pipe (V. S. Sobolev, Sobolev and Larent'ev, 1972). From this crystal, inclusions of garnet, omphacite, and rutile have been extracted, and two inclusions of omphacite have an octahedral shape. The omphacite from one of the inclusions in absolutely fresh, and the other is partly altered, apparently the result of a crack. In the latter, in turn, a rutile crystal has been found. A third octahedral inclusion taken from the same diamond, consists of an intergrowth of garnet and omphacite. Idiomorphic inclusions of rutile and omphacite have also been found in this same diamond, sometimes surrounded by a darkened halo, which evidently developed in a cracked zone around the inclusion (Photo 14a).

The problem of the origin of xenogenic negative crystals has been given little attention in the mineralogical literature. We have noted that they are extremely characteristic of liquid and glassy inclusions, where their origin is explained by redeposition of material of the mineral-host on the walls of the inclusions after their separation (Yermakov, 1950). Such an origin can scarcely apply to solid negative crystals in diamond. The formation of some box-like crystals, especially those of Mariupol zircon, has been regarded as a metasomatic growth in a solid medium. In the present case, such a view is apparently refuted by the fact that the enclosed minerals with idiomorphic shape are frequently present in diamonds in the same crystals, where octahedra have been identified (see Photo 11b).

It is most likely that the monomineralic and polymineralic octahedra were formed as a result of filling of typical octahedral 'growth pits' on the facets of diamond crystals by syngenetic minerals. Evidently, the nuclei of these minerals are deposited in such a pit, and the formation of negative shapes is associated with the fact that the relative velocity of growth of the facets of the enclosed minerals is greater than that of the octahedral diamond facets. As a result, the pits are completely filled with material from the particular mineral with retention of the diamond shape. With further growth of the diamond, such an inclusion may be overlain by layers of diamond growth, and, being deposited in the pits in the later growth phases of the diamond, is sometimes unable to achieve hermetic sealing, as has been observed in an ilmenite in Sample MR-5, which is intergrown with the external zone of a polycrystalline diamond aggregate.

Thus, it may be regarded as proven that inclusions of syngenetic minerals (magnesian and magnesian-iron garnets, olivine, ortho- and clinopyroxenes, chromite, and even ilmenite), occur in the form of negative 'diamond' octahedra, oriented parallel to the outer shape of the diamond, and this phenomenon is extremely widespread. The above-noted features of this morphologic type may probably be recorded in the most of the items studied. An important objective is the detailed study of intermediate cases, where there are complex induction surfaces on the facets of the inclusions, along with 'diamond' or specific facets (Grigor'ev, 1961; Shafranovsky, 1960). A wide-ranging morphologic investigation should help more completely to devise a model of the combined crystallization of the diamond and its accompanying minerals.

P o l y c r y s t a l l i n e syngenetic inclusions have been recorded in very rare cases. Thus, for example, in a study of 50 diamonds from Africa, containing about 200 inclusions of different minerals (Meyer and Boyd, 1972), in only two cases was more than one phase in the form of a single polymineralic inclusion identified. In one case, the inclusion consisted of an intergrowth of olivine and chromite, and in another, the pyrope contained a small (less than 5μ) birefringent inclusion, belonging either to olivine or pyroxene. In a Yakutian diamond, an intergrowth of garnet and chrome-diopside has been identified and studied (Sobolev, Bartoshinsky et al., 1970), and the chrome-diopside may even be regarded as an independent inclusion in the garnet (Photo 14b). A series of intergrowths of garnet and chrome-diopside has been identified in a collection of diamonds from the 'Mir' pipe, specially collected by A. I. Botkunov. Two samples of this collection, representing intergrowth of three minerals in each diamond (Photo 14c, d) are undoubtedly of the rarest type (Sobolev, Botkunov et al., 1976).

Several isolated inclusions of the same mineral are much more common in diamonds than polymineralic intergrowths. For olivine, the number of grains reaches 10 or more. Thus, about 20 colorless inclusions, probably composed of olivine, and 1 inclusion of crimson pyrope are present in a twinned diamond crystal in Sample AV-47 (see Photo 13), and the dimensions of the inclusions, as seen from the photograph, vary widely. A still greater number of colorless inclusions (more than 40) of very different sizes have been identified in the inner zone of diamond AV-91 from the 'Mir' pipe (Photo 15), and the zone of distribution of the inclusions is limited by the planes of the octahedral facets, which are apparently zones of growth.

Similar examples are also known to us for the eclogitic association within a diamond with a marked quantitative predominance of well-facetted inclusions of a light-green pyroxene over those of orange garnet (Photos 16 and 17). Occurrences of a larger number of inclusions in one diamond are more common for chromite.

Of particular interest are the descriptions of a vast number of microcrystals of olivine in one diamond both in the Yakutian and African diamonds (Gnevushev and Futergendler, 1965; Harris, 1968). The dimensions of those crystals do not exceed a few microns, and they are segregations in the form of a 'dusty cloud'. Similar accumulations in diamond have also been identified for garnet microcrystals (Futergendler, 1969). In particular, it is known for garnet that the inclusion is a combination of strictly oriented crystallites, evidently tabular in shape, disseminated throughout the entire sample, and they are the cause of the milk-white color of the diamond. The a_0 value of 11.60 ± 0.005 Å enables us provisionally to assign the garnet to a pyrope-almandine. In our opinion, this inclusion may be assigned to a garnet of the eclogite type.

The orientation of the crystals relatively to one another and to the diamond has been established both for olivine and for garnet.

A further occurrence is known of a single dense white, cloud-like inclusion (Harris, 1968a), for which only its crystalline nature has been established. Accurate descriptions of gas-liquid inclusions in diamonds are not so far available.

2) There is considerable interest in cases of identification of the patterns or orientation of enclosed minerals with respect to the diamond. Following the first description of the orientation of olivine (Mitchell and Giardini, 1953), in which the axis [101] is perpendicular to the plane [111] of the diamond, a considerable amount of factual information has been collected together from the

investigation of the orientation of garnet, olivine, and chrome-spinel
relative to the diamond (Futergendler and Frank-Kamenetsky, 1961;
Frank-Kamenetsky, 1964; Zyuzin, 1969). In a whole series of cases,
a regular arrangement in the diamond has been established for these
minerals, and this has been partly explained by oriented growth on
the surface of an octahedral facet and subsequent overgrowth by the
diamond (Frank-Kamenetsky, 1964). The results of the investigation
of quite numerous isolated inclusions have shown that the orientation
of these inclusions in diamonds is not random. Thus, for example,
during the investigation of 29 isolated inclusions of garnet, in most
cases a regular orientation or tendency towards such has been revealed. Four types of such orientation have been identified, and
the rarest type, the coincidence of the $\underline{L}_4$ axes of the garnet and
diamond, has been found in one sample (Zyuzin, 1969).

Although the very orientation of the inclusions is itself not a
proof of their combined growth with the diamond, repetition of the
regular orientation in many cases and the distortion in shape of many
of the inclusions indicate the active influence of the diamond-host
on the orientation of many of the inclusions. It may be assumed that,
in cases of clear distortion of the shapes of crystals, their orientation is of an epitaxic nature, as has, for example, been shown for
the tabular garnet depicted in Photo 9.

3) A direct morphologic feature of the inclusions is the absence
of crystal fragments within the diamond, which might have been caught
up by it during growth. As is known, there is an extremely widespread
distribution in the kimberlites of fragments of different types both
of diamonds themselves, and also all other early minerals, and in some
cases, it has even been possible to find fragments of one and the same
diamond crystal separated in the pipe.

During the investigation of inclusions, not a single confirmed case
of discovery of fragments of minerals within a diamond has been
reported. During detailed investigations, the possibility of strongly
deformed shapes of inclusions must be allowed for, as these could
sometimes, in error, be mistaken for fragments. Moreover, in some
minerals, especially garnet, well-facetted crystals may only be
observed in inclusions in diamonds.

The absence of mineral fragments in diamonds is one of the significant features of the formation of diamond during the phase preceding
the displacement of the kimberlites, with which the formation of the
fragments is associated.

4) Quite early on, attention was drawn to the radical difference
in the degree of preservation of syngenetic inclusions in the diamond
and those same minerals in kimberlites and xenoliths of various
rocks.

Olivine in kimberlite and in xenoliths has been altered to a
certain degree in the freshest samples; it has been affected by
serpentinization. The pyroxenes have been replaced by various
secondary products. Garnet is the least altered, but it is normally
also surrounded by a kelyphitic rim.

All the minerals in diamonds are distinguished by exceptional
preservation and only in relatively rare cases have the above-noted
secondary changes been noted. In some cases, it may be directly
observed that such alterations are associated with cracks, penetrating
into the diamond as far as the inclusions. However, when the cracks
cannot be directly observed, radical differences in the degree of
distribution of the alterations in the diamonds and outside them,

suggest that they are, in all cases, associated with the penetration of solutions along cracks. Such differences are especially clear for all the pipes where kimberlites have undergone the greatest changes, for example, in the 'Mir' pipe. Here, only rare signs of fresh olivine have been recorded in the kimberlite, and in the ultramafic rocks of the upper horizons (Vasil'ev et al., 1968), but in the diamonds from this same pipe, the olivines have been completely preserved. Even in such pipes as 'Udachnaya East', in which the kimberlite is a fresh variety, the initial phases of serpentinization of olivine, which are absent from the inclusions, are everywhere observed.

5) When describing the composition of the inclusions in diamonds below, it has been shown that, in spite of the clearly typical features of the composition of these minerals, the limits of variation in the amounts of various elements in them are extremely wide in the various samples of diamonds. At the same time, numerous cases of the discovery of several inclusions of minerals of variable composition, such as garnets, chromites, and clinopyroxenes in one diamond are known. Such cases are of interest in explaining whether the inclusions of these minerals are grains of different composition picked up at random by the diamond, or are the result of regular crystallization of a melt in which the diamond also grew: Detailed investigations of the compositions of such inclusions in one diamond suggest (see §11) that they are always practically identical within the limits of analytical accuracy.

6) Even a preliminary study of a large number of diamonds containing inclusions, makes it possible to separate stable combinations of minerals in the diamonds (a detailed account of all the established combinations has been given in §11). In this case, intergrowths of various minerals, along with individual inclusions have been identified in the diamonds, the presence of such intergrowths alone suggesting the clear paragenetic association of the intergrown minerals, such as garnet and olivine or garnet and diopside, which sometimes even form inclusions of one mineral in another inside the diamond.

7) The clear differences between them and the minerals of the same type and parageneses of most of the xenoliths, established as a result of an examination of the composition of the inclusions, have a regular nature (see below) and are the result of a change in the conditions of crystallization of these minerals and parageneses in the direction of increasing pressure.

Epigenetic inclusions. We have already listed above, minerals comprising 11 species, which have been identified as epigenetic inclusions, with the aid of X-ray methods. It is noteworthy that the abundance of secondary inclusions (based on the variety of mineral species) is specific to the diamonds of the oldest deposits, for example, in Ghana and Brazil, which are of Precambrian age (Harris, 1968b).

The most widely distributed epigenetic inclusions include graphite, the later development of which with respect to diamond, as exemplified by the Uralian diamonds, has been demonstrated by Kukharenko (1955). Graphite has been found in the form of individual flakes, arranged along the cracks in the cleavage planes, in the form of rosettes in cracks around syngenetic inclusions (mainly olivine), or powdery in the form of minute flakes through the entire diamond.

In the first description of sulfide inclusions (Sharp, 1966), it was pointed out that pyrrhotite, and even more often, intergrowths

of pyrrhotite and pentlandite, are the most common widely distributed
inclusions in the South African diamonds. It has been assumed that
the pyrrhotite-pentlandite intergrowths were formed as a result of
dissociation of an original high-temperature sulfide phase, syngenetic
with the diamond. In addition to pentlandite, cohenite and pyrite
have been identified in such intergrowths in isolated cases.

A later detailed investigation of inclusions of this type (Harris,
1968b) has shown that the rosette-like inclusions in one of the
diamonds are located in a zone of cracks around four olivine inclusions, one of which was covered by a crust of pentlandite. In another
diamond, iron-nickel sulfides have also been identified in association with olivine. Such a situation, with respect to olivine,
according to Harris (1968b), indicates the secondary nature of these
inclusions. However, in our opinion, the assignment of these iron-
nickel sulfides to epigenetic inclusions is inadequately based.
First, these are high-temperature associations, differing markedly
from all the other epigenetic inclusions. Second, in a number of
xenoliths of pyrope peridotites, similar sulfide minerals have been
identified in fresh samples, which from the nature of their relationships with the silicate minerals, and discordant veins of serpentine and late sulfides, may be assigned to primary formations
(Vakhrushev and Sobolev, 1971).

Quartz has been identified, as a rule, in abundance in diamonds
of certain ancient placer deposits, especially in those from the
Precambrian places of Ghana and Brazil, mostly in the diamond surface
or along cracks.

Garnets

X-ray investigations have established not only the assignment of
crystalline inclusions to a definite group of minerals (olivine,
garnet, clinopyroxene, and chrome-spinels), but have also assisted
in interpreting the most common compositional features.

However, only with the inception of the microprobe, did it become
possible to get a more precise understanding of such features of
the inclusions which could enable a comparison to be made of their
composition with those of analogous minerals, formed under conditions of lower pressure. Although the conduct of the microanalysis
of crystalline inclusions is associated with a number of significant
difficulties, primarily with the very small dimensions of the
inclusions (0.05 - 0.4 mm), the first results of determining the
composition of pyrope inclusions were obtained in 1968 (Meyer, 1968b).
With the subsequent resources of American and Soviet investigators
over a number of years, numerous determinations of their composition
have been made. The characteristics of the minerals, enclosed in
diamonds, presented below, are based on the results of determinations of the chemical composition of more than 90 such inclusions
from the deposits of South Africa, Ghana, Sierra Leone, Venezuela,
Yakutia, and the Urals (Meyer and Boyd, 1972; Sobolev, Lavrent'ev
et al., 1969; Sobolev, Bartoshinsky et al., 1970; Sobolev, Botkunov
et al., 1971; Sobolev, Gnevushev et al., 1971)*.

*Up to 1976, as result of further investigations (Meyer and
Svisero, 1975; Prinz et al., 1975), and new work by the present
author and his colleagues, more than 400 syngenetic inclusions in
diamonds have been analyzed, although the earlier conclusions and
patterns (see §§9-11) have not been revised.

Although garnets are not widely distributed inclusions in diamonds, their composition has been examined in great detail on a large number of samples (about 40) (Meyer and Boyd, 1972; Sobolev, Lavrent'ev et al., 1969; Sobolev, Bartoshinsky et al., 1970; Sobolev, Botkunov et al., 1971; Sobolev, Gnevushev et al., 1971), this being connected with the importance of garnet as an indicator of the composition of the environment and conditions of formation (Sobolev, 1964a).

The possibility of determining the composition of the garnets enclosed in diamonds on the basis of physical properties, appeared after the publication of the first Soviet work listed above. During the interpretation of the compositions, several differences were noted between those of the Uralian and Yakutian diamonds (Futgergendler, 1960). The accumulation of information based on the physical properties of the garnet inclusions suggested a similarity between the compositions of the yellow-orange garnets, enclosed in the Uralian diamonds and those of the diamond-bearing eclogites (Sobolev and Kuznetsova, 1966), and that the garnets of crimson and lilac tints should be assigned to a peridotite paragenesis (Sarsadskikh et al., 1960a; Sobolev and Kuznetsova, 1966). The correctness of such a distinction between garnets of peridotite and eclogite types was confirmed by new information, including data on the physical properties of garnets from South African diamonds (Meyer and Boyd, 1972). The results of a study of the physical properties of garnets from Yakutian and Uralian diamonds (Sobolev, Gnevushev et al., 1971) are shown in Table 26. By recording the presence of two different types of garnets in diamonds, it follows that, when assessing their composition on the basis of their physical properties, we must primarily allow for the possible addition of chromium (crimson and lilac inclusions) or its absence (yellow-orange inclusions), since increased refractive indices will be characteristic both of the pyropes rich in chromium and of the non-chromium ferriferous garnets.

The first determinations of garnet compositions from South African diamonds (Meyer, 1968b) enabled us to recognize a new, earlier unknown variety of chrome pyrope, poor in calcium, with a substantial amount of the knorringite component ($Mg_3Cr_2[SiO_4]_3$), up to 30 mol %. Results of analyses of garnets from the Yakutian diamonds (Sobolev, Lavrent'ev et al., 1969) have underlined the low content of CaO identified by Meyer, along with the significantly broadened hypothesis of the entry of Cr_2O_3 into the garnets. This has been graphically illustrated by the wide variations in the amount of Cr_2O_3 from 6.1 up to 15.9 in Table 27, where the results of analyses of 25 inclusions of chromepyropes with decreased amounts of calcium have been presented. The garnet with maximum amount of Cr_2O_3 in Sample AO-458, contains up to 46% knorringite, which is much greater than the limiting amount so far identified in the magnesian garnets. Below, we present data from a partial analysis of four garnets from Yakutian diamonds, the amount of the individual components in which have been calculated by analogy with garnets of similar composition.

Samples	Cr_2O_3	FeO	CaO	Cr-comp.	Almandine	Ca-comp.	$\underline{f}$
AO-1	8,24	6,88	2,03	23,6	13,9	5,2	14,7
AO-454	8,75	6,27	1,76	25,0	12,6	4,5	13,2
AO-457	9,78	6,71	2,17	28,0	13,5	5,5	14,3
AO-429	10,1	7,28	1,82	29,9	14,8	4,6	15,5

TABLE 26. Physical Properties of Garnets .from the Urals and Yakutia (Sobolev, Gnevushev et al., 1971)

Spec. No.	Sample No.	Color of inclusions n	a_0, $\overset{\circ}{\text{A}}$	N
		Chromium-bearing pyropes		
		Inclusions in Uralian diamonds		
1	8285	dark violet	11,571	1,765
2	913	bright-violet	11,522	1,768
3	931		11,568	1,787
		Inclusions in Yakutian diamonds		
4	1003	pale violet	11,533	1,744
5	1005		11,507	1,746
6	1020	violet	11,511	1,746
7*		violet	11,55	1,748
8	26	pale violet	11,528	1,749
9	3a	"	11,530	1,749
10	1007	"	11,537	1,754
11*		crimson	11,56	1,755
12	1006	violet	11,525	1,756
13	1002	dark-violet	11,543	1,760
14	1		11,565	1,765
15	1010		11,550	1,766
		Pyrope-almandines		
		Inclusions in Yakutian diamonds		
16	13372	pale yellow	11,535	1,754
17	4338		11,540	1,761
18	1a	pale orange	11,547	1,757
19	8815	orange	11,548	1,764
20	3570	yellow-orange	11,549	1,763
21	1964	pale yellow	11,558	1,764
22	6946	"	11,560	1,763
23	6097	yellow	11,560	1,755
24	15422	yellow-orange	11,562	1,766
25	3	dark orange	11,564	1,770
26	15383	pale orange	11,564	1,769
27	138	orange	11,566	1,764
28	7b	pale orange	11,570	1,773
29	850	light orange	11,571	1,763
30	174	orange	11,580	1,766
31	7a	yellow-orange	11,584	1,763
32	7c	pale orange	11,586	1,760
33	865	pale yellow	11,596	1,763
34	4		11,603	1,760
35	6	yellow-orange	11,612	1,760
36	880	pale	11,624	1,768
37	3*	pale yellow	11,636	1,755— 1,760
		Inclusions in Yakutian diamonds		
38	8	pale yellow	11,538	1,756
39*		orange	11,54	1,746
40*		orange-red	11,55	1,755
41	1001	yellow	11,566	1,758
42	1004	"	11,594	1,766
43	1011	orange	11,606	1,771

*Data of É. S. Nikolaeva (Gnevushev and Nikolaeva, 1958, 1961)

In addition to the recorded features in the amount of Cr_2O_3 and CaO, which are clearly typomorphic of these garnets, distinguishing them from pyropes, not only from peridotitic xenoliths, but also from the concentrate of the heavy fraction from kimberlites, a decreased iron index (taken by analogy with other garnets, as the FeO/(FeO + MgO) ratio, where FeO is the amount of total iron), varies for almost all the samples examined within the range of 9.5 to 16%. It is interesting to note that the range of variation of the iron index for the African and Venezuelan garnets is narrower than that for the Yakutian types (9.5 to 13.1%), and consequently, a trend appears in their compositions towards a low iron-index value. In a number of cases, olivine is present, along with garnet, in the diamond crystals.

In addition to the magnesian garnets with low calcium content, a chromium-bearing pyrope has been identified in diamonds with a CaO content exceeding 13% (Sobolev, Bartoshinsky et al., 1970). This unusual garnet composition has been emphasized by two analyses from one and the same diamond (an individual inclusion and a garnet from an intergrowth with chrome-diopside) (see Table 27, Sample 57/9a, b). The insignificant discrepancies between the compositions of the two garnets depend mainly on analytical error. Although garnets, rich both in Cr and Ca have also been found in the kimberlite concentrates, they are also quite rich in iron, and the inclusions examined are distinguished by the markedly lower value of the iron index (14.5%), completely agreeing with that of the low-calcium pyropes from diamonds. Along with the garnet and diopside in diamond Sample 57/9, olivine is present: in essence, a completely new type of association with pyrope has been established (pyrope wehrlite).

In addition to the above two types of garnets, characterized by a high content of Cr_2O_3, magnesian garnets have been found in diamonds, with low (1-2%) contents of Cr_2O_3 and a slightly higher amount of CaO (up to 4%), as compared with the low-calcium garnets. Such garnets have only been discovered in two diamond samples from the 'Mir' pipe along with olivine, enstatite, and diopside (see Table 27, Samples AV-34 and AV-14) and, judging by their external features, especially the crimson-pink color of weak intensity, they are extremely rarely found in diamonds as inclusions. The low Cr_2O_3 content and the amount of calcium, normal for most of the magnesian garnets from kimberlites, seemingly makes it possible to regard them as similar to garnets from lherzolite xenoliths. However, a significant difference, unequivo-cally assigning these garnets to a magnesian group of special type, is their very low iron index. This same compositional feature, as is clearly seen from the analyses in Table 27, and the average iron index of 33 garnets 13.3 ± 1.8), is common for all the magnesian garnets from diamonds in spite of substantial variations in their composition based on the amount of Cr_2O_3 and CaO.

Thus, amongst the magnesian garnets with variable content of Cr_2O_3 and CaO, identified as inclusions in diamonds, three types may be clearly recognized: 1) chrome pyropes, poor in calcium (harzburgite-dunite association); 2) chrome pyropes, rich in calcium (wehrlitic association); and 3) pyropes with low Cr_2O_3 content and intermediate amounts of CaO (4-5%) (lherzolitic association). These three types combine the compositions of the magnesian garnets involved in the ultramafic paragenesis, a substantial portion of which is magnesian olivine.

An independent type constists of inclusions of pyrope-almandine

TABLE 27. Chemical Compositions of Magnesian Garnets, Enclosed in Diamonds

Oxides	Yakutia													
	AV-34	AV-14	AO-434	B-4	AO-460	B-5	AO-438	AO-449	AO-432	AO-90	AO-430	AO-456	AV-79	AO-445
SiO_2	43,2	43,1	42,7	41,7	42,6	42,3	42,7	42,5	41,7	41,4	42,5	41,0	41,5	40,4
TiO_2	0,47	0,68	0,02	0,08	0,02	0,07	0,02	0,02	0,03	0,02	0,03	0,02	0,31	0,17
Al_2O_3	22,1	22,3	18,0	18,1	17,9	17,9	18,4	18,2	17,6	17,0	16,8	16,6	15,2	13,2
Cr_2O_3	1,27	1,65	6,08	6,97	7,09	7,15	7,19	7,57	7,72	8,30	8,80	9,28	10,7	12,8
FeO	6,02	6,21	6,02	8,12	6,93	6,95	6,55	6,34	7,01	7,28	6,85	7,09	5,92	6,15
MnO	0,26	0,23	0,21	0,37	0,28	0,37	0,28	0,21	0,37	0,28	0,37	0,31	0,36	0,30
MgO	22,5	22,5	23,4	21,1	22,6	21,5	23,3	24,0	22,5	21,5	23,3	21,7	21,2	23,0
CaO	3,81	3,96	1,33	2,66	1,24	3,56	1,09	1,94	1,46	2,38	2,22	2,32	3,90	2,33
Total	99,63	100,63	97,8	99,10	98,66	99,80	99,53	100,78	98,39	98,46	100,87	98,32	99,09	98,35
Si	3,038	3,009	3,087	3,027	3,071	3,041	3,052	3,005	3,033	3,035	3,028	3,011	3,036	2,999
Ti	0,025	0,038	0,002	0,004	0,002	0,004	0,002	0,002	0,003	0,002	0,003	0,002	0,017	0,009
Al	1,834	1,838	1,537	1,553	1,525	1,521	1,545	1,522	1,512	1,471	1,427	1,439	1,309	1,151
Cr	0,071	0,091	0,347	0,401	0,407	0,406	0,403	0,425	0,446	0,485	0,496	0,539	0,619	0,750
Fe^{3+}	0,070	0,033	0,114	0,042	0,066	0,069	0,050	0,051	0,039	0,042	0,074	0,020	0,055	0,090
Fe^{2+}	0,285	0,328	0,251	0,451	0,350	0,350	0,341	0,323	0,385	0,407	0,337	0,417	0,305	0,294
Mn	0,016	0,013	0,013	0,022	0,017	0,022	0,017	0,013	0,022	0,018	0,021	0,018	0,022	0,018
Mg	2,358	2,341	2,518	2,281	2,430	2,302	2,481	2,529	2,438	2,339	2,475	2,375	2,311	2,544
Ca	0,287	0,298	0,104	0,205	0,095	0,272	0,086	0,149	0,114	0,189	0,167	0,181	0,308	0,187
Pyrope	80,0	78,6	67,7	61,6	66,9	57,9	64,9	64,9	62,0	59,2	59,5	57,4	54,8	47,5
Almandine	9,7	11,0	8,6	15,2	12,0	11,9	11,6	10,7	12,9	13,7	11,1	13,9	10,4	9,4
Spessartine	0,5	0,4	0,5	0,8	0,6	0,7	0,6	0,4	0,7	0,6	0,7	0,6	0,7	0,6
Grossular	1,3	1,8	—	—	—	—	—	—	—	—	—	—	—	—
Andradite	3,5	1,7	3,6	2,1	3,3	3,5	2,7	2,7	2,1	2,2	3,9	1,1	2,7	5,0
Ti-andradite	1,3	1,9	0,1	0,2	0,1	0,2	0,1	0,1	0,1	0,1	0,1	0,1	0,9	0,5
Uvarovite	3,6	4,6	—	4,6	—	5,5	0,2	2,2	1,8	4,2	1,7	5,0	6,9	1,2
Koharite	—	—	2,2	—	0,1	—	—	—	—	—	—	—	—	—
Knorringite	—	—	17,4	15,5	17,1	14,8	20,0	19,1	20,5	20,1	23,1	22,0	23,6	36,3
f	13,1	13,4	12,7	17,8	14,5	15,4	13,6	12,9	14,8	16,1	14,2	15,5	13,5	13,1
Mg-component	80,0	78,6	85,1	77,1	83,0	78,2	84,9	84,0	82,5	79,3	82,6	79,4	78,4	83,8
Ca-component	9,7	10,0	3,6	6,9	3,3	9,2	2,9	4,9	3,9	6,4	5,6	6,1	10,5	6,2
Cr-component	3,6	4,6	17,4	20,1	20,4	20,3	20,2	21,3	22,3	24,3	24,8	27,0	30,5	37,5
n measured	—	—	1,753	—	—	—	—	1,757	—	—	—	1,767	—	1,780
n calculated	—	—	1,754	—	1,757	—	1,757	1,758	1,760	1,765	1,764	1,765	—	1,778

(table 27)

Oxides	Urals	Yakutia				Africa			Venezuela			Africa			
	931	AO-458	57/9a	57/9b	G9a	G12a	G20a	GL32b	GL24a	GL25a	GL26e	GL1	GL6	GL15e	GL15h
SiO_2	41,4	41,4	41,0	41,0	42,7	42,1	41,0	42,7	41,6	42,5	41,0	42,3	42,8	41,8	42,2
TiO_2	0,04	0,08	0,46	0,42	0,02	0,00	0,02	—	0,02	0,05	0,25	0,02	0,00	0,02	0,02
Al_2O_3	12,0	11,1	16,2	16,3	18,4	16,3	14,3	18,4	16,0	17,3	15,5	17,2	18,2	15,7	15,7
Cr_2O_3	15,6	15,5	8,0	7,82	5,85	8,14	12,6	6,97	10,1	8,77	10,7	8,93	4,75	10,9	10,7
FeO	6,22	6,18	4,66	4,64	5,38	6,31	6,11	5,65	5,92	5,04	6,08	5,36	4,75	5,71	5,57
MnO	0,27	0,23	0,30	0,29	0,18	0,24	0,28	0,29	0,27	0,19	0,29	0,21	0,17	0,20	0,19
MgO	24,1	23,1	15,5	15,5	25,1	24,0	23,6	24,5	24,6	25,1	22,8	25,3	25,5	24,2	24,5
CaO	2,13	2,24	13,5	13,3	2,46	2,79	1,74	1,22	1,18	1,77	3,10	1,09	1,35	2,19	2,22
Total	101,76	100,23	99,32	99,27	100,09	99,88	99,75	99,72	99,69	100,75	99,52	100,41	100,67	100,72	101,10
Si	2,987	3,035	3,031	3,031	2,999	3,036	2,997	3,028	2,989	2,993	2,973	2,998	3,007	2,999	3,004
Ti	0,002	0,004	0,026	0,024	0,001	0,000	0,001	—	0,001	0,003	0,014	0,001	0,000	0,001	0,001
Al	1,023	0,960	1,413	1,422	1,528	1,374	1,228	1,537	1,351	1,437	1,326	1,439	1,512	1,324	1,318
Cr	0,893	0,925	0,470	0,456	0,325	0,460	0,721	0,391	0,571	0,489	0,611	0,503	0,439	0,619	0,559
Fe^{3+}	0,082	0,111	0,091	0,098	0,146	0,164	0,050	0,072	0,077	0,071	0,049	0,057	0,049	0,056	0,082
Fe^{2+}	0,291	0,268	0,198	0,191	0,171	0,213	0,321	0,263	0,279	0,226	0,320	0,320	0,230	0,288	0,252
Mn	0,017	0,013	0,018	0,018	0,011	0,014	0,017	0,017	0,016	0,011	0,018	0,013	0,008	0,013	0,013
Mg	2,592	2,524	1,676	1,711	2,633	2,554	2,554	2,589	2,631	2,639	2,457	2,674	2,580	2,580	2,602
Ca	0,165	0,176	1,071	1,053	0,185	0,213	0,138	0,093	0,031	0,133	0,240	0,085	0,101	0,168	0,167
Pyrope	40,1	38,7	57,1	58,0	70,3	61,2	50,2	67,8	57,8	64,0	55,1	62,9	67,0	56,4	57,1
Almandine	9,5	8,9	5,8	5,7	5,7	7,1	10,6	8,9	9,2	7,5	10,5	8,6	7,6	9,3	8,3
Spessartine	0,6	0,4	0,6	0,6	0,4	0,5	0,6	0,6	0,5	0,4	0,6	0,4	0,3	0,4	0,4
Grossular	—	—	7,0	6,8	—	—	—	—	—	—	—	—	—	—	—
Andradite	4,1	5,8	4,6	4,9	6,2	7,1	2,5	3,1	3,0	3,4	2,5	2,8	2,5	2,8	4,2
Ti-Andradite	0,1	—	1,3	1,2	—	—	—	—	—	0,2	0,7	—	—	—	—
Uvarovite	1,2	0,1	23,6	22,8	—	—	2,1	—	—	0,8	4,7	—	0,9	2,7	1,3
Koharite	—	—	—	—	1,1	1,1	—	0,5	0,8	0,6	—	—	—	—	—
Knorringite	44,4	45,1	—	—	16,3	23,0	34,0	19,1	28,6	23,7	25,9	25,2	21,1	28,4	28,7
f	12,6	13,1	14,6	14,5	10,7	12,9	12,7	11,5	11,9	10,1	13,1	10,7	9,5	11,6	11,3
Mg-Component	84,5	84,8	57,1	57,0	87,7	85,3	84,2	87,4	87,2	87,7	81,0	88,1	88,7	84,8	85,8
Ca-component	5,4	5,9	36,4	35,6	6,2	7,1	4,6	3,1	3,0	4,4	7,9	2,8	3,4	5,5	5,5
Cr-Component	45,7	45,2	23,5	22,8	16,3	23,0	36,1	19,6	28,6	24,5	30,6	25,2	22,0	31,1	30,0
n measured	1,787	1,790	1,769	1,769	1,758	1,768	1,778	—	—	1,768	—	—	—	1,771	1,771
n calculated	1,786	1,789	1,768	1,768	1,752	1,764	1,775	—	—	1,769	—	—	—	1,768	1,768

Note. Analyses of Samples G9a... GL15h, after Meyer & Boyd. Samples AV-14, B-4, B-5, and AV-79 from 'Mir' pipe; Sample 57/9a from 'Udachnaya' pipe; Samples G9a and G12a, from Ghana; Sample GL32b, from Sierra Leone. Samples with AV index in this and later tables are materials of A. I. Botkunov and N. V. Sobolev.

garnets with variable content of calcium and variable iron-index.
These fluctuations in the amount of principal components are well
defined by the increase in the value of the unit-cell parameter, a_0
(increase in amount of CaO) and the refractive index (increased iron
index). From the data given in Table 26, and with the aid of a
composition-properties diagram in the system pyrope-almandine-
grossular (Gnevushev et al., 1956; Giller, 1962), we may estimate
the limits of variation in the amount of calcic component (from 10
to 30-35%) and f (20-25 up to 55-60%). The results of determining
the compositions of 11 garnets of this type with the aid of the microprobe (Meyer and Boyd, 1972; Sobolev, Gnevushev et al., 1971;
Sobolev, Botkunov et al., 1971) support the hypothesis of their
similarity to the garnets from eclogites, especially the diamond-
bearing types (Sobolev and Kuznetsova, 1966). The results of 14
complete analyses of garnets are given in Table 28. In the analyzed
garnets, the amount of calcic component varies from 3.7 to 37.3%,
and the value of the iron index, from 39 to 69%, which passes
beyond the limits of variation in the amount of the corresponding
components in the garnets from the diamond-bearing eclogites studied.

The composition of the overwhelming majority of analyzed inclusions of pyrope-almandine garnets from a comparison with the literature data, raises no doubt about their assignment to the garnets of
the eclogite type (Sobolev, 1964a), the more so since omphacite has
been identified along with such garnets in the diamond. An apparent
exception may be the analysis of garnet No. 20f (Meyer and Boyd, 1972),
shown in Table 28, for which, as compared with the eclogitic garnets,
a much lower calcic component has been recorded. Recently, garnets
with a very low CaO content (down to 3%) have been found in eclogite
xenoliths and even in association with two pyroxenes (Sobolev, 1970),
where the decrease in amount of CaO in the garnet is associated with
an increased amount of Na_2O in the clinopyroxene (Sobolev, Zyuzin
and Kuznetsova, 1966; Banno, 1967). Similar garnet compositions have
also been recorded in eclogite xenoliths from kimberlite pipes in
Arizona (O'Hara and Mercy, 1966). Consequently, garnet No. 20f may
also be assigned to the eclogite paragenesis, assuming that the
clinopyroxene, possible in association with such a garnet, may contain a significant amount of Na_2O, and is also distinguished by a
significantly increased amount of Mg + Fe over Ca.

Emphasis on the special nature of the inclusions of pyrope-
almandine garnets in diamonds and an important feature, distinguishing
them from those of eclogites of the metamorphic complexes, is the
stable amount of Na_2O in these garnets (Sobolev and Lavrent'ev, 1971).
Such a constant amount of Na_2O from 0.10 to 0.22% is typical only of
the garnets of the diamond-bearing eclogites (see §12). The possibility of isomorphous entry of Na_2O into these garnets according to
the scheme $NaSi \rightleftharpoons CaAl$, providing for the initial phases of the
transition $Si^4 \rightleftharpoons Si^6$, is associated with high pressure and features
of the paragenesis, a component part of which is a jadeite-bearing
clinopyroxene. At the same time, the presence of an isomorphous
amount of sodium in the yellow-orange garnets provides an additional
support for the eclogitic nature of their paragenesis. As in the
peridotites, the identified features of the composition of the
minerals in the diamond-bearing paragenesis, enables us in a number
of cases to assign several samples of eclogites, in which both
graphite and diamond are absent to the diamond-bearing facies.

TABLE 28. Chemical Compositions of Pyrope-Almandine Garnets Associated with Diamonds

Column groups: samples 20f through 880 = **Inclusions in diamonds**; samples ME-131 through B-22 = **Intergrowths with diamonds**. ("Mir' pipe over B-19 and over B-18/B-22; Urals over 138 and 880.)

Oxides	Africa 20f	Sierra Leone G16	Africa D15	1a	Urals 138	850	'Mir' pipe B-19	AV-6	1583	Urals 880	ME-131	B-6	'Mir' pipe B-18	B-22
SiO_2	37,8	40,5	40,1	41,8	41,2	40,7	39,2	40,5	40,3	39,9	39,7	40,8	40,4	40,6
TiO_2	0,25	0,74	0,64	0,18	0,21	0,64	0,22	0,93	0,33	1,18	0,33	0,34	0,48	0,50
Al_2O_3	20,3	21,9	21,0	22,2	22,1	21,2	21,6	21,0	21,7	20,7	21,2	21,8	21,7	22,0
Cr_2O_3	0,06	0,11	0,05	0,09	0,07	0,04	0,06	0,11	0,07	0,15	0,06	0,08	0,09	0,18
FeO	29,5	17,2	14,5	16,6	16,1	16,3	20,5	16,4	18,7	15,2	21,7	18,1	18,7	8,74
MnO	0,39	0,47	0,54	0,29	0,24	0,29	0,26	0,29	0,24	0,21	0,50	0,35	0,17	0,14
MgO	7,35	13,7	9,25	14,7	13,2	13,0	9,32	9,38	8,40	11,7	11,3	12,7	12,5	12,0
CaO	1,27	5,35	13,9	4,85	6,98	7,15	7,90	10,4	10,1	11,1	3,46	5,23	5,86	14,7
Na_2O	—	—	—	0,11	0,12	0,19	0,12	0,2	0,22	0,17	0,12	0,09	0,15	0,17
Total	96,9	100,0	100,0	100,86	100,29	99,51	99,18	99,25	100,06	100,31	98,37	99,49	100,05	99,03
Si	3,040	2,995	3,015	3,044	3,035	3,032	2,995	3,044	3,043	2,987	3,041	3,041	3,014	3,003
Ti	0,016	0,042	0,036	0,013	0,013	0,036	0,014	0,052	0,018	0,067	0,019	0,018	0,027	0,027
Al	1,922	1,913	1,862	1,907	1,920	1,863	1,948	1,861	1,932	1,821	1,914	1,917	1,911	1,919
Cr	0,004	0,006	0,005	0,006	0,003	0,003	0,007	0,007	0,003	0,011	0,003	0,007	0,005	0,011
Fe^{3+}	0,054	0,039	0,097	0,074	0,064	0,098	0,034	0,080	0,057	0,103	0,064	0,007	0,057	0,043
Fe^{2+}	1,928	1,027	0,816	0,936	0,927	0,919	1,275	0,950	1,132	0,848	1,326	1,071	1,109	0,495
Mn	0,026	0,029	0,032	0,017	0,013	0,018	0,061	0,018	0,014	0,015	0,032	0,022	0,009	0,009
Mg	0,882	1,509	1,035	1,596	1,447	1,442	1,061	1,052	0,943	1,300	1,288	1,411	1,391	1,324
Ca	0,110	0,424	1,121	0,380	0,553	0,573	0,648	0,836	0,816	0,888	0,285	0,421	0,467	1,164
Na	—	—	—	0,018	0,020	0,031	0,018	0,027	0,036	0,027	0,018	0,013	0,022	0,024
Pyrope	29,9	50,4	34,4	54,1	48,9	48,4	35,1	36,5	32,0	42,3	43,6	48,0	46,4	43,9
Almandine	65,5	34,4	27,2	31,8	31,3	30,8	42,2	33,0	38,5	27,6	45,0	36,5	37,0	16,4
Spessartine	0,9	1,0	1,1	0,6	0,4	0,6	0,6	-0,6	0,5	0,4	1,1	0,7	0,3	0,3
Grossular	—	9,8	30,3	8,8	15,3	13,3	19,5	22,9	25,0	20,6	5,2	10,6	11,8	35,3
Andradite	2,7	2,0	4,9	3,7	3,2	4,9	1,7	4,0	2,9	5,2	3,2	2,9	2,9	2,1
Ti-andradite	0,8	2,1	1,8	0,7	0,7	1,8	0,7	2,6	0,9	3,4	1,0	0,9	1,4	1,4
Uvarovite	0,2	0,3	0,3	0,3	0,2	0,2	0,2	0,4	0,2	0,5	0,2	0,4	0,2	0,6
f	69,3	41,4	46,9	38,8	40,6	41,4	55,2	49,5	55,8	42,2	51,9	44,4	45,6	28,9
Ca-component	3,7	14,2	37,3	13,5	19,4	20,2	22,1	29,9	29,0	29,7	10,3	14,8	16,3	39,4

Note. Analyses of Samples 20f, G15, and D15, after Meyer & Boyd (1972).

Olivine

Olivine is one of the most widely distributed silicate mineral-inclusions in diamonds. It has usually been observed in the form of colorless, well-facetted, elongate or equant grains which, without a special check may be visually mistaken for enstatite or coesite (Harris, 1968a). Very often, several grains (up to 10 or more) of olivine are observed in the one diamond (see Photo 16). In the Yakutian and Uralian diamonds, olivines were first discovered by Orlov (1959) and assigned to forsterite on the basis of a study by optical methods. Their assignment to forsterite has also been confirmed by Futergendler (1958, 1960), using X-ray methods.

Judging by the available results of an analysis of 17* inclusions of olivines from diamonds of different localities (Meyer and Boyd, 1972; Sobolev, Bartoshinsky et al., 1970), presented in Table 29, they are characterized by an exceptionally large amount of forsterite and consequently low iron index. A qualitative analysis on the microprobe and the determination of the unit cell for 31 samples of olivines from South African diamonds (Meyer, 1968a), have shown the very narrow variations in their composition. A comparison of the average composition of the olivines from diamonds studied, on the basis of the forsterite content (inclusions in diamonds) with those of magnesian olivines from different ultramafic rocks, taken from the work of Il'vitsky and Kolbantsev (1968), suggests a trend towards their greatest enrichment in magnesium as compared with the olivines from kimberlites, and ultramafic xenoliths in the kimberlites and basalts, etc. (see below).

Rock type	No. of analyses	$\bar{x}$	s
Inclusions in diamonds	17	92.9	1.0
Ultramafic inclusions in kimberlites	18	92.0	0.9
Alpine-type ultramafic units	55	91.5	1.1
Kimberlites	40	91.4	1.5
Olivines in basalts	34	90.4	0.2
Olivinites of ultramafic alkaline complexes	15	89.8	2.7

Further accumulation of data on the composition of olivines from diamonds should lead to the recognition of a significant difference in their composition from that of olivines from various ultramafic associations. The exceptionally high magnesium content of the olivines enclosed in diamonds, shows that they are apparently the earliest products of crystallization from a medium of ultramafic composition.

One of the most interesting established features of the composition of olivines from diamonds is the isomorphous amount of Cr_2O_3 (from 0.02 to 0.08%) (Meyer and Boyd, 1972), on the basis of the amount of which they are significantly distinguished from the olivines of many types of ultramafic rocks (see §15).

Chromite

Using X-ray methods, the diagnosis of inclusions in diamonds has not only identified the chrome-spinels as the most widely distributed

*Including those for two samples from Yakutian diamonds (57/33 and 41/33) in which iron was determined only as FeO (7.45 and 7.39%). The content of forsterite is respectively 92.4 and 92.4% [sic].

TABLE 29. Chemical Compositions of Olivines Enclosed in Diamonds

Oxides	Africa					Venezuela			Ghana			Yakutia			
	9b	10c	11a	13c	23b	GL24c	GL29b	GL47a	G10b	G16a	G17a	57/9	38/8	AV-14	AV-34
SiO₂	40,7	40,8	40,4	41,0	40,7	41,0	40,2	41,1	41,3	40,8	41,2	41,4	41,2	41,0	41,1
TiO₂	0,00	0,00	0,00	0,02	0,00	0,00	0,00	0,00	0,00	0,00	0,00	0,00	0,00	0,05	0,08
Al₂O₃	0,07	0,02	0,02	0,06	0,02	0,02	0,02	0,00	0,05	0,02	0,02	0,02	0,00	0,02	0,02
Cr₂O₃	–	0,06	0,08	0,06	0,04	0,07	0,07	0,08	0,05	0,02	0,07	0,06	0,03	0,02	0,02
FeO	7,79	7,09	5,52	7,11	7,60	6,65	7,14	6,21	6,91	7,10	8,31	6,00	6,15	6,30	7,01
MnO	0,17	0,12	0,09	0,12	0,12	0,10	0,12	0,10	0,11	0,11	0,11	0,11	0,06	0,08	0,06
MgO	49,6	51,6	52,9	50,4	51,1	52,7	51,9	52,7	51,6	52,5	50,7	52,9	53,2	51,2	50,1
CaO	0,18	0,07	0,01	0,09	0,07	0,01	0,04	0,01	0,04	0,03	0,06	0,04	0,00	0,04	0,02
NiO	Не опр.	Не опр.	Не опр.	Не опр.	Не о пр.	0,38	0,40	0,40	0,43	0,43	0,40	0,34	Не опр.	Не опр.	Не опр.
Total	98,51	99,8	99,0	98,9	99,7	100,9	99,9	100,6	100,5	101,0	100,9	100,89	100,67	98,71	98,41
Si	1,001	0,991	0,982	1,003	0,992	0,986	0,981	0,989	0,998	0,983	0,997	0,991	0,988	1,001	1,009
Ti	0,000	0,000	0,000	0,000	0,000	0,000	0,000	0,000	0,000	0,000	0,000	0,000	0,000	0,001	0,001
Al	0,002	0,001	0,001	0,002	0,001	0,001	0,001	0,000	0,001	0,001	0,001	0,000	0,000	0,000	0,001
Cr	–	0,001	0,001	0,001	0,001	0,001	0,001	0,002	0,001	0,000	0,001	0,001	0,001	0,000	0,000
Fe²⁺	0,161	0,144	0,112	0,145	0,155	0,134	0,146	0,125	0,140	0,143	0,169	0,121	0,123	0,129	0,144
Mn	0,003	0,003	0,002	0,003	0,003	0,002	0,002	0,002	0,002	0,002	0,002	0,002	0,001	0,001	0,001
Mg	1,821	1,868	1,918	1,839	1,887	1,890	1,887	1,892	1,858	1,886	1,831	1,887	1,899	1,864	1,833
Ca	0,005	0,112	0,001	0,002	0,002	0,000	0,001	0,000	0,001	0,001	0,001	0,001	0,000	0,001	0,001
Ni	–	–	–	–	–	0,007	0,008	0,008	0,008	0,008	0,008	0,006	–	–	–
Total of cations	2,993	3,010	3,017	2,995	3,041	3,021	3,027	3,018	3,009	3,024	3,010	3,009	3,011	2,997	2,990
f	8,1	7,2	5,5	7,3	7,6	6,6	7,2	6,0	6,9	7,0	8,5	6,0	6,1	6,5	7,3
Forsterite	91,5	92,6	94,3	92,5	92,2	93,4	92,8	93,8	93,0	93,0	91,6	93,10	93,9	93,4	92,5

types of inclusions, but has also emphasized their assignment to chromium-rich varieties (chrome-picotite) (Futergendler, 1960).

The results of the first determinations of the composition of inclusions of chrome-spinels with the aid of the X-ray microprobe (Meyer and Boyd, 1972) have emphasized and substantially confirmed the assignment of these inclusions to chromites, which are most chromium-rich as compared with those found anywhere in ultramafic rocks (see §18). Their compositions are comparable only with those of the chromites from meteorites (Meyer and Boyd, 1972). Our additional study of chromite inclusions from the Yakutian diamonds has completely supported this conclusion (Table 30).

The total amount of chromium component in these chromites in all cases exceeds 80%, which is a type feature. However, the available information enables us to recognize two separate compositional types. The first, and markedly predominant type is characterized by comparatively narrow variations in the $Fe^{2+}/Fe^{2+} + Mg$ ratio, from 20 to 41% (in the calculation of this ratio, no account has been taken of the portion of Fe transferred to the R^{3+} group, in addition up to 2.000). The ratio $Fe^{3+}/Fe^{3+} + Fe^{2+}$ in these chromites is also relatively small and stands in inverse ratio to their iron index: from 7 to 29%.

The second type, consisting in fact of one sample only (two analyses of chromite inclusions from one diamond) is distinguished by marked enrichment in the iron component (Table 30).

It should be noted that owing to the exceptionally large amount of Cr_2O_3 and some addition of Al_2O_3, almost all the iron enters the chromites assigned to the second type, in the form of FeO. The entire R^{2+} group consists almost exclusively of iron component (97%). In addition, these chromites are characterized by an unusually large amount of ZnO (about 2.0%).

The chromites of the first type may be compared in composition with that of the chromites of some ultramafic rocks, mainly dunites, being distinguished from them by the increased amount of chromium component. The chromites of the second type are significantly more difficult to assign to a possible paragenesis, in which, evidently, an increased iron index and a large amount of Cr_2O_3 ought to coincide in the associated minerals. Some analogy may be found in individual samples of chondrites, where the ferriferous chromite is associated with olivine, containing about 25-30% of fayalite (Bunch et al., 1967).

Enstatite

Enstatite was first identified in two diamonds (one Uralian and one Yakutian (Futergendler, 1960)) on the basis of X-ray data. It was later identified in African diamonds, and its composition was determined in four samples from Sierra Leone and Ghana (Meyer and Boyd, 1972), and also in two samples from the 'Mir' pipe (see Table 31).

The principal difference between these enstatites and those from ultramafic xenoliths is the lower (on average) amount of Al_2O_3, which would be expected, arising from data on the solubility of Al_2O_3, in enstatite in the stability field of pyrope, during further increase in pressure (V. S. Sobolev, 1963a; Boyd and England, 1964); see also Figure 17. Along with the insignificant amount of Al_2O_3, not exceeding 1%, the enstatites examined are characterized by the presence of Cr_2O_3 (from 0.1 to 0.55%). Such extremely wide variations

TABLE 30.   Chemical Compositions of Chromites Enclosed in Diamonds

Oxides	'Mir' pipe			'Udachnaya' pipe	Africa			Ghana		Venezuela
	BM-11	BM-8	BM-6	46/8	GL2	GL3	GL3b	G4	G20b	GL47e
SiO$_2$	0,07	0,12	0,12	0,18	0,13	0,29	0,43	0,26	0,23	0,26
TiO$_2$	0,19	0,16	0,12	0,47	0,12	0,09	0,09	0,05	0,03	0,00
Al$_2$O$_3$	4,82	6,71	6,97	5,24	5,12	3,20	3,26	6,74	5,94	5,81
Cr$_2$O$_3$	65,2	64,1	65,0	64,7	67,2	61,4	62,1	63,3	64,0	65,3
Fe$_2$O$_3$	3,70	3,90	2,70	1,20	—	—	—	—	—	—
FeO	14,8	12,7	13,6	14,4	14,5	31,5	31,7	14,4	15,1	10,3
MnO	0,32	0,29	0,26	0,29	0,00	0,42	0,45	0,01	0,00	0,00
MgO	12,0	13,6	13,0	11,8	14,2	0,54	0,48	14,4	13,8	16,4
CaO	0,00	0,00	0,00	0,00	0,02	0,02	0,04	0,01	0,00	0,05
Total	101,20	101,58	101,77	98,28	101,3	99,4	100,8	99,2	99,1	98,1
Si	0,002	0,004	0,004	0,006	0,004	0,011	0,016	0,009	0,008	0,009
Ti	0,005	0,004	0,002	0,012	0,003	0,003	0,003	0,001	0,001	0,000
Al	0,189	0,256	0,266	0,208	0,195	0,138	0,139	0,260	0,231	0,224
Cr	1,703	1,640	1,652	1,733	1,716	1,781	1,776	1,635	1,667	1,686
Fe^{+3}	0,091	0,092	0,066	0,028	0,080	0,062	0,061	0,092	0,093	0,081
Fe^{+2}	0,404	0,344	0,365	0,407	0,313	0,905	0,900	0,301	0,323	0,199
Mn	0,009	0,008	0,008	0,008	0,000	0,013	0,014	0,000	0,000	0,000
Mg	0,592	0,653	0,641	0,596	0,685	0,029	0,026	0,699	0,677	0,799
f	40,6	34,2	36,2	39,9	31,4	96,9	97,2	30,1	32,3	19,9
Cr-Component	85,2	82,0	82,6	86,7	87,2	89,2	89,0	81,9	83,4	84,4
Al-Component	9,5	12,8	13,3	10,4	8,8	7,6	7,9	13,5	12,0	11,8
Ulvospinel	0,5	0,4	0,2	1,2	0,3	0,3	0,3	0,1	0,1	0,0
K$_0$	18,4	21,7	15,2	7,1	20,4	6,4	6,3	23,4	22,4	28,9

Note. In Samples GL2, GL3, and GL3b, the total includes ZnO, 0.004, 1.93, and 2.20% respectively.

in the amount of chromium indicate a paragenesis of these enstatites with pyropes of variable chromium index, which is especially well emphasized by Sample AV-34, which is associated with a pyrope poor in chromium.

An interesting feature of the composition, which may serve as a diagnostic feature for determining the possible paragenesis, is the amount of Na$_2$O in the enstatites from diamonds, which consists in two samples (AV-34 and AV-42) from the 'Mir' pipe of 0.20 and 0.08% respectively. The increased amount of Na$_2$O here may evidently be regarded, as in the enstatites from peridotites (see §14), as a feature of the association with a Na-bearing clinopyroxene, which is extremely significant for the individual inclusions of enstatite. In Sample AV-34, containing an increased amount of Na$_2$O, such an association has been identified in one diamond, and in Sample AV-42, the harzburgitic association without clinopyroxene is more likely. Unfortunately, in the samples from the African diamonds, the amount of Na$_2$O has not been determined, which limits the possibility of assessing their paragenesis. Their assignment to the harzburgite association is most likely, since the high value of the Cr/Cr + Al ratio indicates a paragenesis with chromium-rich garnets, and the overall similarity to Sample AV-42, which contains almost no Na$_2$O, also suggests that clinopyroxene is absent from the paragenesis.

The amount of CaO, that is, the isomorphous amount of the diopside

TABLE 31. Chemical Analysis of Pyroxenes Enclosed in Diamonds

Oxides	Enstatites						Clinopyroxenes											
	'Mir' pipe		Sierra Leone		Ghana		'Mir' pipe		'Udachnaya' pipe		Urals	'Mir' pipe		Africa		Urals		'Mir' pipe
	AV-34	AV-42	GL27a	GL28a	G93a	G10a	AV-34	AV-14	57/9a	57/9b	927	29 9	56/9	M	D-15	880	7a	AV-6
SiO_2	59,3	59,5	57,7	57,0	56,9	57,3	55,4	54,5	55,4	54,1	54,6	54,7	55,6	52,8	52,9	54,9	54,5	55,7
TiO_2	0,15	0,00	0,00	0,00	0,00	0,00	0,33	0,58	0,07	0,10	0,05	0,06	0,08	0,43	0,16	0,33	0,37	0,27
Al_2O_3	0,65	0,62	0,78	0,97	0,67	0,44	3,14	2,75	1,75	1,50	1,61	2,04	2,86	0,86	5,11	5,76	6,34	12,4
Cr_2O_3	0,11	0,37	0,31	0,48	0,55	0,33	0,69	1,01	1,65	1,62	1,88	2,10	3,54	0,09	0,02	0,06	0,09	0,11
FeO	4,22	4,30	4,36	4,48	4,43	4,29	2,05	1,56	1,36	1,11	1,75	1,77	1,92	5,89	7,84	5,87	6,68	5,26
MnO	0,09	0,07	0,09	0,12	0,12	0,12	0,08	0,10	0,03	0,03	0,07	0,09	0,07	0,71	0,12	0,02	0,10	0,09
MgO	36,0	36,2	36,2	35,8	35,9	37,0	16,1	16,5	16,6	16,8	17,3	16,4	15,3	16,1	12,6	12,3	11,5	7,52
CaO	0,32	0,24	0,42	0,45	0,49	0,35	19,0	20,0	21,4	22,1	20,9	19,7	17,5	20,9	19,1	16,6	15,4	12,2
Na_2O	0,17	0,08	—	—	—	—	2,44	2,12	1,40	1,31	1,49	2,25	3,33	1,38	2,30	3,93	3,93	6,89
K_2O	—	—	—	—	—	—	—	—	0,15	0,14	0,00	0,05	0,02	0,04	0,01	0,00	0,00	0,33
Total	101,01	101,38	99,86	99,30	99,1	99,8	99,31	99,12	99,81	98,81	99,65	99,16	100,22	99,2	100,2	99,7	98,9	100,77
Si	2,000	2,000	1,976	1,966	1,969	1,966	2,001	1,977	1,999	1,979	1,980	1,990	1,998	1,970	1,947	1,991	1,993	1,972
Al^{IV}	—	—	0,032	0,040	0,027	0,018	—	0,023	0,001	0,021	0,020	0,010	0,002	0,018	0,053	0,009	0,007	0,028
Ti	0,004	—	—	—	—	—	0,009	0,016	0,002	0,002	0,001	0,001	0,002	0,012	0,004	0,009	0,011	0,007
Al^{VI}	0,026	0,025	0,008	0,013	0,015	0,009	0,134	0,095	0,073	0,045	0,050	0,077	0,119	0,020	0,168	0,229	0,265	0,491
Cr	0,003	0,010	—	—	—	—	0,020	0,029	0,048	0,048	0,052	0,061	0,101	0,003	—	0,001	0,003	0,003
Fe^{3+}	—	—	—	—	—	—	0,062	0,047	0,041	0,033	0,054	0,055	0,056	0,083	0,039	0,039	0,009	0,012
Fe^{2+}	0,119	0,121	0,125	0,129	0,128	0,123	0,002	0,003	0,001	0,001	0,002	0,002	0,002	0,100	0,202	0,140	0,195	0,143
Mn	0,003	0,002	0,003	0,003	0,003	0,003	0,002	0,003	0,001	0,001	0,002	0,002	0,002	0,002	0,002	0,000	0,002	0,002
Mg	1,810	1,814	1,845	1,839	1,849	1,889	0,866	0,891	0,893	0,917	0,934	0,890	0,819	0,895	0,690	0,664	0,626	0,396
Ca	0,012	0,008	0,016	0,017	0,018	0,013	0,736	0,778	0,828	0,865	0,812	0,767	0,674	0,834	0,751	0,645	0,604	0,464
Na	0,011	0,005	—	—	—	—	0,171	0,150	0,100	0,092	0,105	0,159	0,232	0,100	0,164	0,279	0,281	0,472
K	—	—	—	—	—	—	—	—	0,007	0,007	0,003	0,002	0,001	0,000	0,000	0,000	0,000	0,013
Total of cations	3,988	3,985	4,005	4,007	4,009	4,021	4,001	4,009	3,992	4,010	4,010	4,014	4,006	4,037	4,020	4,016	3,996	4,003
l'	6,2	6,3	6,3	6,5	6,5	6,1	6,7	5,0	4,4	3,5	5,5	5,8	6,4	10,0	22,8	17,4	23,8	26,5
f	6,2	6,3	6,3	6,5	6,5	6,1	6,7	5,0	4,4	3,5	5,5	5,8	6,4	17,0	25,9	21,2	25,2	28,1
Ca/Ca+Mg	0,7	0,4	0,9	0,9	0,9	0,7	45,9	46,6	48,1	48,6	46,5	46,3	45,1	48,2	52,1	49,3	49,1	54,0
Ca/Ca+Mg+Fe	0,7	0,4	0,8	0,9	0,9	0,6	44,2	45,3	47,1	47,7	45,1	44,8	43,5	45,6	45,7	44,5	42,4	46,3
Cr/Cr+Al	10,3	28,6	20,0	24,5	35,7	33,3	13,0	19,7	39,3	42,1	42,6	41,2	45,5	—	—	—	—	—

Note. Analyses of Samples GL27a, GL28a, G93a, G10a, M, and D-15, after Meyer & Boyd (1972).

Photo 1. Shape and dimensions of grospydite xenoliths from the
'Zagadochnaya' pipe, Yakutia.

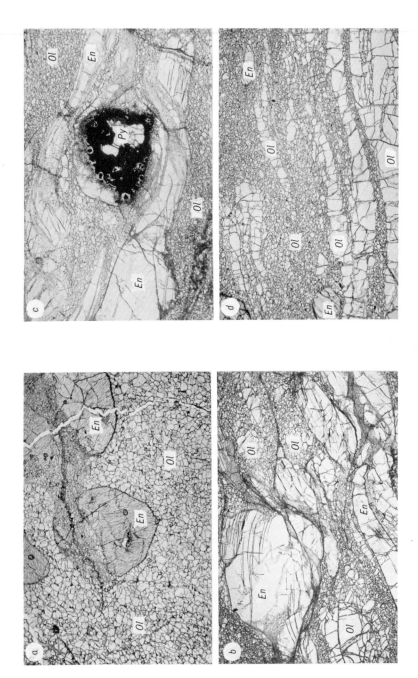

Photo 2. Some specific features of the textures of sheared pyrope
peridotites from Udachnaya pipe, Yakutia. a, b, c, - peculiarities
in deformation of enstatite and olivine grains; d - interbanding
of fine mosaic olivine with relics of coarse olivine grains. Scale:
field width = 0.5 cm. Transmitted light.

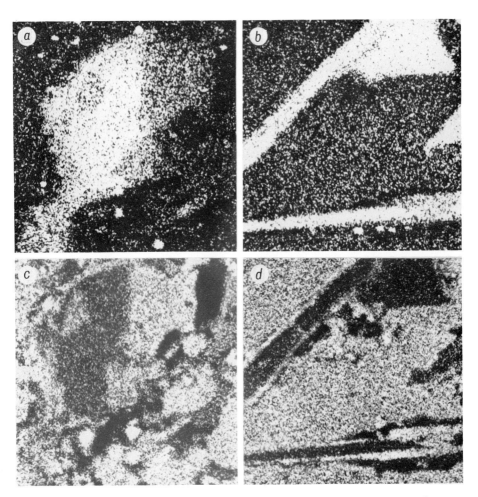

Photo 3. Ingrowths of chrome-spinels in zoned garnet M-49 (a, b -- in X-ray Al K_d-radiation; c, d -- X-ray Ti K_d-radiation).

Photo 4a. Eclogite xenolith of complex composition from 'Obnaz-hĕnnaya' pipe. (0.7 of natural size)

Photo 4b. Banded texture of grospydite, resulting from alternation of bands consisting of kyanite and garnet with pyroxene (garnet -- dark sectors, kyanite and pyroxene -- light sectors). (0.7 of natural size)

Photo 5. Poikiloblastic texture of grospydite (light idiomorphic grains, sometimes with cleavage -- kyanite; light pyroxene is present only in the form of relicts of large altered grains; dark sectors -- alteration products of pyroxene; ordinary light, magnification X 16).

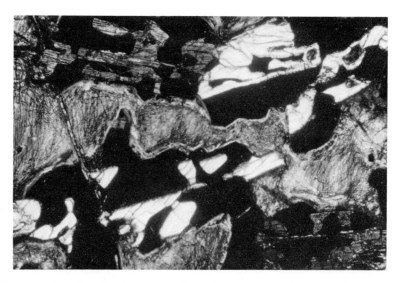

Photo 6. Diablastic intergrowths of garnet (dark sectors) and
kyanite (light) in grospydite (a rim, consisting of secondary products is developed around xenomorphic pyroxene grains). magnification X 20.

0 5cm

Photo 7. Fragment of xenolity of diamond-bearing corundum eclogite
from 'Mir' pipe with impression of one of the apices and facets of
a diamond octahedron below the surface of the xenolith (magnification X 4)

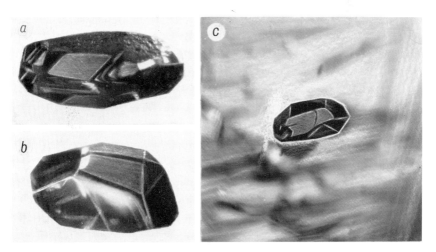

Photo 8. Distorted forms of chrome-pyrope inclusions in diamonds
(*a*, *b* -- elongate crystals, magnification X 30; *c* -- distorted inclusion, magnification X 15)

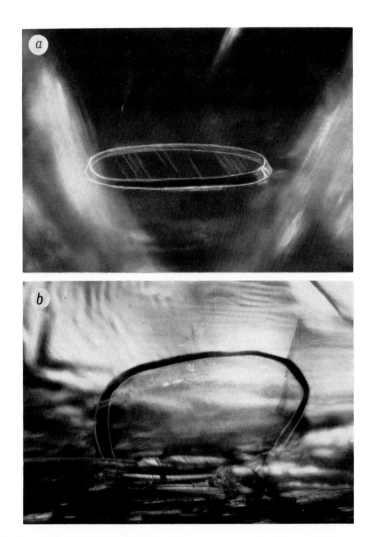

Photo 9. Laminar inclusion of chrome-pyrope in diamond (*a* -- appearance of lamina in cross-section; *b* -- anomalously developed facet, probably (110), magnification X 50)

Photo 10 a. Laminar inclusions of pyrope in diamond (magnification X 50): the laminae are oriented parallel to the octahedral facets; in this same plane is seen a transverse section of a lamina, oriented parallel to another octahedral facet (indicated by arrow).

Photo 10 b. Inclusions of yellow-orange garnet and light-green clinopyroxene with octahedral facets in diamond, magnification X 30.

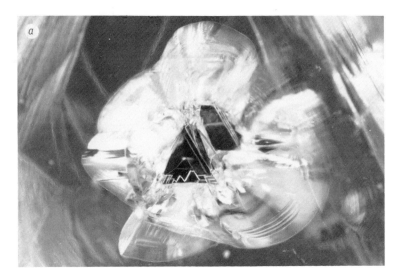

Photo 11 a. Inclusion of chrome-diopside with octahedral facets in diamond, sample AV-7, magnification X 20.

Photo 11 b. Inclusions of pyrope and chrome-diopside with octahedral facets in diamond, along with euhedral inclusion of olivine sample AV-10, magnification X 20.

Photo 12. Intergrowth of chromium-bearing garnet (dark) and enstatite (light) with octahedral form, in diamond, sample AV-93 (seen through adjacent facets of diamond octahedron, magnification X 20.

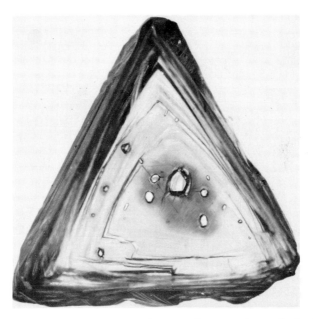

Photo 13. Series of olivine with well developed octahedral facets in twinned diamond crystal (sample AV-37); dark inclusion on left above is euhedral garnet grain; the most developed olivine facets are parallel to facet of diamond octahedron; most of the edges of the inclusions are parallel to those of the outer limit of the diamond (magnification X 15).

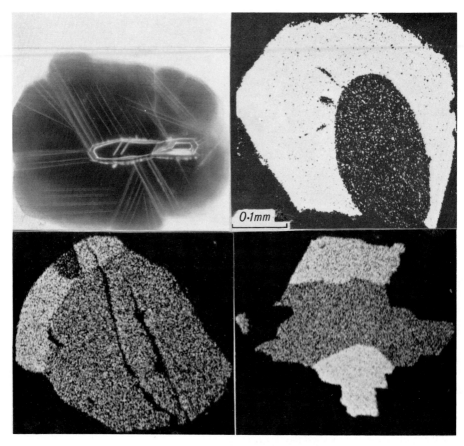

Photo 14. Examples of one-two and three-phase inclusions in diamonds
(a - elongated inclusion of clinopyroxene, surrounded by graphite
halo, in diamond, extracted from diamond-bearing eclogite, sample
M-46 from Mir pipe, magnification X 50; b - intergrowth of garnet
(light) and chrome-diopside (dark) from diamond, sample 57/9, in
X-ray Al K -radiation; c - three-phase intergrowth of inclusions
from diamond of Mir pipe, (sample AV-75), main dark area-garnet,
light rim-enstatite, dark spot within the rim -chromite, in X-ray
MG K -radiation; d - three phase intergrowth of inclusions, sample
AV-25 from Mir pipe; from the top: enstatite, diopside, olivine -
in X-ray Al K -radiation. For c and d scale field width = 0.4 cm.

Photo 15. Segregation of colorless inclusions of olivine in central zone of diamond crystal (*a* -- field of inclusions bounded by octahedral planes, magnification X 20; *b* -- the same inclusions at magnification X 50).

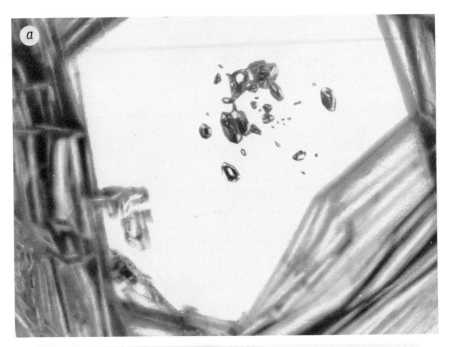

Photo 16. Group of inclusions of yellow-orange garnets in light-green clinopyroxenes in central zone of diamond (a -- general appearance at magnification X 20, b -- detail at magnification X 50).

Photo 17. Inclusion of yellow-orange garnet (left upper) and series
of light-green, almost colorless inclusions of clinopyroxenes in
diamond from 'Mir' pipe (magnification X 25).

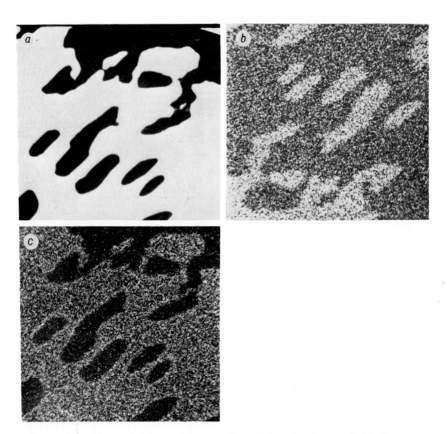

Photo 18. Dissociation texture of solid solutions of Al_2O_3 -
Cr_2O_3 (a -- in absorbed electrons; b -- in X-ray Cr K -radiation;
c -- in X-ray Al K -radiation).

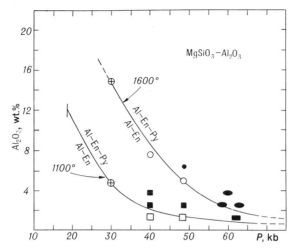

Fig. 17. Isotherms for the enstatite-pyrope solvus in the system $MgSiO_3$-$Mg_3Al_2Si_3O_{12}$, according to Boyd (1970)

component, in the enstaites from diamonds is low and is approximately the same for the six analyzed samples. The value of the iron index is also approximately identical and low for all for the samples, reflecting the feature of the magnesian associations with the diamond.

Clinopyroxenes

Clinopyroxene was first identified in diamond in a sample from the Urals (Futergendler, 1960). The inclusion of greenish tint, was assigned on X-ray data, to diopside. However, inclusions of a bright-green chrome-diopside as shown in a hand-colored plate had been visually recorded earlier in African diamonds (Williams, 1932, Pl. 148, Nos. 2, 3, 4), including several tens of small inclusions in one diamond, and also in company with a crimson garnet, in another. It must be stressed that with chrome-diopside, visual diagnosis is almost unquestionable, since no other bright-green mineral with such a specific color as chrome-diopside has been observed either in diamond or in the xenoliths of deep-seated rocks. In the Yakutian diamonds, two tabular inclusions of bright-green color have been visually recorded and provisionally assigned to chrome-diopside (Bobrievich et al., 1959).

In contrast to other minerals, found in the form of inclusions, not only the X-ray methods of diagnosis but also the study of the optical constants are inadequate for deciding on the compositional features of the clinopyroxenes. Data on the physical properties may serve only for the most general definition of the clinopyroxenes because of complexities in their compositions. Therefore, the direct determination of the composition is particularly important for inclusions of clinopyroxenes. The first such analysis was carried out for a diopside from an African diamond (Meyer and Boyd, 1972). The almost complete absence of Cr_2O_3, the low content of Al^{IV}, and the high content of iron suggested that this diopside be assigned to a paragenesis of the eclogitic type (see Table 31, Sample $\underline{M}$). A difference

from the typical eclogitic pyroxenes is the low content of Na_2O and
the corresponding low content of Al_2O_3. However, pyroxenes of such
composition may also be present in the eclogitic paragenesis, where
the low content of sodium in the pyroxene may be dependent only on
its low amount in the original composition. The garnet, possible in
association with such a pyroxene, may be characterized by an intermediate value for the iron index (about 40-45%). Pyroxene No. D-15
from an African diamond (see Table 31) has a composition intermediate
between diopside and a typical omphacite. Pyroxenes of similar
composition have also been recorded in some xenoliths of eclogites
from the South African kimberlites (Kushiro and Aoki, 1968).

The first discovery of typical omphacites in diamonds was made in
material from the Uralian diamonds (Sobolev, Gnevushev et al., 1971).
In these diamonds, two pyroxenes have been analyzed (see Table 31,
Samples 880 and 7a), the compositions of which are distinguished
almost only by the amount of iron. Calculation of the formulae of
these pyroxenes demonstrates the complete absence of Al^{IV} and the
very low possible amount of the aegirine component. A significantly
greater amount of Fe^{3+} may be assumed for the pyroxene of Sample D-15
and especially for Sample M, if we, of course, assume the complete
absence of errors in determining the amount of silica, which is
unlikely (see Chapter 1).

A substantially larger amount of Na_2O, equal to 6.9%, has been
found for a light-green pyroxene, extracted from a diamond in Sample
AV-6 in the 'Mir' pipe (see Table 31). Along with the analyzed
idiomorphic crystal, there is also an intergrowth in this diamond of
two crystals of pyroxene, measuring about 0.05 mm. The composition
of this pyroxene is practically indistinguishable from the average
composition of the pyroxenes from the diamond-bearing eclogites
(see §13). Additional emphasis for the eclogitic nature of the
paragenesis comes from the combined occurrence of two crystals of
orange garnet and the indicated pyroxene crystals in the diamond
from Sample AV-6.

Thus, the five pyroxene inclusions examined from diamonds may
primarily, on the basis of absence of chromium and increased iron
index, be assigned to a basic type of paragenesis, and on the basis
of the sodium content, to the eclogitic type. The average composition of these pyroxenes (see §13) is characterized by comparatively
small fluctuations in the iron index and the Ca/Ca + Mg + Fe ratio,
although the amount of sodium and also Al, Ca, and Mg vary extremely
widely. The difference in the amount of sodium, as recorded above,
may be associated with the difference in its concentration in the
diamond-forming medium.

In addition to these pyroxenes, which form an independent type,
inclusions of chrome-bearing clinopyroxenes have been studied in the
diamonds from Yakutia and the Urals (see Table 31), the composition of
which is of interest in comprasion with pyroxenes, widely distributed
in ultramafic parageneses (Sobolev, Bartoshinsky et al., 1970).
The amount of Cr_2O_3 in them varies over quite a wide range, exceeding
the maximum figure for the pyroxenes of the kimberlites and the
pyrope peridotites (Boyd, 1970). The same must also be noted for
Na_2O. However, these anomalously large amounts of Na_2O and Cr_2O_3 as
compared with the typical pyroxenes of the peridotites apply only to
one sample of Yakutian diamond (56/9), which is best termed a chrome-
omphacite, rather than chrome-diopside.

Of the seven analyses of chrome-bearing pyroxenes considered, two inclusions have been extracted from one diamond and consist of an individual inclusion (57/9a) and a pyroxene from an intergrowth with pyrope (57/9b). Their compositions are distinguished mainly within the limits of analytical error (Sobolev, Bartoshinsky et al., 1970). In the paragenesis with these pyroxenes, olivine has also been identified.

A common feature of all the chrome-bearing pyroxenes from diamonds examined is the increased value of the Ca/Ca + Mg ratio, the low iron index, and the almost complete absence of Al^{IV}.

Other Inclusions

R u t i l e was first identified in diamond by X-ray methods (Harris, 1968a). There has been great interest in the identification of rutile inclusions in diamonds, extracted directly from a xenolith of a diamond-bearing eclogite (Gurney et al., 1969).

C o e s i t e is perhaps the rarest of inclusions and has been found previously in only two African diamonds (Harris, 1968a). Its rarity has prevented the achievement of any completely unequivocal conclusion on its syngenetic relations with the diamond. In one case, the coesite consists of a well-facetted crystal, and in the other, it consists of a series of grains of irregular shape. The frequency of occurrence of the coesite is very difficult to estimate statistically, since without the use of X-ray techniques it is almost indistinguishable from colorless inclusions of the olivine type which are widely distributed in the diamond. Final assessment of the nature of coesite in diamond requires further careful search for it and the identification of the paragenesis. Since coesite (if it be syngenetic with diamond) cannot occur together with olivine, it most probably belongs to a specifically eclogitic paragenesis. In such a case, with the object of extending the possibility of diagnosing it visually, we may recommend the search for and consequent later careful checking of colorless inclusions, associated with orange garnet in one diamond. First discovery of the paragenesis of coesite in nature was made in two diamonds from the places of the northern part of the Siberian platform: coesite + garnet + omphacite in each diamond sample (Sobolev, Yefimova et al., 1976).

M a g n e t i t e has also been noted amongst the identified syngenetic inclusions (Harris, 1968a), and confirmed by Prinz et al., (1975) as a mineral of eclogitic paragenesis.

I l m e n i t e has been recorded as a possible syngenetic inclusion on the basis of its wide distribution in kimberlite (Harris, 1968a). However, it is extremely clear that magnesian ilmenite has not so far been successfully identified as an inclusion in diamond, which is most likely associated with the particular rarity of such inclusions. Magnesium-free ilmenites were identified by Meyer and Svisero (1975) in Brazilian diamonds. Cases of intergrowth of ilmenite with diamond are alone known, and it has been emphasized (Harris, 1968a) that ingrowths of ilmenite on the outer portions of diamond crystals are quite frequent.

D i a m o n d is also typical in the form of inclusions, and the diamond-captive in most cases consists of plane-facetted octahedra which are independent of the shape of the crystal enclosing it.

Probable Syngenetic Inclusions

In addition to the series of minerals, widely distributed in kimberlites, but not found in the form of inclusions in diamonds, such as phlogopite, zircon, and moissanite, attention must, in our opinion be focussed on the possibility of finding as syngenetic inclusions in diamonds, certain minerals which are extremely rarely found in the concentrates of the heavy fraction of the kimberlites. These minerals may include kyanite* and corundum, and a suitable basis for such a proposal may be the discoveries of kyanite and corundum diamond-bearing eclogites (see §8). It is especially important to dwell on the possibility of finding in the diamond an exceptionally rare variety of kyanite, containing Cr_2O_3, the presence of which has been demonstrated in the grospydites for the specific conditions of simultaneous increased concentrations of Al_2O_3 and Cr_2O_3 (Sobolev, Kuznetsova and Zyuzin, 1966). Thus, in the diamond we may find a chrome-kyanite, containing the largest amount of Cr_2O_3 as compared with those found already in the concentrate from the kimberlite from the 'Zagadochnaya' pipe and even compositionally approaching Cr_2OSiO_4, the chromium analogue of kyanite. Such an hypothesis may also be proposed relative to the discovery of compositions in the diamond, belonging to the system $Al_2O_3-Cr_2O_3$ corundum-eskolaite). Discoveries of such inclusions are most likely in the 'Roberts Victor' pipe, which contains diamond-bearing kyanite eclogites. There is no doubt that a further detailed study of the inclusions in diamonds with the aid of all the available methods should produce many new precise facts, so necessary for unequivocal conclusions about the conditions of formation of the diamond and the composition of the diamond-forming medium, and consequently, about the substrate of significant depths of the upper mantle.

Problematic Syngenetic Inclusions

In a detailed consideration of the inclusions in diamonds, the problem may arise as to whether all the possible syngenetic inclusions have been found in them. In actual fact, the list of minerals identified in diamonds is quite small and does not include many minerals that occur in the kimberlite pipes together with the diamonds. Such minerals, not found in diamonds, may include m o i s s a n i t e, first identified under terrestrial conditions as an accessory mineral in a xenolith of pyrope peridotite (Bobrievich et al., 1957) and later in abundance found in a concentrate of the heavy fraction of several kimberlite pipes (Marshintsev et al., 1967). The presence of moissanite in kimberlite has even served as a basis for assigning this mineral to the principal satellites of the diamond (Vasil'ev et al., 1968). However, all X-ray investigations so far carried out on inclusions, which have produced conclusions about the distribution of particular minerals in diamonds, have given a negative answer to the question of the presence of moissanite inclusions. So far not a single confirmed moissanite inclusion has been found (Harris, 1968a, b).
P h l o g o p i t e, which is one of the most widely distributed minerals in kimberlites, has also not been found in diamonds. If we

*According to Prinz et al., (1975) kyanite has been found in the form of an inclusion in an African diamond.

could succeed in finding a confirmed syngenetic phlogopite in diamond, this would immediately answer the question as to the source of potassium and water in the stability field of the diamond. However, such a discovery is unlikely. If phlogopite had actually crystal- lized in the stability field of the diamond, it would be amongst the most widely distributed inclusions in diamonds. So far, only isolated occurrences of phlogopite are known in intergrowths with diamond in South Africa (Boyd, 1969) and the 'Mir' pipe.

Ilmenite, or more strictly picroilmenite, one of the commonest minerals in kimberlite, has not been identified in the form of inclu- sions, although it has frequently been found in intergrowths with diamonds (Harris, 1968a). Judging by the peculiar composition of ilmenite from intergrowths with diamonds, almost devoid of Fe^{3+}, the ilmenite possible as inclusions in diamonds must comprise an insig- nificant portion of the total ilmenite in the kimberlite pipes. The visual diagnosis of possible ilmenite inclusions in diamonds is extremely difficult, owing to the similarity of its small grains to those of chromite, which markedly predominates in the inclusions. However, there is absolutely no doubt that true inclusions of picroilmenite will be found in diamonds, most probably in a para- genesis with chromite.

Z i r c o n was recorded as an inclusion in diamond on the basis of a preliminary diagnosis, but as a result of more careful X-ray investigations, this conclusion has not been supported (Harris, 1968a). The reported visual determinations of zircon in diamonds (Williams, 1932; Kukharenko, 1955) have clearly been due to incorrect diagnosis of inclusions of a yellow-orange garnet and olibine. A zircon inclu- sion has been identified recently in a Brazilian diamond (Meyer, Svisero, 1975).

In general, when considering the position of zircon in kimberlites, where it constantly accompanies diamond, there is a predominant view that the zircon occurs in the kimberlite as a result of crushing of xenoliths of metamorphic rocks. The discovery of a xenolith of an unusual zircon-bearing pyroxenite in the South African kimberlites (Williams, 1932) has stimulated a series of investigations to con- sider the part played by zircon in the pipes as a syngenetic mineral with diamond or, in any case, genetically associated with the kimber- lite (Ilupin and Kozlov, 1970)*. A search for such zircon inclusions in diamond has so far yielded only one positive result (Meyer, Svisero, 1975). In this respect, with the object of comparison with the composition of other inclusions, an attempt to determine the composition of the garnet enclosed in a grain of zircon from the South African kimberlites, may be of interest. This sample was kindly presented to us for examination by Dr. H. Milledge (England). The microprobe analysis of the garnet gave: SiO_2, 36.3%; Al_2O_3, 20.0%; FeO, 29.8%; MnO, 9.77%; MgO, 1.46%; CaO, 0.71%; total, 98.04%; $\underline{f}$ = 92.0%.

Along with predominance of the almandine component, this garnet has a substantial amount of the spessartine component (23%), which dis- tinguishes it markedly from all the garnets of the deep-seated xenoliths. A comparison of the composition of the garnet with

---

*The ultramafic type of paragenesis of certain zircons from the South African kimberlites is indicated by discoveries of phlogopite, and also picroilmenite (MgO = 8.77%) in the form of inclusions in individual zircon grains (Kresten, 1975).

those of the known paragenetic types of garnets (Sobolev, 1964a)
demonstrates its complete analogy to those from the muscovite pegmatites. Thus, in this case, we have not only succeeded in proving
the alien nature of the zircon grain studied in relation to the kimberlite, but also in providing a definite answer as to its assignment
to the muscovite pegmatites.

By the same method, we have obtained another unique result, sup-
porting the possible association of zircon with the ultramafic
paragenesis. Thus, in one of the zircon grains from the Yakutian
kimberlites, a birefringent inclusion of emerald-green color has been
found (Khar'kiv et al., 1972). A preliminary investigation of this
inclusion suggested that it is a chrome-diopside, and this has
subsequently been confirmed by a X-ray microprobe analysis: SiO_2,
52.7%; TiO_2, 0.08%; Al_2O_3, 1.49%; Cr_2O_3, 2.64%; FeO, 20.9%; MnO,
0.10%; MgO, 16.3%; CaO, 20.6%; Na_2O, 2.24%; total, 98.24%; Ca/Ca +
Mg = 47.6%; Cr/Cr + Al = 54.2%; $\underline{f}$ = 6.7%.

In interpreting this result, we must take account that, so far, no
inclusions of zircon have been found in xenoliths of deep-seated
peridotites in the Yakutian pipes, although many of the samples have
been carefully checked for the possible presence of moissanite after
its first discovery in a xenolith of peridotite (Bobrievich et al.,
1957). Therefore, if zircon had been present in the heavy fractions
of the rocks, it would certainly have been recorded. However zircon
has been found in garnet lherzolite from Bultfontein (Boyd, written
communication).

In some peridotites with pyrope, known in particular as xenoliths
in the Minusinsk basaltoid pipes, zircon has been found in quantities,
even sufficient for determining the absolute age of these rock fragments (Kryukob, 1968). Finally, zircon has been identified with
the aid of X-ray methods as inclusions in Czech pyropes (Bauer, 1966).
These cases, and also the absence of zircon in xenoliths of especially
deep-seated peridotites and diamonds, in our opinion, indicate the
assignment of zircon mainly to certain types of shallower pyrope-
bearing ultramafic rocks. It is possible that the presence of zircon
may serve as a supplementary feature, distinguishing the shallower
pyrope peridotites of the Czech Massif and the Minusinsk pipe types
from those with similar mineral composition in the 'Udachnaya' pipe,
which contain chrome-rich garnets. The possibility of finding
zircon-bearing peridotites is also not excluded in the Yakutian
kimberlites; support for this is the discovery of chrome-diopside
inclusions in a zircon, enabling us with confidence to assign this
zircon grain to the ultramafic association.

Thus, on the basis of the above information, there seems little
likelihood of finding syngenetic inclusions of moissanite, phlogopite,
and zircon* in diamonds. The wide distribution of these minerals in
kimberlite and the nature of the accumulated data on the precise
diagnosis of inclusions in diamonds are the most favorable basis for
such an assumption.

The facts at our disposal about the discoveries of ilmenite in
intergrowths with diamond (see §10), based on the composition of
markedly different ilmenites occurring in the pipes, suggest that a
small portion of it crystallized in equilibrium with diamond and

---

*The first confirmed inclusion of zircon in diamond has been
found and studied in a crystal of Brazilian diamond (Meyer and
Svizero, 1975).

chromite in an ultramafic association. Only such an ilmenite with
a maximum content of Fe^{3+} may be found as a typical inclusion in
diamond.

§10. MINERAL INTERGROWTHS WITH DIAMONDS

A study of the minerals, occurring in intergrowth with diamonds,
is of great interest, along with an investigation of inclusions in
diamonds and diamond-bearing xenoliths, because such intergrowths
are a valuable additional source of information about the composition
of the medium and the conditions of formation of diamonds. Intergrowths of minerals with diamonds may be separated into two types:
1) fragments of diamond-bearing eclogites and other diamond-bearing
rocks with entire crystals of diamond, growing into other minerals;
2) polycrystalline diamond aggregates, on the surface of which are
intergrowths of other minerals.

In the intergrowths of the first type, diamond, certainly in the
form of individual crystals, comprises a minor portion, and sometimes
has seemingly been enclosed in garnet or another mineral. Such
intergrowths of diamond and yellow-orange garnets are evidently also
fragments of diamond-bearing eclogites, for example, Sample ME-131
from the 'Mir' pipe. In the intergrowths with diamonds, both garnet
and chromite must commonly predominate over other minerals, including
clinopyroxene, as a result of their greater stability towards
secondary alterations.

Intergrowths are known, in which a chromium-bearing garnet or
pyroxene predominates over diamond. Thus, in Williams' monograph
(1932), clear figures are presented (plates 148, 149) from which it
may be seen that the diamond crystal occupies a subordinate position
with respect to the rounded grain of lilac garnet (pl. 148, No. 6),
the diamond octahedron is intergrown on one facet with a lilac garnet
(pl. 149, No. 7), and is also involved in the composition of the
intergrowths of lilac garnet and chrome-diopside (pl. 149, Nos. 8
and 9).

In the collection of Yakutian diamonds available to us for examination, there is a sample with an ingrowth of a diamond octahedron
into a rounded grain of crimson pyrope (Sample No. ME-137), and also
an ingrowth of diamond into a red-orange garnet, in which we found
an idiomorphic inclusion of clinopyroxene. These samples may
evidently also be regarded as fragments of diamond-bearing xenoliths,
differing in composition both from eclogites and peridotites
(dunites) with diamonds. Such fragments may also include Sample
No. ME-155, the main portion of which is a pyrope of red color.

The polycrystalline diamond aggregates of complex construction
comprise a separate group; on their external surface, minerals are
often found in cavities intergrowing with them.

When considering the features of minerals, associated with diamonds
in intergrowths, it follows, in our opinion, to use examples those
minerals or groups of minerals, the syngenetic nature of which with
respect to diamond, may be regarded as established in the basis of
a study of typical inclusions in diamonds. Such a standpoint is necessary, because on the diamonds and even more so on the diamond
aggregates, in the surface of which there are numerous cavities,
later minerals may be superimposed, which crystallized in the
kimberlite significantly later than the diamond itself. Such minerals,
which probably crystallized during replacement of other silicates,

may include the phlogopite from the polycrystalline diamond aggregate described below.

The minerals, possibly in equilibrium with diamond, and in any case with the outer zones of diamond, may (by analogy with the syngenetic inclusions) in the intergrowths include olivine, chromite, garnets (pyropes and pyrope-almandines), pyroxenes, and also ilmenite, rutile, and possibly kyanite and corundum.

The compositions of the corresponding mineral-inclusions may serve as a standard. A special case here is that of ilmenite, for which numerous intergrowths with diamonds have been identified, but not a single ilminite crystal in the form of a definite inclusion in diamond is known (Harris, 1968a).

When considering the features of intergrowths of minerals with polycrystalline diamond aggregates, assigned by us to the second type, it should be noted that the garnets, pyroxenes, and chromites do not simply fill the cavities in the surfaces of the diamond aggregates, but in all the samples examined, they are clearly intergrown with the diamond, being present not only in the outermost portion of the aggregates, but overgrowing the diamond also. Such a feature seemingly suggests that those minerals, whose syngenetic nature with diamonds has been proved on the basis of the examination of typical inclusions, crystallized in equilibrium with the diamond aggregates, and in any case, with their external zones.

Garnets

The garnets intergrowing with diamonds, like the inclusions, may be separated into pyropes and pyrope-almandines. During the study of intergrowths of garnets and polycrystalline diamond aggregates from the 'Mir' pipe (Sobolev, Botkunov et al., 1971), calcium-poor chrome-pyropes have been recorded which are completely analogous in composition with the pyropes enclosed in the diamonds (Table 32, Samples B-2, B-3, B-14, and B-12). These garnets are characterized by variations in the amount of Cr_2O_3 from 6.17 to 10.8%, and CaO, from 0.62 to 3.05%. Such garnets comprise about 50% of all the pyrope garnets from the intergrowths examined.

Along with such garnets, already well-known in the form of inclusions, pyropes with considerably smaller amounts of Cr_2O_3 (from 1.24 to 3.59%), characterized at the same time by a larger amount of CaO (2.45 - 4.67%) (see Table 32, Samples ME-155, B-10, 59/12, B-1, B-20, B-13, and ME-137), have been identified in intergrowths with polycrystalline diamond aggregates. Their compositions in individual cases are quite similar to those of normal pyropes, found in kimberlite concentrates, although a clear difference from the magnesian garnets of these concentrates is the lower iron index of the garnets from the intergrowths with diamonds, which comprises 12-16% for the magnesian garnets examined from these intergrowths, with some isolated exceptions (e.g. Sample 59/12).

In regard to the established variation in the garnet compositions, it is interesting to compare the quantitative relationships between the pyropes of various types, associated with diamond. Whereas, in most cases, only pyropes rich in chromium and poor in calcium, forming a special type of magnesian garnet (Fig. 18), occur in the form of inclusions in diamonds, those in the intergrowths in half of the cases consist of chromium-poor types, and gradual transitions between the two types of magnesian garnets have not so far been found.

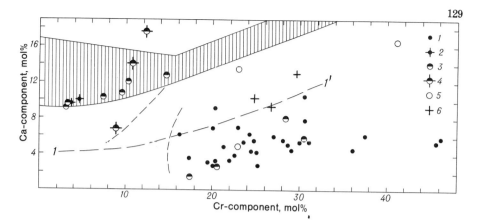

Fig. 18. Compositional features of garnets of ultramafic parageneses based on amount of calcium and chromium. 1) from diamonds harzburgite-dunite paragenesis); 2) from diamonds (lherzolite paragenesis); 3) from intergrowths with diamonds; 4) from intergrowths with diamonds (along with clinopyroxene); 5) from diamond-bearing peridotites (dunites); 6) from chromite-pyroxene-garnet intergrowths in kimberlite; 1 - 1') lower possible boundary of amount of calcium in garnets from paragenesis with Na-clinopyroxenes.

This relationship remains practically unchanged in a visual estimate of the number of samples containing chromium-rich and -poor pyropes, and separately for the inclusions and intergrowths with polycrystalline diamond aggregates. Such a visual estimate is quite accurate, because the garnets, markedly differing in the amount of chromium, can be separated by the intensity of color. The chrome-poor pyropes have a predominantly light, pinkish tint.

Amongst the 24 samples of polycrystalline diamond aggregates available for examination, which contain garnets (together with the analyzed types), chrome-pyropes have been identified in 10 samples and the remainder are light-tinted chromium-poor pyropes.

In regard to the pyrope-almandine garnets with variable amounts of calcium (see Table 28, Samples ME-131, B-6, B-18, and B-22), they are in every way, including the increased amount of Na_2O (Sobolev and Lavrent'ev, 1971), similar to the corresponding garnets from inclusions in diamonds and diamond-bearing eclogites. It is interesting that in spite of the extremely limited number of garnet samples of such type from intergrowths, the limits of variation in their composition, based on the calcium content (from 9 to 40% calcic component), characterize the entire known range for the eclogitic garnets. In this respect, it must be emphasized that the garnet from Sample B-22 is the most calcic amongst those known in association with diamond, both from the inclusions, and also from the diamond-bearing eclogites, including the kyanite and corundum types. The variations in iron index in these garnets are also extremely significant (from 30 to 52%).

Clinopyroxenes

The pyroxenes from intergrowths consist of clinopyroxenes only, and enstatite in intergrowth with diamond has not so far been

TABLE 32. Chemical Compositions of Pyropes from Intergrowths with Diamonds from the 'Mir' Pipe

Oxides	ME-155	B-10	B-1	B-20	B-13	59/12	MR-4	ME-137	B-2	B-3	B-14	B-12
SiO_2	43,2	43,7	43,2	42,6	42,9	42,0	41,1	42,9	42,6	43,0	40,7	42,9
TiO_2	0,31	0,23	0,12	0,24	0,39	0,32	0,41	0,25	0,01	0,00	0,03	0,05
Al_2O_3	22,1	22,5	21,3	20,7	20,5	19,3	19,7	19,4	18,9	18,5	15,9	15,6
Cr_2O_3	1,00	1,24	3,14	3,31	3,59	3,74	4,21	5,24	6,17	7,28	9,81	10,8
FeO	7,23	7,30	6,94	6,84	5,42	9,14	5,75	6,98	5,41	5,81	6,51	5,65
MnO	0,38	0,31	0,34	0,23	0,21	0,37	0,26	0,19	0,22	0,26	0,29	0,25
MgO	21,9	21,8	23,3	20,6	21,8	19,4	19,7	20,9	24,2	24,2	21,8	23,2
CaO	3,40	3,67	2,45	3,88	4,67	5,30	6,79	4,90	0,62	1,10	3,05	2,21
Na_2O	0,08	0,05	0,13	0,07	—	0,04	—	0,06	0,01	0,01	0,02	0,01
Total	99,60	100,80	100,92	98,47	99,48	99,61	97,92	100,82	98,17	100,17	98,11	100,69
Si	3,048	3,049	3,021	3,062	3,043	3,045	3,001	3,045	3,055	3,042	3,004	3,062
Ti	0,017	0,013	0,004	0,013	0,021	0,017	0,022	0,013	0,000	0,000	0,001	0,002
Al	1,840	1,854	1,756	1,753	1,713	1,646	1,694	1,621	1,594	1,538	1,385	1,312
Cr	0,059	0,067	0,176	0,190	0,205	0,215	0,246	0,294	0,353	0,408	0,569	0,609
Fe^{3+}	0,084	0,066	0,064	0,044	0,061	0,122	0,038	0,072	0,053	0,054	0,046	0,077
Fe^{2+}	0,344	0,362	0,339	0,366	0,259	0,431	0,313	0,342	0,291	0,290	0,353	0,262
Mn	0,025	0,017	0,021	0,013	0,013	0,022	0,018	0,013	0,013	0,017	0,018	0,017
Mg	2,302	2,269	2,428	2,207	2,305	2,095	2,145	2,209	2,585	2,550	2,401	2,466
Ca	0,259	0,273	0,185	0,302	0,354	0,414	0,531	0,371	0,047	0,085	0,244	0,167
Na	0,012	0,007	0,022	0,010	—	0,006	—	0,009	0,001	0,001	0,002	0,001
Pyrope	78,3	77,4	75,8	74,6	76,3	67,0	71,3	69,0	70,2	66,1	56,9	55,9
Almandine	11,7	12,4	11,3	12,6	8,8	14,5	10,4	11,6	9,9	9,9	11,7	9,0
Spessartine	0,8	0,6	0,7	0,4	0,4	0,7	0,6	0,4	0,4	0,6	0,6	0,6
Grossular	1,1	2,2	—	—	—	—	2,4	—	—	—	—	—
Andradite	4,2	3,3	3,2	2,2	3,1	6,1	1,9	3,6	1,6	2,7	2,3	3,9
Ti-andradite	0,9	0,7	0,2	0,7	1,1	0,9	1,1	0,7	—	—	0,1	0,1

Uvarovite	3,0	3,4	3,5	7,9	7,9	7,2	12,3	8,6	—	0,2	5,8	1,8
Knorringite	—	—	5,3	1,6	2,4	3,6	—	6,1	17,8	20,2	22,6	28,7
f	15,7	15,9	14,2	15,7	12,2	20,9	14,1	15,8	11,1	11,9	14,3	12,1
Mg-component	78,3	77,4	81,1	76,2	78,7	70,6	71,3	75,1	88,1	86,6	79,5	84,6
Ca-component	9,2	9,6	6,9	10,8	12,1	14,2	17,7	12,9	1,6	2,9	8,2	5,8
Cr-component	3,0	3,4	8,8	9,5	10,3	10,8	12,3	14,7	17,7	20,4	28,4	30,5

identified. All the known clinopyroxenes from intergrowths with
diamonds belong to the chromium-bearing varieties (Table 33) with the
lower limit of Cr_2O_3 at 0.6%. We must take particular note of the
difference, for the most part, between the clinopyroxenes examined
from intergrowths and analogous minerals, enclosed in diamond, based
on the wider variation in the Cr_2O_3 and Na_2O content. From these
features, they are distinguished from the pyroxenes of the peridotite
xenoliths in the kimberlite pipes. In the intergrowths with polycrystalline diamond aggregates, pyroxenes with a large amount of the
ureyite component $(NaCrSi_2O_6)$ (more than 10 mol %) were first
described under natural conditions (Sobolev, Botkunov et al., 1971).
Moreover, a study of the pyroxenes from intergrowths has revealed
compositions in which the amount of the ureyite component exceeds 30
and even 40 mol %. These are Samples B-7 and MR-9 from the 'Mir' pipe,
also characterized by similar paragenetic features. In intergrowths
with polycrystalline diamond aggregates, these pyroxenes form, in
turn, intergrowths with chromite with marked quantitative predominance
of the latter. The chromites of such intergrowths are almost completely devoid of aluminum (Table 34).

Other Minerals

C h r o m i t e is the most widely distributed mineral in intergrowths with diamonds. It is usually present in the form of numerous
grains clearly overgrowing the diamond aggregates. Determination
of the composition of the chromite in samples from the 'Mir' pipe
(see Table 34) has shown complete analogy with that of the chromite
enclosed in the diamonds. These data have suggested some extension
of the limits of variation of the composition of the chromites,
associated with diamonds, in the direction of a larger amount of
Cr_2O_3 up to 68.6%, and consequently, a decrease in the amount of
Al_2O_3 to 1.5 - 2.0%. The values of the iron index and the coefficient
of oxidation (K_0) for the chromites from intergrowths are analogous
to those for the chromites, enclosed in diamonds. The complete
identity of the composition of the chromite in the intergrowths with
that in the inclusions is of great significance in assessing the
equilibrium nature of the minerals of the intergrowths and diamond.

There is substantial interest in the investigation carried out on
the composition of i l m e n i t e, associated with diamond in the
intergrowths. Ilmenite compositions have been determined in samples
from the 'Mir' pipe (BM-30 and BM-10), representing monocrystals of
diamond with ilmenite, occurring on the outer surface. Such a relationship with the diamond does not allow us to consider these
ilmenites as typical inclusions in the diamond. In a third sample
(MR-5), ilmenite has been identified in the outer zone of a polycrystalline diamond aggregate.

The compositions of these ilmenites (Table 35) are sharply
distinguished from almost all those studied so far from the kimberlite concentrates with increased content of TiO_2, and consequently,
a decreased amount of Fe_2O_3. It is interesting that amongst more
than 90 analyzed ilmenites from the Yakutian kimberlites (Milashev
et al., 1963; Bobrievich et al., 1964; Frantsesson, 1968), only
isolated samples contain more than 53% TiO_2.

Comparison of the composition of the ilmenites studied from the
intergrowths with diamonds with all those from the kimberlites (see
§19), based on the content of MgO and TiO_2 (wt %), graphically

TABLE 33. Chemical Composition of Clinopyroxenes from Intergrowths with Diamonds

Oxides	South Africa GL-50	'Mir' Pipe 59/12	B-23	B-8	B-1	MR-4	MR-10	B-9	B-7	MR-9
SiO_2	54,7	55,2	55,5	55,0	56,9	55,1	54,8	55,0	55,5	54,7
TiO_2	0,2	0,05	0,20	0,14	0,12	0,42	0,17	0,15	0,19	0,17
Al_2O_3	2,64	1,01	3,05	4,37	8,00	4,87	4,89	3,23	3,14	4,56
Cr_2O_3	1,36	0,63	1,34	1,62	3,04	3,45	4,41	6,15	11,8	15,6
FeO	2,37	3,41	1,76	2,23	2,12	1,52	2,12	1,76	1,68	1,73
MnO	0,1	0,13	0,08	0,04	0,10	0,26	0,07	0,07	0,05	0,09
MgO	17,5	19,0	15,2	14,5	12,0	13,1	12,0	12,9	9,27	6,25
CaO	18,9	19,2	19,1	17,1	11,0	16,9	13,6	14,7	10,6	6,21
Na_2O	2,1	0,87	2,45	3,83	6,58	4,33	5,83	4,93	7,07	10,1
K_2O	0,05	0,00	0,07	0,04	0,00	0,03	0,02	0,00	0,05	0,00
Total	99,9	99,50	98,70	98,87	99,86	99,68	97,91	98,89	99,35	99,41
Si	1,973	1,999	2,013	1,993	2,014	1,987	2,009	2,005	2,023	2,007
Al^{IV}	0,027	0,001	—	0,007	—	0,013	—	—	—	—
Ti	0,004	0,001	0,007	0,004	0,002	0,011	0,004	0,004	0,007	0,004
Al^{VI}	0,085	0,043	0,131	0,180	0,334	0,182	0,211	0,136	0,136	0,199
Cr	0,039	0,017	0,038	0,048	0,085	0,100	0,128	0,180	0,342	0,435
Fe	0,072	0,102	0,054	0,068	0,064	0,046	0,064	0,055	0,053	0,053
Mn	0,003	0,004	0,001	0,001	0,002	0,009	0,002	0,002	0,002	0,002
Mg	0,941	1,025	0,821	0,784	0,634	0,704	0,656	0,701	0,504	0,342
Ca	0,730	0,744	0,743	0,664	0,417	0,652	0,535	0,574	0,414	0,245
Na	0,145	0,065	0,172	0,274	0,451	0,303	0,414	0,351	0,504	0,718
K	0,002	0,000	0,003	0,002	0,000	0,001	0,001	0,000	0,002	0,000
Total of cations	4,021	4,001	3,983	4,025	4,003	4,008	4,024	4,008	3,987	4,023
f	7,1	9,1	6,2	8,0	9,2	6,1	9,0	7,3	9,5	13,4
Ca/Ca + Mg	43,7	42,1	47,5	45,9	39,7	48,0	45,0	45,0	45,0	41,7
Ca/Ca + Mg + Fe	41,9	39,8	45,9	43,8	37,4	46,5	42,6	43,2	42,6	38,3
Cr/Cr + Al	43,7	27,9	22,5	20,4	20,3	34,0	37,8	57,0	71,5	69,5

Note. Analysis of Sample GL-50, after Boyd & Nixon, 1970.
demonstrates their similarity to the theoretical composition of $FeTiO_3$–$MgTiO_3$ solid solutions, almost without addition of Fe_2O_3. This similarity will be even closer if we take account of the amount of Al_2O_3 and Cr_2O_3 in the composition of the ilmenites from intergrowths and diamonds, reaching in total 0.5%.

Although analysis with the aid of the microprobe does not make it possible directly to determine Fe_2O_3, recalculation of analyses allowing for the amount of Fe^{3+} both in the group of divalent and that of trivalent cations (Mikheev and Kalinin, 1961), permits an objective allowance for Fe^{3+} in the composition of ilmenite.

The features of the composition of ilmenite Samples BM-10 and BM-30 suggest their paragenesis with chromite and, possibly, pyrope and a diamond-bearing medium of ultramafic composition. It is evident that ilmenite in such a medium belongs amongst the latest minerals

TABLE 34. Chemical Composition of Chromites and Phlogopite from Intergrowths with Diamonds

Oxides	Chromites						Phlogopite
	BM-13	A-3a	B-7	MR-9a	BM-7	MR-9b	MR-9
SiO_2	0,08	0,16	0,10	0,10	0,13	0,24	43,0
TiO_2	0,55	0,74	0,25	0,15	0,32	1,89	2,38
Al_2O_3	5,59	5,78	2,31	1,98	2,24	7,06	11,9
Cr_2O_3	64,4	65,3	65,8	67,0	68,6	50,9	2,54
Fe_2O_3	4,00	2,2	2,0	1,3	1,83	5,6	—
FeO	13,7	16,4	19,2	18,3	17,7	19,6	3,10
MnO	0,27	0,27	0,28	0,27	0,36	0,26	0,0
MgO	13,1	11,2	8,73	8,97	9,77	10,2	24,2
Total	101,8	102,05	98,67	98,07	100,95	95,75	95,1
Si	0,003	0,006	0,004	0,004	0,004	0,009	3,003
Ti	0,014	0,018	0,007	0,004	0,008	0,051	0,125
Al	0,215	0,224	0,096	0,082	0,090	0,297	0,981
Cr	1,656	1,689	1,833	1,872	1,844	1,435	0,141
Fe^{3+}	0,098	0,055	0,053	0,034	0,045	0,157	—
Fe^{2+}	0,373	0,448	0,565	0,539	0,503	0,616	0,181
Mn	0,008	0,008	0,008	0,008	0,010	0,008	—
Mg	0,635	0,546	0,459	0,472	0,495	0,542	2,516
f	36,1	44,1	54,9	53,1	50,0	51,0	6,7
Cr-component	82,8	84,5	91,7	93,6	92,2	71,8	—
Al-component	10,8	11,2	4,8	1,7	4,5	14,9	—
Ulvöspinel	1,4	1,8	0,7	0,4	0,8	5,1	—
K_0	20,8	11,0	8,6	5,9	8,6	20,3	—

Note. For phlogopite, $Na_2O = 0.17$, $K_2O = 7.90$, $Na = 0.023$, $K = 0.705$

and therefore it has not been found enclosed in diamond. The chromite, possible in association with such an ilmenite, from compositional features must be analogous to all the known chromites, associated with diamonds, characterized, as emphasized by the available data, by a very large amount of the chromium component and a low amount of Fe^{3+}. The fundamental possibility of the association of chromite with ilmenite has been demonstrated for certain samples from the Bushveld (Legg, 1969), in meteorites (Buseck and Keil, 1966), and also lunar rocks of specific composition, rich in titanium and chromium (Haggerty et al., 1970). We have also identified an actual case of association between chromite and ilmenite and pyrope in a peridotite xenolith from the 'Aikhal' pipe (see §18).

R u t i l e is also known in intergrowths with diamonds (see Table 35), and cases has even been described of overgrowth of rutile by an aggregate of diamond crystals (Futergendler, 1965). Rutile, associated with diamond in inclusions (like, it seems, as in the intergrowths), probably belongs to the eclogite association, since rutile, associated with chrome-bearing pyrope and ilmenite, contains

TABLE 35. Chemical Composition of Ilmenites and Rutiles from Intergrowths with Diamonds and Xenoliths of Ultramafic Rocks (without diamonds)

Oxides	Ilmenites								Rutiles			
	xenoliths of ultramafic composition					intergrowths with diamonds			intergrowths with diamonds			
	A-287	I-140	Ush-3	Ud-112	A-70	BM-30	BM-10	MR-5	BM-9	BM-3	BM-4	BM-2
SiO$_2$	0,59	0,15	0,36	0,38	0,36	0,02	0,03	0,32	0,04	0,00	0,06	0,00
TiO$_2$	34,6	39,5	47,2	51,2	53,0	55,0	55,6	57,8	95,0	97,4	95,6	95,1
Al$_2$O$_3$	0,42	0,67	0,46	0,57	0,19	0,20	0,22	0,25	0,53	0,53	0,51	0,46
Cr$_2$O$_3$	5,67	1,74	2,63	0,36	3,33	0,13	0,23	0,04	0,09	0,07	0,13	0,07
Fe$_2$O$_3$	22,1	17,3	8,6	6,5	3,1	2,8	1,20	—	1,62	1,33	1,74	1,63
FeO	28,0	33,4	31,5	26,8	24,2	32,1	29,6	31,0	—	—	—	—
MnO	0,00	0,22	—	0,34	0,33	0,25	0,28	0,25	0,00	0,00	0,00	0,00
MgO	8,23	6,12	8,13	12,1	12,8	9,62	11,3	9,70	0,00	0,02	0,06	0,02
CaO	0,06	0,00	0,00	0,00	0,00	0,00	0,00	0,00	0,00	0,00	0,00	0,00
Total	99,67	99,1	98,88	98,2	97,31	100,12	97,86	99,36	97,36	99,33	97,90	97,35
Si	0,016	0,004	0,009	0,009	0,009	0,000	0,001	0,007	0,001	0,000	0,001	0,000
Ti	0,701	0,797	0,883	0,925	0,944	0,970	0,984	1,013	0,979	0,983	0,983	0,980
Al	0,013	0,021	0,013	0,016	0,005	0,006	0,006	0,007	0,008	0,008	0,008	0,008
Cr	0,121	0,036	0,052	0,007	0,063	0,003	0,004	0,001	0,001	0,001	0,002	0,001
Fe^{3+}	0,224	0,172	0,080	0,059	0,028	0,026	0,010	—	0,017	0,013	0,018	0,017
Fe^{3+}	0,224	0,173	0,080	0,059	0,028	0,026	0,011	—	—	—	—	—
Fe^{2+}	0,633	0,741	0,656	0,539	0,479	0,630	0,583	0,604	—	—	—	—
Mn	—	0,005	—	0,007	0,007	0,004	0,006	0,005	—	—	—	—
Mg	0,330	0,242	0,301	0,433	0,451	0,337	0,396	0,338	—	—	—	—
f	65,7	75,4	68,5	55,5	51,5	65,0	59,5	64,1	—	—	—	—
Cr-component	6,1	1,8	2,6	0,4	3,2	0,1	0,4	0	—	—	—	—
Hematite	22,4	17,3	8,0	5,9	2,8	2,4	1,0	0	—	—	—	—
K_0	41,4	31,8	19,6	18,0	10,5	7,1	3,5	0	—	—	—	—

Note. Samples A-287, I-140, BM-30, BM-10, MR-5, BM-9, BM-3, BM-4, and BM-2, from 'Mir' pipe; Ush-3, and Ud-112, from 'Udachnaya' pipe; A-70, from 'Aikhal' pipe.

a significant amount of Cr_2O_3 (1-5%), according to MacGetchin and
Silver (1970).

P h l o g o p i t e as an intergrowth with diamond and chrome-
diopside has been recorded in the South African diamond (Meyer and
Boyd, 1970). We have found phlogopite in Sample MR-9 from the 'Mir'
pipe in intergrowth with a clinopyroxene, rich in Na and Cr (ureyite
component). The results of an analysis of such a phlogopite, carried out on the microprobe, are presented in Table 34. We may note
that the phlogopite in this intergrowth has been developed in the
peripheral part of the pyroxene, creating the impression of its
being a later mineral. In the sample with phlogopite studied, there
are associated small segregations of a later chromite (see Table 34),
enriched in titanium and aluminum as compared with the chromite
of the first generation in this same intergrowth, which is almost
devoid of titanium and aluminum.

Thus, individual consideration of the features of the composition
of the minerals in intergrowths with diamonds, suggests a similarity
in composition in such important minerals as garnets and chromite
to that of the corresponding minerals, enclosed in the diamonds.
In drawing attention to the typical features of the composition of
these minerals in the paragenesis with diamond, we may concede the
equilibrium nature of the mineral-intergrowths with the diamond
aggregates, on the surface of which they are developed.

The absence from intergrowths of one of the most widely distributed minerals in the inclusions (olivine) may be explained by
its complete serpentinization on the outer surface of the intergrowths, in contrast to the isolated inclusions in diamonds, for
which serpentinization of the enclosed olivines has been observed
only in rare cases, where the diamond is cracked.

The discovery of certain new compositional features of minerals
in the intergrowths with diamonds, especially the clinopyroxenes
exceptionally rich in Na and Cr, and also chromites containing almost
no Al_2O_3, may be explained by the wide representative nature of the
material.

An individual examination of the mineral intergrowths with diamonds
suggests certain, so far preliminary, conclusions about the quantitative distribution of the various types of magnesian garnets in the
inclusions and intergrowths. This applies to the pyropes, which
contain a decreased amount of Cr_2O_3 and a somewhat increased amount
of CaO, as compared with the typical chrome-pyropes which we have
assigned to the harzburgite-dunite paragenesis. Whereas chrome-
poor pyropes are found in diamonds in isolated cases in the lherzolite
paragenesis, even some predominance of such garnets over chrome-
pyropes has been found in the intergrowths. This feature may indicate
the lower-temperature nature of most of the intergrowths as compared
with most of the single diamond crystals.

§11. PARAGENETIC FEATURES OF MINERALS ASSOCIATED WITH DIAMONDS

The data on the composition of minerals from the various diamond-
bearing associations set out in the preceding sections, definitely
indicate their assignment of two most general types: the ultramafic
type, which includes olivine, enstatite, and all the chromium-
bearing minerals, and the basic (eclogitic) type, equivalent in
chemical composition to some kinds of basalts with variable amounts
of Al_2O_3 and Fe/Fe + Mg (Table 36). We may cite as an example of

xenoliths of rocks, belonging to the first type of association, the diamond-bearing pyrope peridotites (dunites or altered harzburgites) of the 'Aikhal' pipe (V. S. Sobolev, Nai et al., 1969), and as an example of the second type, the diamond-bearing eclogites.

As has been shown when describing the minerals that intergrow with the polycrystalline diamond aggregates, their composition in the great majority of samples is similar to that of the inclusions in diamonds, at the same time differing from that of the equivalent minerals from xenoliths of the non-diamond-bearing peridotites. The substantial number of inclusions studied has revealed certain variations in the iron index in the garnets, the average value of which (13.1 ± 1.8) differs significantly even from that of the garnets from the most deep-seated peridotites from the 'Udachnaya' pipe (17.0 ± 0.7).

As a result of an investigation of the mineral-intergrowths it has been possible to identify some wider variations in the compositions of such minerals as chromites, almost devoid of aluminum, and also unusual clinopyroxenes, which contain up to 45% of the ureyite component and 20% of the jadeite component. Examination of the intergrowths has also significantly broadened our ideas about the composition of the magnesian garnets in association with diamonds. In this respect, it becomes necessary to analyze the already available factual information about the inclusions in diamonds, and about the minerals of the diamond-bearing xenoliths and intergrowths with diamonds, with the object of establishing their assignment to definite associations or the possibility of grouping them into independent types. A careful diagnosis of the inclusions in diamonds by X-ray methods and the direct determination of the composition of about 90 syngenetic inclusions in diamonds have created the prerequisites for a well-founded recognition of the stable associations of the syngenetic mineral-inclusions, associated with diamonds.

This fact, which concerns not only one, but several associations in which diamond has crystallized, raises no doubts at present. However, when attempting to interpret the results, the question immediately arises as to the possibility of considering the minerals, which occur in a single diamond, as equilibrium phases. Since such minerals as garnets and also chromites, enclosed in different diamonds, differ substantially in composition, verification of the composition of some of the grains of garnet or chromite, enclosed in a single diamond crystal, may provide unequivocal answers to such a question.

Support for the hypothesis of their equilibrium nature may come from the identity in composition of isolated inclusions of a particular mineral in a single diamond. Such data have been obtained for two chromite grains, GL-3 and GL-3b (Meyer and Boyd, 1972) (see Table 30). Our results from determining the refractive index of two garnet grains from a Yakutian diamond (n = 1.757 for each grain), also emphasize the closeness in their composition. Equivalence in composition has also been established for two inclusions of pyrope-almandine garnet, rich in calcium, from Sample AV-6 of a Yakutian diamond. A still more convincing example may be the multi-mineral association from Sample 57/9 of a Yakutian diamond. Isolated inclusions of pyrope and chrome-diopside and an intergrowth of these minerals of the intergrowth are distinguished mainly only within the limits of analytical error.

Observations on the inclusions have established a whole series of combinations of minerals in individual diamond crystals. These

TABLE 36. Types of Parageneses of Minerals Coexisted with Diamonds

Paragenesis	Minerals	Principal compositional features of garnets and other minerals	Xenoliths	Inclusions	Intergrowths
		I Ultramafic (with olivine)			
Harzburgitic (Dunitic-without En)	Ga + Ol + En + Chr	Ga (Cr$_2$O$_3$ = 5-20%; CaO = 0, 6-4%; f = 9-16% Ol (Fo$_{92-94}$); En (Al$_2$O$_3$ = 0,1-1,0%; Cr$_2$O$_3$ < 0.5 Chr (Cr$_2$O$_3$ > 62%; Cr-comp. > 80%)	(+)	(+)	+
Lherzolitic	Ga + Ol + En + Cpx (± Chr ± Ilm ± Ru)	Ga (Cr$_2$O$_3$ = 1-7%; CaO = 4-5%; f = 13-16%) Ol (Fo$_{98-98}$; En (Al$_2$O$_3$ = 0.5-0.6)	(+)	(+)	+
Wehrlitic	Ga + Ol + Cpx (±Chr)	Cpx (Ca/Ca Mg = 46-47%) or inclusions, 45,38%-for xenoliths) Ga (Cr$_2$O$_3$ = 8%; CaO = 13.5% (>6-7%); f = 13-15%) Ol (Fo$_{93}$)	–	(+)	+
?	Cpx + Chr ± Ol	Cpx (Cr$_2$O$_3$-up to 15.6%; Na$_2$O-up to 10.1%) Chr (Cr$_2$O$_3$ > 67%; Cr-comp. > 90%)	–	–	+
		II Eclogitic (without olivine)			
Eclogites +	Ga + Cpx + Ru (± Mt ± Gr)	Ga (CaO = 3.1-14.7%; Ca-comp = 8.7-40%; f = 34-69%) Cpx (Na$_2$O = 1.4-9%; Na-comp = 10-61%)	(+)	(+)	–
Coesite eclogites	Ga + Cpx + Cs	Ga (CaO = 7-11.6%; Ca-comp = 20-33%; f = 45-57%) Cpx (Na$_2$O = 4.1-4.5%; Na-comp = 29-31%)	–	(+)	–
Kyanite eclogites	Ga + Cpx + Ky + Ru (±Gr)	Ga (CaO = 12.3-18.6%; Ca-comp = 33-48.5%; f = 25-41%) Cpx (Na$_2$O = 6.8-9.6%; Na-comp = 47-66%)	(+)	+	–
Corundum eclogites	Ga + Cpx + Cor + Ru + Gr	Ga (CaO = 13%; Ca-comp = 35%; f = 35%) Cpx (Na$_2$O = 10.7%; Na-comp = 71%)	(+)	+	–
Grospydites	Ga + Cpx + Ky (± Cor ± Ru)	Ga (CaO = 20.3%; Ca-comp = 54%; f = 42%) Cpx (Na$_2$O = 8.88%; Na-comp = 60%)	(+)	–	–

Note (+ indicates that all minerals occur in one sample; + indicates that individual minerals have been found with clear features of the given paragenesis; – indicates that the paragenesis and its features have not been found.

Abbreviations: Ga - garnet; Ol - olivine; En - enstatite; Cpx - clinopyroxene; Chr - chromite; Ilm - ilmenite; Ru - rutile; Cs - coesite; Ky - kyanite; Cor - corundum; Mt - magnetite, Gr - graphite

combinations (Harris, 1968a; Meyer and Boyd, 1972) may include
bimineralic types, which form olivine with chromite, pyrope, enstatite,
diopside, and diamond (in diamond) separately, and also pyrope +
chromite and garnet + omphacite. The trimineralic associations
include olivine + pyrope + enstatite (Harris, 1968a) and olivine +
pyrope + chrome-diopside (Sobolev, Bartoshinsky et al., 1970).

In addition to the associations listed, which have been identified
by X-ray methods and X-ray spectral microanalysis, a significant
number of associated minerals have been identified visually in a
series of diamonds. For a number of minerals, such as garnets and
chromium-bearing clinopyroxenes, visual diagnosis is completely
reliable. Colorless inclusions, if devoid of a light-greenish tint
(which is specific to enstatite), may with reasonable probability be
assigned to olivine (Harris, 1968a).

In the collection of diamonds with inclusions, specially selected
by A. I. Botkunov from the 'Mir' pipe, the following parageneses have
been visually identified: olivine + chromite, olivine + chrome-
diopside, pyrope + chrome-diopside (intergrowths of pyrope and chrome-
diopside have been identified in two samples of diamond), pyrope +
chromite, and garnet + omphacite. The trimineralic associations are
of special interest, being exceptionally rare in diamonds: olivine +
pyrope + chromite, and olivine + pyrope + chrome-diopside; and even
more so the tetramineralic associations: olivine + enstatite +
Cr-poor pyrope and chromite and olivine + enstatite + chrome-diopside +
garnet (Sobolev, Botkunov et al., 1976). Colorless olivine comprises
quite large inclusions, of the order of 0.5 - 0.6 mm, which facili-
tates its diagnosis. The largest inclusions of olivine (more than
1 mm) have a light-green color.

As a fundamental feature for the recognition of stable parageneses
of minerals, associated with diamond, we may use the composition of
the garnet, which, as shown by a whole series of investigations, is
a sensitive indicator of the composition of the medium.

The Paragenesis of Calcium-poor Chrome-pyrope
(harzburgite-dunitic)

As demonstrated when describing the inclusions of pyrope garnets
from diamonds, the great majority of the compositions studied belong
to an unusual variety of pyrope, poor in calcium. The previous
investigations of garnet compositions from ultramafic rocks and
kimberlites suggested that their calcium content was exceptionally
constant (Sobolev, 1964a). A paragenesis is known for the overwhelming majority of garnets examined from ultramafic xenoliths,
which almost always includes a bipyroxene association. Even in very
rare xenoliths, in the original descriptions termed pyrope olivinites
or dunites, the amount of calcium (in mol %) in the garnet was also
close to the average figures for peridotites: 13.4 ± 1.9% (Sobolev,
1964a). This feature of the composition of the garnet in paragenesis
with two pyroxenes (± olivine) is also well supported by the results
of experimental investigations in the system $CaSiO_3$-$MgSiO_3$-Al_2O_3,
both on the basis of a X-ray spectrographic study of the garnets in
the end products (MacGregor, 1965), and by direct determination of
their composition using the microprobe (Boyd, 1970).

Some corrections add an amount of Cr_2O_3 to the system, the increase
in which contributes to the increase in the amount of CaO in the
garnet (see §6), and also an amount of Na_2O, causing the opposite

effect (Sobolev, Zyuzin and Kuznetsova, 1966; Banno, 1967).

If we make no allowance for the approximately constant 'background' amounts of Cr_2O_3, FeO, and Na_2O, which complicate the composition of the ultramafic system and are involved in it in a clearly minor way, then the compositions of the garnets from the various parageneses, both natural (Sobolev, 1964a; Sobolev and Kuznetsova, 1965), and also obtained experimentally (Boyd, 1970), are similar to one another and correspond well with the characteristic point on the paragenetic diagram for ultramafic rocks (Bobrievich et al., 1960) (see Fig. 15).

Starting with this diagram, the compositions of the calcium-poor pyropes must be located in the sector with an amount of CaO from zero up to the peridotite point in the paragenesis olivine ± enstatite. Consequently, in the system $MgSiO_3-CaSiO_3-Al_2O_3$, garnet with a small amount of calcium may crystallize in a paragenesis with olivine and/or enstatite without clinopyroxene. However, as noted above, the presence of Na_2O in the system may markedly influence the amount of CaO in the garnet towards a decrease. In such a case, the garnet with decreased amount of CaO, as compared with the peridotite point, may be associated with two pyroxenes (Sobolev, 1970).

The possibility of a more precise determination of the nature of the paragenesis, based on the composition of single inclusions of garnet in diamond has appeared as a result of investigations of the isomorphous amount of Na_2O in garnets, associated with the Na-bearing pyroxenes and formed in the area of stability of the diamond (Sobolev and Lavrent'ev, 1971). The data obtained as a result of a study of the disseminated amounts of sodium in garnet, have been discussed in Chapter V, and here we shall only dwell briefly on their importance in establishing the paragenesis of pyrope in the inclusions.

On the basis of the example of the coexisting pair, garnet + pyroxene, in the inclusions and intergrowths with diamonds, it has been shown that even if the amount of Na_2O in the pyroxene is close to the minimum (about 1%), then a seemingly notable amount of sodium is fixed in the garnet. In those cases when garnet with decreased content of CaO is associated a sodium-rich pyroxene, the amount of Na_2O in the garnet markedly increases. An example is Sample B-1, the garnet of which, in association with pyroxene rich in Na_2O (6.5%), contains 2.4% CaO and 0.13% Na_2O. However, the typical chrome-pyropes from diamonds with small amounts of CaO, comparable with that in Sample B-1, are almost completely devoid of sodium. This feature may indicate the absence of clinopyroxene in the possible paragenesis. A convincing argument in favor of recognizing a paragenesis without clinopyroxene is the identification of a trimineralic association, garnet + enstatite + olivine, in one diamond (Harris, 1968a), and also the associations, garnet + olivine + chromite and garnet + olivine + enstatite + chromite (Sobolev, Botkunov et al., 1976).

Thus, in the stability field of diamond, the associations olivine + chrome-pyrope (calcium-poor) + enstatite + chromite may be recognized, being emphasized both by the features of the paragenesis of the inclusions in one diamond, and by the composition of the pyrope with low content of calcium and absence of sodium. An example of the occurrence of almost all the listed minerals together with diamond (without enstatite) in xenoliths may be the well-known xenoliths from the 'Aikhal' pipe (V. S. Sobolev, Nai et al., 1969).

Certain features of distribution of the principal components in the minerals, which belong to the recognized association, may be demonstrated from the example of pairs of minerals extracted from one diamond (Meyer and Boyd, 1972), namely: olivine + garnet (Sample GL24); olivine + enstatite (G10); olivine + chromite (GL47); and garnet + chromite (G20) (see Tables 27, 29, 30, 31). Data on the iron index of these minerals and the amount of Cr_2O_3 present in them (and also the Cr-component) are presented below (Meyer and Boyd, 1972).

Amount of component	GL24 Olivine	GL24 Garnet	G10 Olivine	G10 Ensta-tite	GL47 Olivine	GL47 Chrom-ite	G20 Garnet	G20 Chrom-ite
Cr_2O_3	0,07	10,1	0,06	0,33	0,08	65,3	12,6	64,0
CaO (Ca-component)	--	1,118(3,0)	--	--	--	--	1,74(4,6)	--
Cr-component	--	28,6	--	33,0	--	84,3	36,1	83,9
f	6,6	11,9	6,9	6,1	6,0	19,9	12,4	32,3

The enstatite, assigned to this association is characterized by an increased Cr/Cr + Al ratio, which agrees closely with the amount of Cr_2O_3 in the garnet, and also the almost complete absence of Na_2O, the amount of which (0.2 and above) is typical only of enstatites, associated with clinopyroxenes and crystallized under conditions of maximum pressure (see §14).

A comparison of the compositional features of the associated minerals, represented in the table above, strongly emphasizes the similarity of the olivines, associated with garnet and enstatite in individual samples, and also the similarity of the garnet compositions in Samples GL24 and G20, and those of the chromites in Samples GL47 and G20. All the silicate minerals are characterized by minimum values of the iron index as compared with that of analogous minerals of the inclusions of ultramafic rocks in kimberlites and peridotites, known in situ (see Chapter III), which is also typical of all the analyzed individual inclusions of the corresponding minerals.

A comparison between the compositions of the numerous garnets with decreased calcium content and those of garnets from lherzolites (see Fig. 18) demonstrates that the position of the compositional points for the calcium-poor magnesian garnets, associated with diamonds in xenoliths, inclusions, and intergrowths, is clearly defined well below the compositional field for the garnets of the bipyroxene paragenesis. However, in individual cases the compositional field for garnets in paragenesis with clinopyroxene may extend up to the boundary 1 - 1' (see Fig. 18). This is associated with the possibility of an exceptionally large content of Na_2O in the associated pyroxene (up to 6.5%), which significantly affects the amount of CaO in the garnet towards a decrease (Sobolev, Zyuzin and Kuznetsova, 1966).

The Lherzolitic Paragenesis

The lherzolitic paragenesis of inclusions in diamond is amongst the rarest known, as indicated by such features as the marked predominance of calcium-poor chrome-pyropes, and also the exceptional rarity of the enstatite-diopside pair in association with diamond. This feature, after establishing the widespread distribution of garnet lherzolites in kimberlites in general, and the theoretical conclusions arrived

at on this basis, was somewhat unexpected. In truth, the lilac pyropes
from diamonds were for a long time identified with the pyropes of the
garnet peridotites from the pipes, and the similarity in compositions
has been stressed for all the minerals of the peridotites and inclusions in diamonds (Sarsadskikh et al., 1960).

The reliability established lherzolitic paragenesis is limited to
three samples of diamonds from the 'Mir' pipe (AV-34, AV-14 and AV-93),
which consist of plane-facetted octahedra. In all cases, a unique
tetramineralic assemblage of inclusions has been identified: olivine,
enstatite, diopside, and pyrope (Sobolev, Botkunov et al., 1976).
A characteristic feature of the pyrope and diopside in samples AV-34
and AV-14 is their pale color, respectively reddish-pink and light-
green, indicating a low chromium content. We must stress once more
that light-colored pyropes are very rare in diamonds, although in the
present case, there is great interest not only in this feature, but
also in the possibility of identifying their paragenesis. Garnet of
the AV-93 sample is lilac but its calcium content is typical of all
lherzolitic garnets.

The unusual morphology of the inclusions of garnet, diopside, and
enstatite from the diamonds under consideration has been recorded
above; they occur in the form of octahedral shapes, totally atypical
of these minerals, and have the outline of negative crystals of
diamond. In Sample AV-14, one such octahedron consisted of an intergrowth of garnet and diopside, and another, or an intergrowth of
diopside and enstatite. In Sample AV-34, the negative diamond shape
has been defined for an individual pyrope inclusion, and the enstatite consists of an intergrowth of two individuals of octahedral
habit. In Sample AV-93, the intergrowth of garnet and enstatite
has an octahedral shape. The olivine in Samples AV-34 and AV-14
consists of several idiomorphic crystals; in Sample AV-14, nine
olivine inclusions have been identified in all, and in Sample AV-34,
four inclusions; in Sample AV-93, the olivine has an octrahedral habit.

The compositional features of the minerals, coexisting in a single
diamond in both Samples AV-34 and AV-14, emphasize their similarity
to those of the lherzolite xenoliths, along with certain clear
differences, the principal of which is the decreased iron index of
all the minerals, including the garnet. Below are given the principal compositional features of the minerals based on the amount of
certain components from the tetramineralic paragenesis of inclusions in diamonds.

	Garnets		Diopsides		Ensta-tite	Olivines	
	AV-34	AV-14	AV-34	AV-14	AV-34	AV-34	AV-14
TiO_2	0,47	0,68	0,33	0,58	0,15	0,08	0,05
Cr_2O_3	1,27	1,65	0,69	1,01	0,11	0,02	0,02
CaO	3,81	3,96	--	--	0,32	--	--
Na_2O	--	--	2,44	2,12	0,17	--	--
Ca-component	9,7	10,0	45,9	46,0	--	--	--
Cr-component	3,6	4,6	13	20	10	--	--
f	13,1	13,4	6,7	5,0	6,2	7,3	6,5

The clinopyroxenes, associated in both samples with enstatite,
are characterized by an increased value of the Ca/Ca + Mg ratio, and
consequently, a decreased amount of CaO in the enstatite, which must

apparently indicate an extremely moderate equilibrium temperature
for this association, if we begin by comparing the solvus curve for
pyroxenes at a pressure of 30 kb (Davis and Boyd, 1966). However,
our assumption (see §30), based on available experimental data is
that with further increase in pressure (of the order of 50-60 kb
and above), the calcium index of the clinopyroxenes, which crystallized in equilibrium with garnet, must increase in the isothermic
sector. Allowing for these assumptions, the equilibrium temperature
of the inclusions under consideration in the diamonds of Samples AV-34
and AV-14, is determined at 1150-1200°C. Taking account that the
lherzolite paragenesis is eutectic, that is, lowest-temperature in
the stability field of the diamond, we may estimate the lower boundary for the temperatures of equilibrium crystallization of the
inclusions in diamond (and possibly, for diamond itself).

On the basis of the position of the compositional points for the
chromium-poor magnesian garnets from intergrowths with diamonds on
the diagram (see Fig. 18), we may conclude that in contrast to the
minerals considered above, enclosed in diamond, the majority of the
garnets, on the basis of calcium content, fall within the limits of
the compositional field for garnets from lherzolites.

Thus, the lherzolitic type of paragenesis, according to our
observations, is more typical of the polycrystalline diamond aggregates than the plane-facetted monocrystals, which may indicate the
relatively lower temperature nature of part of such intergrowths as
compared with the monocrystals.

The Paragenesis of Calcium-rich Chrome-pyrope (wehrlitic)

Calcium-rich chrome-pyrope has so far been identified in diamond
only in one case (Sobolev et al., 1970). The aim in establishing
its paragenesis has been facilitated by the unique nature of diamond
Sample 57/9, in which olivine, chrome-diopside, and also an intergrowth of pyrope and chrome-diopside were present simultaneously
with the pyrope. Analyses of these minerals have been given in
Tables 27, 29, and 31. Here, we shall only consider the features of
distribution of the individual components in the coexisting minerals.
For convenience, we shall set out the appropriate data separately.

				Intergrowth			
Components	Olivine	Garnet	Diopside	garnet	diopside	Garnet	Diopside
Cr-Component	0,06*	24,9	39,3*	24,3	42,1**	24,6	40,7
Ca-Component		36,4	47,0	35,0	47,7	36,0	47,4
f	6,0	14,6	4,4	14,5	3,5	14,6	4,0

*Cr_2O_3 **$Cr/Cr + al$.

In discussing the features of this unusual paragenesis, defined
only in isolated xenoliths of deep-seated rocks (Sobolev, Lavrent'ev
et al., 1973a), we must primarily draw attention to the large amount
of the calcic component in the garnet, which is neither comparable
with its amount in the garnets from the harzburgite-dunitic paragenesis
with two pyroxenes (websterite-lherzolitic). The presence of olivine
in the present paragenesis excludes the possibility of assigning the
garnet to the specific chrome-bearing eclogite association, such as

that identified in the chrome-bearing grospydites and kyanite eclogites (Sobolev, Zyuzin and Kuznetsova, 1966).

Data from two analyses (see Table 31) enable us to assess the
compositional features of the clinopyroxene from diamond Sample 57/9.
The compositions of the pyroxenes are mainly distinguishable only
within the limits of analytical error, and the results of the analyses
coincide with those on the composition of the pyroxenes from kimberlites (Boyd, 1970), and especially with those of the calcic diopsides.
In addition, it must be emphasized that three further analyzed
pyroxenes from diamonds available to us (see Table 31) also belong
to the calcic group, and on the basis of this feature, together with
their low iron index, they are similar to the pyroxenes from Sample
57/9. It is possible that they also belong to the present paragenesis.

The low iron index and other compositional features of all the
coexisting minerals (see below; Sobolev et al., 1970) agree closely
with the hypothesis of their equilibrium nature. The feature of the
paragenesis with very magnesian olivine is most clearly manifested
in the composition of the garnet, which is distinguished by its very
large content of Ca-component along with a low iron index.

Although pyrope garnets with increased amounts of calcium and
chromium are occasionally found in the heavy-fraction concentrates
from kimberlites (Sobolev et al., 1968; Nixon and Hornung, 1969;
Sobolev, Lavrent'ev et al., 1969; Sarsadskikh, 1970; Khar'kiv and
Makovskaya, 1970; Sobolev, Lavrent'ev et al., 1973a), their iron
index fluctuates over a very wide range (20-45%), the lower limit
of which is significantly above that for the calcium-rich chrome-
bearing pyropes, extracted from diamond (Sample 57/9).

Of considerable interest is the composition of the small (about 5μ)
inclusions, discovered in an individual inclusion of clinopyroxene in
Sample 57/9. Analysis of one such inclusion has given the following
composition: SiO_2, 62.0%; Al_2O_3, 0.7%; Cr_2O_3, 0.7%; FeO, 1.6%; MgO,
27.0%; CaO, 9.3%; Na_2O, 0.4%; total, 101.7%. This corresponds to
a pyroxene of unusual composition with a Ca/Ca + Mg + Fe ratio of
20%, intermediate between pigeonite and subcalcic diopside. Its iron
index is 3.0%. The formula for the inclusion is:

$$(Mg_{1.36}Fe_{0.04}Ca_{0.34}Al_{0.03}Cr_{0.02})_{1.79}Si_{2.09}$$

There is no doubt that the dimensions of the inclusions strongly
affect the quality of the analysis, especially the determination of
SiO_2. Nevertheless, it is patently clear that the inclusion should
be assigned to the pyroxene group with a composition differing significantly from that of the chrome-diopside which encloses it.

Thus, on the basis of the unusual Sample 57/9, the presence has
been demonstrated of an earlier-unknown ultramafic association in the
stability field of the diamond. The composition of the garnet in
this association is distinguished from that of both the pyropes poor
in calcium, and also those from the peridotite xenoliths.

In connection with the reliable identification of a paragenesis
(wehrlitic) new to the deep-seated ultramafic compositions, it is
necessary to define in greater detail the features that will permit
us to recognize the wehrlitic paragenesis of minerals, in cases when
only magnesian garnet with clinopyroxene together or separately are
known. The principal feature of the wehrlitic paragenesis is the
increased amount of CaO in the garnet. Whereas in the lherzolitic
or websteritic paragenesis, the constancy of the amount of CaO in

the garnets has been demonstrated and an estimate given of the effect
of the addition of other components (see §1), an extremely weighty
argument for assignment of the wehrlitic paragenesis may be the
greater amount of calcium as compared with that in the lherzolitic
garnets.

On the basis of a comparison with the lherzolitic garnets (see
Fig. 18 and §12), Sample MR-4 (see Table 32) may be assigned to the
wehrlitic paragenesis; its characteristic feature is the large amount
of Na_2O in the pyroxene (see Table 32). Since the amount of Na_2O
in the pyroxenes can only lower the calcium index of the garnets
(Sobolev, Zyuzin and Kuznetsova, 1966; Banno, 1967), the garnet of
the wehrlitic of lherzolitic paragenesis in association with a sodium-
rich pyroxene may even be calcium-poor. This may apply especially
to Sample B-1, in which the pyrope is associated with a sodium-
bearing clinopyroxene in a polycrystalline diamond aggregate.

The wehrlitic paragenesis may evidently include some chrome-
pyroxene-garnet intergrowths with garnets having an increased calcium
index, as defined in §2.

The high calcium index of the green chrome-bearing garnets from
the kimberlite concentrate and their similarity, on the basis of this
feature, to the garnets of the above-mentioned intergrowths and to
the garnet from the reliable wehrlitic paragenesis (Sample 57/9),
also enable us to regard their paragenesis as wehrlitic in nature.
A weighty argument in favor of such a view is the discovery of xenoliths containing green calcium-rich garnets in a paragenesis with
olivine, clinopyroxene, and chromite (Sobolev et al., 1973).

Thus, there is now a good deal of evidence of the distribution of
an earlier-unknown type of paragenesis of deep-seated ultramafic
rocks (wehrlitic), with a characteristic association of chrome-
bearing, calcium-rich garnet, clinopyroxene, olivine, and chromite.
Some features of this paragenesis have been explained earlier on
the basis of experimental data. Thus, during the study of the spinel-
pyrope peridotite transition in experiments with compositions containing no enstatite, increase in the calcium index of the garnet
was obtained with increase in pressure (MacGregor, 1965). On the
basis of these data, the value of the pressure for the wehrlitic
association with garnet, the calcium index of which reaches 35-40%
is about 40 kb (Sobolev et al., 1970).

The Eclogitic Paragenesis

In addition to the above-defined ultramafic associations, which
occur in diamonds, an eclogitic type of association is also clearly
recognized as a result of the study of inclusions. This type may
include garnets of pyrope-almandine composition with a variable content of calcium and iron index, and also clinopyroxenes devoid of
chromium and distinguished from the chrome-diopsides from diamonds
by the markedly increased iron index (see §13). It includes both
the minerals of the normal bimineralic eclogites, as well as those
of the kyanite and corundum types. Of the syngenetic inclusions in
diamonds, this association may with complete confidence include
rutile also. Finally, coesite, which has been identified in diamond
in the form of an inclusion, may also belong to the eclogite association only, which is confirmed now by discovery of coesite-garnet-
omphacite association in Yakutian diamonds (Sobolev, Yefimova et al.,
1976).

Although inclusions of non-chrome garnets with a calcic component
up to 40% have also been recorded in diamonds, which makes them
similar to the garnets of the kyanite and corundum eclogites (see
Chapter III), neither kyanite nor corundum has been found as inclusions in diamond. However, this fact does not of itself deny the
possibility of finding syngenetic inclusions of kyanite and corundum
in diamonds, the more so since the presence of kyanite and corundum
eclogites, containing diamonds, has already been demonstrated (see
§8).

The principal prerequisite for recognizing the stable eclogite
association, which involves diamond in its composition, is the
numerous discoveries of various diamond-bearing eclogites and the
identified (with difficulty) combined occurrences of pyrope-almandine
garnet and sodium-bearing clinopyroxene in a single diamond from
Samples 880 and 7a (Sobolev et al., 1970) (see Table 31). In
addition to the samples of Uralian diamonds noted, for which the
compositions of the associated garnets and clinopyroxenes are known,
we must take note of Sample D15 from South Africa (Meyer and Boyd,
1972).

There is special interest in the results of an investigation into
the association of yellow-orange garnet and light-green clinopyroxene,
extracted from a diamond from the 'Mir' pipe (Sample AV-6). The
pyroxene of this sample (see Table 31) contains about 7% Na_2O, and
the garnet has about 10% CaO, with an iron index of 50%. In the
pyroxene of this sample, there is an unusual amount of $K_2O = 0.27\%$.
In all their features, the minerals of this association have a
composition, typical of the diamond-bearing eclogites.

The principal features of the compositions of the associated
garnets and clinopyroxenes from diamonds, on the basis of amount of
Na_2O, iron index, and the value of the Ca/Ca + Mg + Fe ratio are
given below.

	Garnets			Omphacites		
	D-15	880	AV-6	D-15	880	AV-6
TiO_2	0,64	1,18	0,93	0,16	0,33	0,27
CaO	13,9	11,1	10,4	19,1	16,6	12,2
Na_2O	0,09	0,17	0,20	2,30	3,93	6,89
Ca-component	37,3	29,7	29,9	45,7	44,5	46,3
f	46,9	42,2	49,5	25,9	21,2	28,1

In emphasizing once again the similarity of the pyrope-almandine
garnets, associated with diamonds in the form of inclusions, and in
intergrowths and eclogites, not only in the amount of calcium and the
iron index, but also in the amount of sodium present (see §20), we
may note that the clinopyroxenes, associated with these garnets,
are characterized by an exceptionally wide range of the amount of
Na_2O (from 2.30 to 10.7%), along with closely similar iron indices.
This may, apparently, be associated with the difference in the concentration of sodium in the diamond-forming medium of basic composition. However, the clinopyroxenes from all known xenoliths of
diamond-bearing eclogites in the 'Mir' pipe are characterized by an
increased amount of sodium. Such enrichment in sodium may have a
local character, since all the diamond-bearing eclogites from the
'Mir' pipe have evidently been formed under closely similar conditions

and from a medium of basic composition. With further accumulation of factual information it is necessary to compare the composition of the omphacitic pyroxenes in diamonds from the 'Mir' pipe with that of the pyroxenes from the diamond-bearing eclogites. Whereas the eclogitic diamond-bearing associations of the 'Mir' pipe are characterized by a general enrichment in sodium, the omphacitic pyroxenes, enclosed in the diamonds from this pipe, must also be distinguished by the similar amount of sodium, support for which may come from the Sample AV-6 from this pipe.

Although the range of chemical composition of the minerals in the diamond-bearing eclogites appears extremely wide, and the parageneses appear different, it is evident that still not all the possible parageneses have been identified. Thus, on the basis of discoveries of coesite and its paragenesis, the establishment of independent coesite-bearing eclogites is possible (see Table 36). Enstatite diamond-bearing eclogites (an association, widely developed in the complex of rocks that do not contain diamonds) have not so far been found. Diamond-bearing grospydite is found now in the 'Udochnaya' pipe (Ponomarenko et al., 1976) (see Table 36), and such paragenesis is also completely possible for the Roberts Victor mine.

Inclusions in Diamond from a Diamond-bearing Eclogite

Since the commencement of detailed investigations into the inclusions in diamonds, special attention has been focussed on the search for inclusions in crystals, directly occurring in the diamond-bearing rocks (eclogites and peridotites). In this respect, a number of investigators, especially Meyer and Boyd, have stressed that such inclusions (garnets and pyroxenes) must be distinguished on the basis of composition from the minerals of the enclosing xenolith by a somewhat lesser iron index, since they must represent the products of the earliest crystallization. Such a proposition, however, is immediately cast in doubt in view of the absence of sufficiently clear zonation in the garnets and pyroxenes of the deep-seated rocks.

The author, along with A. I. Botkunov, first succeeded in finding inclusions in a crystal of diamond from a diamond-bearing eclogite in the 'Mir' pipe (Sample M-46). With more detailed investigations (V. S. Sobolev, Sobolev and Lavrent'ev, 1972), it appeared that the diamond crystal, weighing 0.3 carats contains a whole series of inclusions: two inclusions of clinopyroxene, of which one has been partially altered; an inclusion, consisting of an intergrowth of garnet and clinopyroxene, and also several idiomorphic crystallites of rutile. Along with these inclusions having an octahedral aspect (see §9) with dimensions of the order of 0.1 - 0.3 mm, there are also small idiomorphic inclusions of clinopyroxene, surrounded by a graphite halo (see Photo 14a). Their dimensions (3 such inclusions have been identified) do not exceed 0.05 mm along the long axis.

The investigation of the composition of the inclusions has shown that, being typical minerals of eclogites and close in composition to the minerals of eclogite containing the diamond (M-46), they deviate markedly at the same time in other features. These features have shown up clearly in a concurrent analysis of both mineral-inclusions and the minerals from the eclogite (Table 37).

The garnet from the diamond is distinguished from the minerals of the parent rock by its increased iron index. The discrepancy in the iron index, although of little significance (according to

TABLE 37. Composition of Garnet and Omphacite from Diamond and Eclogite from Sample M-46 (V. S. Sobolev & Lavrent'ev, 1972)

Oxides	Garnet in dia-mond	Garnet in eclo-gite	Omphacite	Omphacite in eclo-gite
SiO_2	40,0	39,7	54,8	55,5
TiO_2	0,46	0,43	0,48	0,56
Al_2O_3	22,0	21,5	9,79	9,39
Cr_2O_3	0,04	0,07	0,05	0,06
FeO	20,9	18,7	4,94	4,48
MnO	0,52	0,39	0,07	0,04
MgO	9,02	9,79	8,97	8,96
CaO	8,18	8,53	13,1	12,7
Na_2O	0,17	0,17	6,70	6,82
K_2O	—	—	0,30	0,08
Total	101,29	99,28	99,20	98,59
Si	3,001	3,014	1,981	2,007
Al^{IV}	—	—	0,019	—
Ti	0,027	0,025	0,013	0,015
Al^{VI}	1,947	1,924	0,398	0,400
Cr	0,002	0,004	0,001	0,002
Fe^{3+}	0,024	0,047	0,071	0,065
Fe^{2+}	1,287	1,139	0,079	0,072
Mn	0,032	0,027	0,002	0,001
Mg	1,005	1,108	0,482	0,482
Ca	0,658	0,693	0,508	0,491
Na	0,027	0,027	0,469	0,478
K	—	—	0,014	0,004
Total of cations	8,010	8,008	4,037	4,017
Pyrope	33,4	37,0	—	—
Almandine	42,7	38,0	—	—
Spessartine	1,1	0,9	—	—
Grossular	20,1	20,2	—	—
Andradite	1,2	2,3	—	—
Ti-andradite	1,4	1,3	—	—
Uvarovite	0,1	0,2	—	—
f	56,6	51,7	23,7	22,1
Ca-component	22,8	24,0	51,3*	50,5*

+Ca/Ca + Mg .

the distribution curve), has also been identified in the composition of the pyroxene, but the principal difference concerns the amount of K_2O, which in the pyroxene from the diamond is four times that found in the pyroxene from the eclogite.

In both cases, completely fresh pyroxenes were examined, and the evenness of distribution of potassium in them is not in doubt. The difference noted in the amount of potassium has been emphasized by an examination of two different inclusions from diamond M-46.

Thus, as a result of new investigations, the hypothesis that the differences have been associated with the sequence of crystallization collapses, since a trend towards an opposite change in the iron index has been identified. One conclusion that may be reached from the data obtained is the assumption that, after the formation of diamond, the minerals of the eclogite underwent some change in composition. If such a change were associated with simple melting out of a small quantity of liquid, then there would most probably be a lowering in the amount of sodium in the pyroxene (and garnet). It may be assumed that such a change in composition has been associated with degassing, accompanied by extraction of a certain amount of potassium and iron from the garnet and pyroxene of the diamond-bearing eclogites. It is interesting to note that in such a pyroxene assigned to the eclogitic paragenesis of a diamond from the 'Mir' pipe, we have identified a maximum amount of K_2O up to 0.27% (Sample AV-6).

Of course, the available data may serve only as a preliminary exposition of an hypothesis, which requires further factual evidence. However, on the whole the features noted reflect a general trend in the behavior of K_2O in the mantle (see §26).

The Clinopyroxene + Chromite Paragenesis

Most of the minerals considered by us, which are associated with diamonds, have been assigned on the basis of a whole series of features to definite parageneses: harzburgite-dunitic, wehrlitic, lherzolitic, and eclogitic. In some cases, doubts arise about the assignment of an actual garnet or garnet + diopside intergrowth to the wehrlitic or lherzolitic paragenesis, although this detail is essential within the well-known peridotite type of association.

Among the minerals, associated with diamonds, individual cases of a completely new and unusual type of association have been identified. This applies to intergrowths, consisting of chromite, which quantitatively predominates markedly over clinopyroxene, found in two samples of a polycrystalline diamond aggregate (B-7 and MR-9) from the 'Mir' pipe.

The unique nature of this association has been expressed in the unusual composition of the clinopyroxene, which in both samples contains an exceptionally large amount of Cr_2O_3 and Na_2O (respectively 11.7 and 15.6%, and 7 and 10%). The chromites of both intergrowths contain very little Al_2O_3 (see Table 34).

Comparison with the compositions of pyroxenes, known in association with diamonds and in the chromite-pyroxene-garnet intergrowths in kimberlite (see §2) demonstrates that the amounts of both Cr_2O_3 and Na_2O in the pyroxenes from Samples B-7 and MR-9 are much greater than that identified for the intergrowths in association with the chrome-rich pyroxenes. Data on the distribution of chromium and aluminum in the coexisting minerals from deep-seated rocks, given in §23, indicate the absence of garnet in paragenesis with the most chrome-rich pyroxene and the chromite.

Features of Crystallization of Minerals Associated with Diamonds

When interpreting the possible temperature conditions of equilibrium of minerals, associated with diamonds, it is important to compare

the compositions of the minerals, enclosed within the diamond, which
have seemingly been protected from the effects of external conditions
after the capture of the inclusions, with those of the minerals
occurring in the outer portions of the diamond aggregates and in
the diamond-bearing xenoliths.

In discussing the conditions of equilibrium of the crystalline
inclusions in diamonds, Meyer and Boyd (1972) believe that these
inclusions represent the earliest crystals, precipitated from the
melt, and that they indicate considerably higher-temperature conditions than the association of the xenoliths, including the diamond-
bearing types, which reflect the subsolidus temperature. Support
for this is, in their view, the idiomorphic nature of the crystalline
inclusions, their isolation in the diamond from subsequent reaction
with the melt, and as a result of this, the lower iron index of the
mineral-inclusions in the diamonds as compared with that of the
minerals in the xenoliths. Th last feature, according to F. R. Boyd,
must also be maintained for analogous minerals, especially the garnets,
in diamond from the diamond-bearing xenoliths.

In agreeing completely with the view that both the diamond and its
enclosed minerals are the earliest products that crystallized from
the kimberlite melt (and especially, those captured from peridotites
and eclogites) (V. S. Sobolev, 1960), we cannot accept the proposition
that the composition of the minerals, associated with diamond, can
be substantially altered with lowering of temperature (Meyer and Boyd,
1972). The factual information available to us on the composition of
the minerals, associated with diamonds, contradicts such an interpretation. Since in the present investigation, we have clearly
demarcated minerals enclosed in diamonds, and those in intergrowths
with polycrystalline diamond aggregates, it is convenient to compare
the compositions of the minerals with highest iron index, that is,
the garnets in both types. As such a comparison demonstrates (see
§11), the magnesian garnets of the inclusions and in the intergrowths
have a completely similar composition. Moreover, garnets of analogous
composition, amounting to 5-15% of the total content of pyrope, have
been found by us in the concentrates of the diamond-bearing kimberlites from Yakutia. Several grains of such garnets even retain
relicts of a kelyphitic rim (Sobolev, Lavrent'ev et al., 1973a).
They have also been found in the kimberlites from South Africa
(Gurney and Switzer, 1973). In this respect, we note that some
variations in the value of the iron index are specific also to the
minerals enclosed in diamonds. This is clearly seen not only in the
garnets, but also in the olivines, in which the amount of the forsterite component varies from 91.5 to 94.3%, and the iron index, from
5.5 up to 8.5%, although the olivines of the magnesian parageneses
belong amongst the minerals with the most constant iron index.

The garnets from the intergrowths are characterized by limits in
the variation of the iron index, identical to those of the garnet-
inclusions (respectively 9.5 - 16.1 and 11.1 - 15.9). The only
exception here is Sample 59/12, which is evidently a fragment of an
unusual garnet-pyroxene diamond-bearing rock, in which an octahedron
of diamond has been enclosed in the garnet.

In emphasizing that in the case of intergrowths with diamond
aggregates, the garnet must constantly react with the melt, we note
that the chrome-pyropes, poor in calcium, have a completely identical
composition both in the inclusions and the intergrowths. Consequently,
we may conclude, on the basis of factual information, that the

features of the composition of the chrome-pyropes, which coexist with diamond, are not associated with their protection within the diamond crystal, but are determined for a definite type of paragenesis, the harzburgite-dunite type. In a well-known fragment of diamond-bearing peridotite (V. S. Sobolev, Nai et al., 1969), the garnet is also close in composition to the garnet-inclusions and intergrowths with diamonds.

A unique example, which emphasizes the equilibrium nature of the minerals of all the diamond-bearing associations and the diamond, is the composition of the inclusions of garnet and omphacite, discovered in a diamond from a diamond-bearing eclogite from the 'Mir' pipe (Sample M-46). A careful analysis and comparison of the composition of the mineral-inclusions and the enclosing eclogite have shown their close analogy, even with a certain excess of the iron index of the mineral-inclusions over that of the minerals in the enclosing eclogite, that is, this established fact completely contradicts F. R. Boyd's hypothesis. Nevertheless, within the stability field of the diamond, associations are defined in which equilibrium has been achieved at different temperatures. For the ultramafic type of paragenesis, the lherzolitic association may be regarded as the lowest-temperature type, having been identified in three samples of diamond (AV-34, AV-14 and AV-93), because it is eutectic for a medium of ultramafic composition. The temperature of equilibrium of the lherzolitic paragenesis may be estimated from the composition of the diopside, associated with enstatite (Davis and Boyd, 1966). For the samples indicated, after introducing certain corrections, set out in §22, the temperature has been estimated as of the order of 1150-1200°C. It is evident that such temperature values are close to the lower limit in the stability field of diamond. It is considerably more difficult to estimate the upper limit of the equilibrium temperature for the diamond-bearing parageneses. In making such an estimate, we must first take note of the high-temperature nature of the pyrope-olivine paragenesis and the low iron index of the coexisting minerals, which overall indicates an extemely high temperature, evidently slightly exceeding 1400°C.

The Situation in Diamond Crystallization

A study of the diamond-bearing xenoliths unequivocally demonstrates that there are diamonds that have been caught up by the kimberlite along with rock from the deep-seated horizons of the mantle. An examination of the inclusions in diamonds emphasizes that a whole series of diamonds, now occurring in the kimberlite, have been derived from such rocks, that is, from eclogites and peridotites.

It remains for us to consider whether the diamonds crystallized in the kimberlite magma itself, and if so, what is their relative abundance to the xenogenic diamonds.

At the present time, there are essentially no unequivocal proofs as an answer to this question, since the early phases of crystallization of the kimberlite magma are not known in the depths of the Earth. However, a number of features (the uniqueness of the diamonds from each kimberlite pipe, including morphologic and color features (Williams, 1932; Bobrievich et al., 1959), the existence of inclusions of diamond in diamond, and zoned diamonds), suggest that portion of the diamonds have not been caught up from the mantle, but have crystallized directly in a deep-seated focus, in which melting did not proceed to completion and was replaced by recrystallization.

If we accept such an hypothesis, it may be stressed that the diamonds, which crystallize in the kimberlite, must belong to the ultramafic paragenesis, that is, they occur in equilibrium with olivine, the crystallization of which has been achieved in all, even the late, phases of crystallization of kimberlite (V. S. Sobolev, 1960; Bobrievich et al., 1959, 1964). In this respect, all the diamonds from eclogites, and the diamonds that contain an eclogitic paragenesis of inclusions, must be assigned to those caught up from the upper mantle during the eruption of the kimberlites.

If we assume that crystallization of the diamond in the kimberlite continued for some time during the movement of the magma, then it is likely that the polycrystalline diamond aggregates of the bort type (although their presence has also been established in eclogites) should primarily be assigned to this type of diamond. In this case, it may be assumed that a complex change in the composition of the silicates during the initial phase took place in the kimberlite melt. Although zoned garnets of an early generation have not so far been found, there is every ground for assuming that the calcium-poor, chromium-rich garnets of the harzburgite-dunitic paragenesis are replaced by crystallization of garnets of a lherzolitic affinity, because this paragenesis is a lower-temperature one and markedly predominantes in the polycrystalline diamond intergrowths as compared with the single crystals.

The first data on zoned garnets of the later phase (V. S. Sobolev, Khar'kiv et al., 1972) suggest that the wehrlitic association for rocks enriched in sodium, is not a higher-temperature type, as believed by O'Hara (1967), but on the contrary, a comparatively low-temperature association, and incongruently replaces the association of the lherzolitic type, with the outer selvedges of garnet enriched in calcium, depending on the degree of crystallization. Such a course of crystallization has been defined in an actual sample of a zoned garnet from the 'Mir' pipe (see §2), the inner zones of which are similar in composition to the garnets of the lherzolites (see Fig. 7), and the outer zones, to the green garnets, rich in calcium and chromium. The latter have been found in the form of separate grains, surrounded by kelyphitic rims, in a number of kimberlite pipes. A direct paragenesis with olivine has been identified for such garnets in xenoliths (Sobolev, Lavrent'ev et al., 1973a). Judging by the composition of the garnet and by analogy with the unique garnet + olivine + chrome-diopside paragenesis, identified in a single diamond (Sample 57/9), the wehrlitic paragenesis of such garnets is quite the most likely.

Thus, the uncertainty in the quantitative relationships for diamonds, which have crystallized in the kimberlite focus or outside it, is maintained precisely for the ultramafic type of paragenesis, and we are inclined to believe that the wehrlitic paragenesis is more widely distributed specifically in the kimberlites than in the typical xenoliths.

The diamonds that crystallize in the harzburgite-dunite paragenesis, belong to the highest-temperature types and are the products of the earliest crystallization. This conclusion agrees well with data based on the examination of kimberlites (V. S. Sobolev, 1960a; Bobrievich and Sobolev, 1962; Bobrievich et al., 1959, 1964; Smirnov, 1970). New factual information on intergrowths of garnet-pyroxene composition with diamonds provides better grounds for suggesting the decreased temperature of formation of most of the polycrystalline diamond aggregates as compared with the single crystals. A comparison of the compositions of garnets, pyroxenes, and chromites from specific non-diamond intergrowths from the 'Mir' pipe, has shown that these associations have probably crystallized under similar pressure conditions (the high-chromite parageneses), but with substantial falls in temperature and change in the activity of oxygen (the variable amount of Fe^{3+} in the associated chromites and ilmenites).

Thus, although we have succeeded in recording the sequence in the crystallization of certain diamond-bearing parageneses, the sequence itself is much more complex as compared with the separation of two main generations of protominerals (Smirnov, 1970), and depends in many ways not only on $\underline{P}$ and $\underline{T}$, but also on $\underline{P}_{O_2}$.

The results of a study of the diamond-bearing parageneses from the kimberlite pipes of Yakutia and the Uralian placers support the conclusions about the common nature of the geochemical conditions of crystallization of the natural diamonds from all the known deposits, arrived at both on the basis of a study of the isotope composition of the carbon in various diamonds (Vinogradov et al., 1965), and on the basis of an examination of the composition of the crystalline inclusions in the diamonds of South Africa, Ghana, Sierra Leone, and Venezuela (Meyer and Boyd, 1972). This conclusion applies equally to the placer deposits (the Urals and Venezuela) with unknown rock sources, and has been distributed in diamonds of different age, including the Precambrian (Ghana).

In connection with the above information, it must be stressed that all the information on the composition of the minerals associated with diamonds, convincingly supports the viewpoint that the diamonds are of deep-seated origin (Wagner, 1909; Williams, 1932; V. S. Sobolev, 1960a), which has been underlined in recent times by most investigators. The clear type features and the nature of the composition of the minerals, associated with diamonds, and their marked differences from the equivalent in the kimberlitic concentrates, are completely inexplicable from any hypothesis that diamonds were formed at shallow depths, within the crust (Davidson, 1966, 1967), in particular in intermediate foci (Trofimov, 1966), and also in explosion chambers at the boundary between the sedimentary cover and the basement (Vasil'ev et al., 1968).

It has been shown that the supporters of the most extreme view on the formation of diamonds under near-surface conditions (Petrov, 1959; Menyailov, 1962; Leont'ev and Kadensky, 1957; Botkunov, 1964; Mikheenko et al., 1970) have completely overlooked or have incorrectly interpreted the data on inclusions in diamonds and diamond-bearing xenoliths, which are the most reliable sources of information about the conditions of formation of natural diamonds. This factual information must be given special attention, and any hypothesis on the formation of diamonds cannot be well-founded unless it explains the specific features of the minerals directly associated with diamonds.

CHAPTER V

MINERAL DATA

The summary of data presented on the compositional features of the various minerals from different depth facies will enable us to estimate the general limits of variation in their composition and to clarify some of the patterns of isomorphous substitutions. Therefore, in this chapter, we shall consider features of distribution of the trace-elements: sodium in garnets and enstatites, potassium in clinopyroxenes, chromium in olivines, and titanium in garnets.

The accepted order of description of the minerals is dependent on the importance of their composition in deep-seated parageneses with respect to the possibility of estimating the composition of the environment and the conditions of formation. From this aspect, the most important minerals are the garnets and clinopyroxenes, although the marked predominance in the rocks of the upper mantle, not of these minerals but of olivine, is not in question.

§12. GARNET

The Composition of the Magnesian and
Magnesian-calcium Chromium-bearing Garnets

The quite recent description of the compositions of magnesian garnets from deep-seated xenoliths and kimberlite has emphasized the complete similarity between the compositions of all garnets from ultramafic rocks, based on iron index and content of calcium, with small differences only in the content of chromium (Sobolev, 1964a). The maximum large amount of Cr_2O_3 in such garnets has reached 7.0% in only one isolated case (Nixon et al., 1963). The continuation of a purposeful study of the mineralogy of the kimberlites of Yakutia and South Africa, and also a study of garnet inclusions in diamonds, has led to a significant broadening of ideas on the limits of variation in the composition of these garnets based on the amount of both CaO and Cr_2O_3.

Figure 19 presents data on the composition of the magnesian garnets (with variations in the iron-index value from 10 to 40%), containing a variable amount of calcium and chromium. The compositions of the most widely distributed kimberlitic garnets have been omitted from this figure; variations in the chromium and calcium components fall within quite a narrow range, respectively from 0 to 10 and from 8 to 18%.

Compositions typical of the parageneses with diamonds, with less than 10% calcium component may be assigned to the strictly magnesian, calcium-poor garnets. Garnets (with variable chromium-index), containing 10-50% calcium component, are conveniently referred to as magnesian-calcium types, and those containing more than 50% calcium component, and representing, in essence, a new variety of the garnet group, as calcium garnets proper. The great majority of the compositions, referred to in Figure 19, are from results of original analyses (Sobolev, Lavrent'ev et al., 1973a).

The data given in Figure 19, convincingly indicate the significant role of chromium in the form of the knorringite component in magnesian garnets from kimberlites. About 100 analyzed garnets contain this component, that is, the amount of chromium exceeds that of calcium in

154

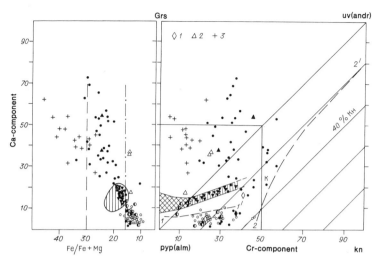

Fig. 19. Compositional features of chromium-rich magnesian-
calcium garnets (Sobolev, Lavrent'ev *et al.*, 1973 *a*). From xenoliths of diamond-bearing dunites, (1) of wehrlitic paragenesis of
minerals, enclosed in diamonds (2), chromium-bearing grospydites
and kyanite eclogites (3) (for remaining symbols, see Figure 5).
Field of garnets from paragenesis with two pyroxenes is shown by
vertical lines and cross-hatching; position of compositions of most
common garnets from kimberlites is shown by cross-hatching; the line
2 - 2' indicates the position of the assumed boundary of solubility
of chromium in natural garnets from kimberlites; the boundary of the
iron index for most garnets, coexisting with diamonds, is shown by
the dash-dot line; the limiting iron index for most garnets in paragenesis with olivine is shown by the broken line; K) compositional
point for garnet from Kao pipe (Nixon & Hornung, 1968); pyr) pyrope;
alm) almandine; grs) grossular; uv) uvarovite; and) andradite;
kn) knorringite.

their composition. Most of such compositions, specifically containing
more than 10% knorringite, belong to calcium-poor garnets, but in
individual cases, a substantial amount of uvarovite has also been
identified in them.

The varilable data on the composition of magnesian and calcium
chromium-bearing garnets may evidently be regarded as quite representative, since most of the rare types of these garnets from inclusions in diamonds, and from the pyrope peridotites, and the green
garnets, rich in chromium and calcium, from the kimberlite concentrates,
having already been examined in detail. Special collections and
the study of green garnets in concentrates from various pipes in
Yakutia have enabled us to determine the limiting compositions on
the basis of chromium content (Sobolev, Lavrent'ev et al., 1973a).
From determinations of the amount of chromium and calcium in 215
grains of such garnets, we have calculated the average amounts of
the appropriate oxides in the magnesian–calcium chromium-bearing
garnets (Cr_2O_3 = 11.2 ± 2.7%; and CaO = 13.9 ± 3.7%). The search
for such garnets has been carried out according to a plan based on
their color and their high refractive-index values.

Particular attention has been paid to the search for low-calcium

chrome-pyropes in kimberlite concentrates, based on color features
and the manifestation of the alexandrite effect (Sobolev, 1971a). This
search has led to the discovery of more than 150 grains of such garnets
in concentrates from diamond-bearing pipes. We also record that our
investigations of average garnets from various pipes (based on 200-
300 grains of magnesian garnets, selected at random) have enabled
us to reconsider the view that chromium-rich garnets are rare in
kimberlites, and to acknowledge that garnets in many of the pipes,
which contain more than 7% Cr_2O_3, and which were not found earlier,
occur in the garnet concentrates in amounts of from 10 to 30%, that
is, they are quite widely distributed.

Data on the parageneses of the magnesian garnets enable us in
Figure 19 to recognize a clear field of garnet compositions from the
paragenesis with two pyroxenes ($\pm$ olivine). In §1, it has been shown
that the amount of calcium in such garnets depends both on the amount
of Cr_2O_3 in them, and on the amount of Na_2O in the associated
pyroxenes.

The value of the CaO content as an indicator of the paragenesis
of the garnets enables us, on the basis of having identified the
position of the compositional field for garnets from the two-pyroxene
paragenesis (the lined area on Figure 19), to recognize two additional fields of garnet compositions (in paragenesis without clinopyroxene (harzburgitic or dunitic) and without enstatite (wehrlitic
or eclogitic)).

Only the garnets in paragenesis without enstatite are richer in
calcium as compared with those from the lherzolites and websterites.
This applies to both the reliably known cases of non-olivine parageneses (specific chromium-bearing eclogites, most commonly with
kyanite), and also the parageneses with olivine (wehrlitic). The
factual information is more in favor of the predominant distribution
of the wehrlitic paragenesis, as compared with the eclogitic type,
in the kimberlites.

Judging by the general tendency towards increase in the iron index
with increased amount of calcium in the magnesian garnets (see Fig. 19),
it may be assumed that portion of the calcium-rich garnets crystallized
later than the calcium-poor garnets, that is, the sequence of crystallization of the parageneses has the following form: harzburgite-
lherzolite-wehrlite. Clear evidence for the trend towards enrichment
in calcium in the garnet of later generation concurrently with increase
in the iron index may be seen in the unique zoned garnet M-49, described in §2 (see Figure 7).

The iron index of the garnets under consideration, like their
calcium index, is a diagnostic feature, clearly distinguished, in
particular, the garnets associated with diamond, that is, the deepest
and less deep types. The average iron index for all magnesian garnets,
associated with diamonds in inclusions, intergrowths, and xenoliths,
is significantly lower than that of the garnets from the deepest
xenoliths (among the non-diamond types) of pyrope peridotites from
the 'Udachnaya' pipe (see §1). This difference is also clearly
defined by the position of the points on the diagram. On the basis
of the iron index (trending towards increase), many garnets from
websterites and most garnets from kimberlite concentrates are clearly
separated. With increase in the calcium index in the chromium-
bearing garnets, a trend appears towards increase in the iron index,
which varies in the garnets from the kimberlite concentrates approximately from 20 up to 35%.

Thus, the calcium and iron indices of the magnesian garnets with variable content of chromium clearly reflect the features of their paragenesis, and in a number of cases, the sequence of crystallization also. The identified significant amounts of Cr_2O_3 are a typical feature of most of the garnets from diamonds, especially from the deep-seated peridotites and the specific garnets from the kimberlite concentrates, which probably belong to the wehrlitic paragenesis. It should also be noted that the average chromium index of garnets from the Yakutian kimberlite pipes seems significantly greater (by 1 1/2 - 2 times) than earlier estimates, given in the works of Bobrievich et al., (1964) and Khar'kiv and Makovskaya (1970).

The Determination of the Composition of Magnesian Garnets from Their Physical Properties

With the increasing interest in the mineralogy of the rocks formed under conditions of high pressures, and also the detailed study of the mineralogy of the kimberlites and inclusions in diamonds, numerous new data on the composition of magnesian garnets have appeared.

Almost in parallel with conformation of a new end member in the garnet series (knorringite, $Mg_3Cr_2(SiO_4)_3$ (Nixon and Hornung, 1969)), a substantial number of magnesian garnets, containing this component, have been studied, including garnets extracted directly from diamonds (Meyer, 1968; Sobolev, Lavrent'ev et al., 1969; Meyer and Boyd, 1972).

It has been shown that magnesian garnets contain a significant amount of calcium- and chromium-components, and only in quite rare cases do calcium and chromium in equal amount form uvarovite. A series of compositions of magnesian garnets is known, beginning with non-chromium types, containing 40% calcium component (Sobolev and Kuznetsova, 1965), passing through garnets simultaneously rich in calcium and chromium (Sobolev and Kuznetsova, 1968; Sobolev et al., 1970; Khar'kiv and Makovskaya, 1970), up to garnets with a predominance of chromium over calcium, reaching up to 46% of knorringite in an inclusion of garnet from a Yakutian diamond (Sample AO-458; Sobolev, Lavrent'ev et al., 1969). The data in Figure 19 convincingly indicate significant variations in the amounts of calcium and chromium in magnesian garnets.

Magnesian garnets with increased content of chromium are distinguished from non-chromium natural pyropes by the markedly increased refractive index, that is, the effect of the amount of chromium on n in them is analogous to that of Fe^{2+} in the pyrope-almandine garnets and Fe^{3+} in the grossular-andradite garnets. However, in the magnesian garnets, the amount of Fe^{2+}, and even more so Fe^{3+} is very small, and the effect of chromium markedly predominates. Increase in the value of a_0, in all the varieties of garnets listed is associated with the effect of the calcium content, which forms grossular, andradite, or uvarovite separately, or all three components together. Therefore, the attempt to determine the composition of magnesian garnets with high refractive index, based on the values of n and a_0, using the pyrope-almandine-grossular diagram (Rickwood et al., 1968), leads to an erroneous interpretation. Insufficiently accurate results are also obtained during the application of the composition triangle, pyrope-almandine-uvarovite, which distort the true compositions of the garnets in a number of cases, especially those with high $\underline{n}$ values (Rickwood $\underline{et\ al}$., 1968), in association with the impossibility of calculating the amount of knorringite.

Thus, it is essential to calculate the amount of the knorringite component when considering the composition of magnesian garnets on the basis of physical properties. Since information is not available on the value of $\underline{n}$ and $\underline{a}_0$ for synthetic knorringite (Coes, 1955), we have used the results of an analysis of garnet AO-458 from a Yakutian diamond with a maximum amount of 46% of this component and $\underline{n} = 1.789$ (Sobolev, Lavrent'ev $\underline{et\ al}$., 1969), and data on the end members of the garnet group (Skinner, 1956), for calculating the value of $\underline{n}$ for pure knorringite.

The data obtained for theoretically pure knorringite ($\underline{n} = 1.830$; Sovolev, Lavrent'ev $\underline{et\ al}$., 1969) have been checked against Sample K-47 (Nixon and Hornung, 1968), for which a good correlation has been established between the measured and calculated values of $\underline{n}$ on the basis of our data. Sample K-47 was unsatisfactory for use as a basic standard, because in comparison with Sample AO-458, it contained a markedly increased amount of CaO, as a result of which an ambiguity in the treatment of its composition during calculation into components was assumed.

The value of $\underline{n}$ for knorringite found by calculation was significantly lower than that calculated earlier (McConnell, 1966): respectively 1.875 (according to McConnell, 1966) and 1.830 (according to our data). At the same time, the value of $\underline{a}_0 = 11.622$, calculated for knorringite (McConnell, 1966), corresponds well with the value obtained from the calculation of some compositions (rich in knorringite), for which the parameters of $\underline{a}_0$ are known (Nixon and Hornung, 1969; and see also Table 31).

The series of composition-properties diagrams for garnets available in the literature make it possible to determine the three components (Gnevushev $\underline{et\ al}$., 1956; Biller, 1962; Sriramadas, 1959). Some more complex diagrams enable us to determine four components, using density, in addition to data on the values of $\underline{a}_0$ and $\underline{n}$ (Winchell, 1958).

All these diagrams make possible an approximate determination of the composition of garnets from an unknown paragenesis, which contain all possible combinations of the principal end members: pyrope, almandine, spessartine, grossular, and andradite. In addition, simplified diagrams are also recommended, which on the basis of data on the values of $\underline{a}_0$ and $\underline{n}$ for random garnets, permit an approximate assessment of their composition (Sastry, 1962; Vasil'ev, 1969).

In this connection, it must be emphasized that the recognition of paragenetic types of garnets (Sobolev, 1964a) and the definition of the limits of variation in composition for each paragenetic type provides the possibility of a much more precise determination of garnet composition based on physical properties within a paragenetic type or group of types. The clearest examples are the magnesian garnets, belonging to the ultramafic type of association.

An attempt to construct composition-properties diagrams, which could be universal for all magnesian garnets, encountered difficulties owing to the multi-component composition. Here, it is convenient to carry out certain simplifications, which have no substantial effect on the results. In particular, we may ignore the effect of a separate andradite component, without, however, excluding it from consideration, but combining it, in view of the similarity in the physical properties of andradite and uvarovite, with the latter and with Ti-andradite.

The total content of Ti-andradite and andradite almost never exceeds 10% and for most analyses, comprises only 3-5%.

The principal condition in representing these garnet compositions on a plane is the possibility of calculating the amount of the almandine component. From general works on the garnet group (Sobolev, 1964a), it is known that the iron index of the pyrope garnets from deep-seated ultramafic rocks varies within a relatively narrow range, mainly from 15 to 25%, and rarely exceeding this figure. This is evidenced by the results of calculating the average compositions of garnets from ultramafic rocks with $\underline{f}$ = 16 $\pm$ 4% for the garnets for peridotites (21 analyses) and $\underline{f}$ = 19 $\pm$ 4% for garnets from ultramafic rocks (89 analyses) (Sobolev, 1964a). In this case, the iron index only implies the ratio Fe^{2+}/Fe^{2+} + Mg, without allowance for the andradite component, which has been grouped together with uvarovite. Therefore, it is necessary in constructing the composition-properties diagram to select some value of the iron index, which would permit determination of the composition of the garnets not only from the ultramafic xenoliths, but also from diamonds, for which the iron-index values are sometimes lowered to 10%. Taking a constant iron index ($\underline{f}$) for the magnesian garnets at 15%, we may thus reduce the possibility of a change in their composition to a change in amount of two variables: the calcic component, defined by a change in the value of $\underline{a}_0$, and the chromium component, defined by a change in $\underline{n}$.

Figure 20 is a diagram, constructed for values of $\underline{a}_0$ and $\underline{n}$ for garnets of constant iron index ($\underline{f}$ = 15%). The compositional point for the corresponding non-calcic pyrope lies on the line of pyrope-almandine compositions. This point is joined by straight lines with compositional points for pure grossular, uvarovite (together with andradite), and knorringite, for which $\underline{f}$ = 15% is also accepted. Starting from the addition of the properties and composition in the pyrope-grossular series, we have plotted the lines of equal amounts of calcic component. The compositional points for 21 garnets (from the literature and our own data), for which there are chemical analyses and known values of $\underline{a}_0$ and $\underline{n}$ (Table 68), have been plotted on the diagram (Fig. 20). The values of the iron index of these garnets vary from 8 to 24%.

As is seen from the results in Table 38, the difference in the determination of the amount of the additive-components in the magnesian garnets (grossular, andradite + uvarovite, and knorringite), based on the values of $\underline{a}_0$ and $\underline{n}$, and from the chemical analyses, very rarely exceeds 3 mol %, and does not in fact pass beyond a limit of 5 mol %. Since the diagram provides an unequivocal solution, its application only requires the determination of the point of intersection of the straight lines corresponding to definite values of $\underline{a}_0$ and $\underline{n}$ (see Fig. 20).

The chemical analyses of garnets (21) selected for verifying the diagram embrace almost all the known compositions of magnesian garnets with variable amounts of Ca and Cr. The only exceptions are garnets, rich in Ca and Cr, from kimberlitic concentrates with greater iron-index value (25% and more) (Sobolev et al., 1968), for which the amount of Cr, taken from the diagram, will clearly be overestimated. In order to determine the composition of such garnets, it is recommended that a similar diagram be constructed, taking, however, the value of the iron index for pyrope and knorringite at 25-30%. It should be noted that amongst the magnesian garnets,

160

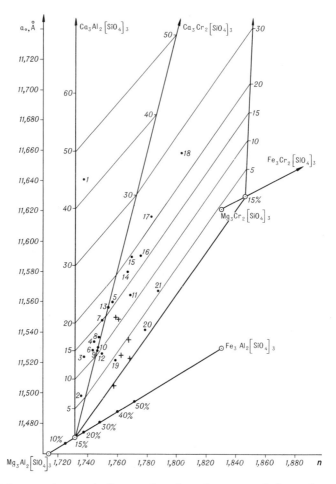

Fig. 20. Diagram ($n - a_0$) for estimating the composition of magnesian-iron garnets with variable amounts of Ca and Cr, based on physical properties: points numbered 1 - 21) compositions of analyzed garnets (Table 38); crosses) compositions of garnets, enclosed in diamonds from Africa (Meyer & Boyd, 1972); lines joined between compositions of Mg-Ca garnets with and without Cr are those for equal amounts of Ca-component (in mol %).

containing Cr_2O_3, those with increased iron index are in a minority as compared with the more widely distributed low-iron types.

In spite of a general satisfactory correlation between the analytical data and results obtained with the aid of the diagram based on the amount of even insignificant amounts of andradite + uvarovite and knorringite, the analyses of individual samples demonstrate significant discrepancies. Such samples include E-15 (Nixon et al., 1963), which is apparently characterized by a certain underestimate in the value of a_0 (11.526 Å), since on the diagram this gives 6% knorringite, whereas in this sample, according to the analysis, there is little Cr and half the amount of Ca is involved in grossular. Such a discrepancy

Table 38

Spec. No.	Samples	Analytical Data						Data taken from Diagram					n	a_0 (Å)	Difference From Analytical Data				Source
		grossular	andradite	uvarovite	andradite + uvarovite	Ca-component	knorringite		grossular	andradite + uvarovite	Ca-component	knorringite			grossular	andradite + uvarovite	Ca-component	knorringite	
1	O-160₁	42	—	—	—	42	—	24	42	3	45	—	1,737	11,641	0	+3	+3	0	Sobolev and Kuznetsova, 1965
2	O-145	9	1	1	2	11	—	21	5	3	8	—	1,734	11,498	−4	+1	−3	0	This volume (table 1)
3	A-6	4	—	6	6	10	—	13	9	4	13	—	1,737	11,524	+5	−2	+3	0	
4	A-3	5	1	7	7	12	2	19	4	10	14	3	1,744	11,534	−1	+3	+2	0	O'Hara and Mercy, 1963
5	A-4	—	1	12	13	13	—	15	—	16	16	—	1,756	11,559	0	+3	+3	+1	
6	E-11	4	3	6	9	13	—	16	3	9	12	—	1,743	11,528	−1	0	−1	0	
7	E-1	4	—	9	9	13	—	17	2	14	16	—	1,749	11,548	−2	+5	+3	0	
8	E-3	5	2	6	8	13	—	13	2	12	14	—	1,747	11,537	−3	+4	+1	0	
9	A-79	1	2	8	10	13	—	15	1	11	11	—	1,746	11,528	−1	+1	0	0	
10	G 12	6	2	5	7	11	—	19	—	11	12	—	1,746	11,531	−5	+4	0	0	Nixon et al, 1963
11	E-10	—	3	13	16	12	9	17	—	16	16	6	1,758	11,565	0	0	0	−3	
12	E-15	5	5	1	6	16	—	17	—	9	9	6	1,749	11,526	−5	+3	−2	+6	Sobolev, 1970
13	Ud-3	—	2	13	15	11	2	17	—	15	15	3	1,755	11,557	0	0	0	+1	
14	Ud-5	—	6	9	15	15	12	13	—	18	18	11	1,766	11,579	0	+3	+3	−1	
15	Ud-4	—	2	15	17	15	10	14	—	20	20	10	1,768	11,590	0	+3	+3	0	This volume (table 2)
16	Ud-18	—	5	13	18	17	17	14	—	17	17	19	1,774	11,591	0	−1	−1	+2	
17	Ud-92	—	7	13	20	18	19	13	—	22	22	19	1,782	11,617	0	+2	+2	0	Nixon and Hornung, 1969
18	K-47	—	3	19	22	20	34	17	—	26	26	34	1,803	11,659	0	+4	+4	0	
19	Gl 25a	—	4	1	5	22	24	8	—	5	5	19	1,758	11,522	0	0	0	−5	Meyer and Boyd, 1972
20	G 20a	—	4	1	5	5	35	11	—	3	3	36	1,778	11,542	0	−2	−2	+1	
21	931	—	4	1	5	5	43	10	—	6	6	42	1,787	11,568	0	+1	+1	−1	Sobolev et al, 1971b

has only been recorded in the one case and may be associated either
with a certain underestimate in the value of a_0, or with the complex
nature of the garnet composition in the samples. In some xenoliths
of ultramafic rocks from African kimberlites, a variation in the
garnet composition has been recorded (Rickwood et al., 1968), and the
difference in the extreme values of a_0 reaches 0.02 Å, which corresponds to approximately 5 mol % in the amount of the calcium component.

For convenience in determining the compositions of these garnets
from their physical properties, a transformed variant of the diagram
is proposed (Fig. 21). The compositional points for the magnesian
garnets, based on values of a_0 and n, graphically and quite precisely
enable us to assess their features, namely, the amounts of calcium
and knorringite components. On this diagram, for example, differences
are clearly seen in the compositions of the garnets from peridotites
and those enclosed in diamonds. Good correlations between the results
of analyses and data obtained from the diagram have been achieved
for three garnets from diamonds (Analyses 19-21).

In order to emphasize the necessity to use the described method
for estimating the composition of garnets, let us consider the compositional features of some garnets from ultramafic xenoliths in the
South African pipes, for which there are data on their physical
properties and for which even an attempt has been made to assess
these compositional features on the basis of these data (Rickwood
et al., 1968). It is expedient to dwell on the compositional features
of the garnets from peridotites, described in the above-mentioned
work with the largest n values (Samples 34/551 and 368), for which
respectively, a_0 = 11.5895 and 11.577 Å, and n = 1.7733 and 1.7625.
The olivine, associated with these garnets, contains (optical data),
8% fa, that is, it is the most magnesian type (Mathias et al., 1970).
In clear contrast to such an olivine composition, there are garnet
compositions, taken by the authors from the pyrope-almandine-uvarovite
diagram, with iron indices respectively of 30 and 25.

In using our proposed diagram (see Fig. 21), an increased amount of
knorringite (15% for Sample 34/351 and 10% for 368) has been found in
the garnets mentioned, which indicates their unusual composition,
in which respect both samples deserve more detailed examination.

We have made a special comparison of the compositions of garnets
with n greater than 1.760, since in the normal pyropes from the
lherzolite paragenesis with smaller n values, the knorringite
component is clearly absent, and their composition, without special
corrections, may be determined from the pyrope-almandine-uvarovite
diagram. This also applies to Sample XK-1 with a_0 = 11.5626 Å
and n = 1.7635 in paragenesis with olivine containing 8% fa (Mathias
et al., 1970).

The suggested method provides a rapid means of determining the
composition of the pyrope garnets and of estimating the amount of
excess Cr-component with respect to calcium (knorringite) in them.
Without resorting to analysis, it is possible by using the proposed
diagram, to assess the composition features of magnesian garnets.

The grossular-uvarovite-pyrope-knorringite system of garnets,
as is known, is three-component, and 'mutual' (Zavaritsky and Sobolev,
1961), and may be illustrated in the form of a square (see Fig. 19).
Resolution into components may be achieved by two methods, depending
on which diagonal divides the square. In one variant, grossular is
impossible with knorringite, and in the other, pyrope is impossible
with uvarovite. In this system, the use of both variants as equal

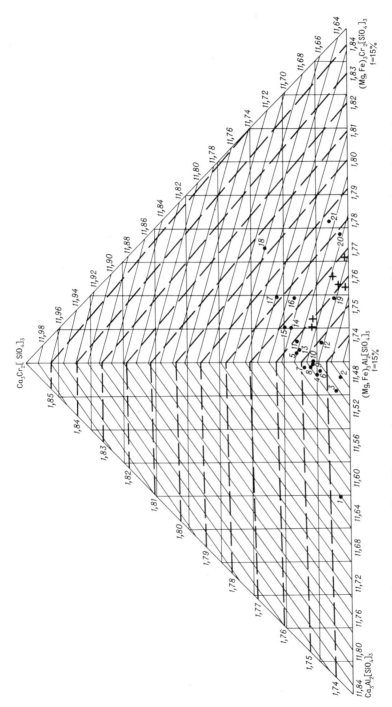

Fig. 21. Composition-properties diagram for magnesian garnets with constant iron index (f = 15%), with variable amounts of Ca and Cr: 1-21) compositional points for garnets, plotted on the basis of analytical data; crosses denote six compositional points for garnet inclusions in African diamonds (Meyer & Boyd, 1972), plotted on the basis of a_0 and n data.

in value is known (Nixon and Hornung, 1969), and also the complete failure of recalculation into components owing to the indeterminacy of the results (Boyd, 1970).

The existence of two types of mutual systems is possible in principle: one, when we know the diagonal, based on the existence of actual parageneses or a continuous series of solid solutions along the diagonal with a break in a perpendicular direction, and the solution of the variant for recalculation is unequivocal; the other, when there are continuous solid solutions in the area of the entire variant. In the present system, both variants are of equal value, and the selection is of an arbitrary nature, in order to retain non-ambiguity in the comparisons.

This and other systems seem relatively complex for the garnets and do not fall directly into the first type; nevertheless, it is possible here to begin with actual existing end members and the relative distribution of solid solutions along the diagonal. On this basis, a scheme of recalculations has been devised, accepted by the present author, which excludes the possibility of appearance of problematic components of the koharite type in the presence of calcium, etc. The correctness of such a variant is also emphasized by the existance of a wide-ranging series of solid solutions along the pyrope–uvarovite diagonal with more than 50% uvarovite (see Fig. 19).

In this scheme of recalculation, the largest divalent cation of Ca is first combined with the larger trivalent cations, Fe^{3+} and Cr^{3+}, and then with Al (Sobolev, 1944). Extending this scheme to the present system of garnets, we find it convenient to accept a definite variant, in which the coexistence of knorringite and grossular is excluded, and to construct a diagram of the rectangular type and one of the Winchell type. We note that, knowing the values of $\underline{n}$ and $\underline{a_0}$, we always obtain an unequivocal solution here (see Fig. 21). The final solution to the problem of selecting components during recalculation may be obtained only after experimental investigation of the system at pressures in which pure knorringite will be stable, that is, about 100 kb (Sobolev, Lavrent'ev $\underline{et\ al.}$, 1973$\underline{a}$).

The Composition of the Pyrope–almandine Garnets

Apart from the chromium-bearing garnets, let us consider garnet compositions, almost devoid of Cr, from websterites, eclogites, kyanite eclogites, and grospydites. These include the yellow-orange garnets, enclosed in diamonds and intergrowing with them. The amount of Cr_2O_3 in the garnets from eclogites extremely rarely exceeds 0.2% (Fig. 22) and only in the websterites are certain variations observed as far as samples of complex composition with a markedly differing amount of chromium in different zones.

Both in regard to the amount of Cr, and the iron index, the garnets from websterites represent a transitional type between the ultramafic and basic associations, which is emphasized by the position of their compositional points on the diagrams for the chromium-bearing garnets (in the field of higher iron index), where they are grouped with the garnets from lherzolites (see Fig. 19), and for the non-chrome garnets (see Fig. 22), where they are denoted by a separate symbol. The compositions of garnets from websterites, on the basis of calcium, define the lower limit of compositions of the eclogite garnets, with some deviation as a result of the differing amounts of sodium in the pyroxenes.

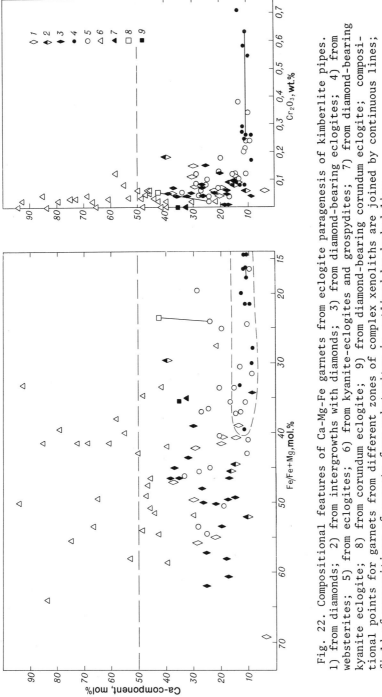

Fig. 22. Compositional features of Ca-Mg-Fe garnets from eclogite paragenesis of kimberlite pipes.
1) from diamonds; 2) from intergrowths with diamonds; 3) from diamond-bearing eclogites; 4) from
websterites; 5) from eclogites; 6) from kyanite-eclogites and grospydites; 7) from diamond-bearing
kyanite eclogite; 8) from corundum eclogite; 9) from diamond-bearing corundum eclogite; compositional points for garnets from different zones of complex xenoliths are joined by continuous lines;
field of compositions of garnets from websterites is outlined by dashed line.

The principal difference from the chromium-bearing garnets of the olivine-bearing and specific eclogite parageneses, is the wider variation in iron index with extreme limits of 15 and 70% and the absence of any tendency towards a positive correlation between iron index and calcium index.

The compositions of all the garnets from deep-seated xenoliths and kimberlite concentrates are presented in a compositional triangular diagram, (Fe + Mn)-Ca-Mg (Fig. 23). On this composite diagram, the general features of all the garnets examined are more clearly shown: the presence of a continuous series of magnesium-calcium garnets with wide variations in iron index from 30 to 65%, and a general wide range of variations in iron index from 9 to 70%.

Garnets of established compositions have crystallized in the pressure range corresponding to the two facies of the upper mantle under consideration (see Chapters III and IV). The data presented contradict the conclusions reached during the classification of all the known garnets from eclogites (Coleman et al., 1965; Smulikowski, 1964, 1968), in the fact that the garnets from kimberlites and associated eclogites are the most magnesian as compared with those from eclogites of the metamorphic complexes. As is seen from Figure 23, the compositions of garnets from deep-seated eclogites overlap the entire field of those from eclogites of gneissic complexes, and the boundaries of the identified garnets, on the basis of iron index, reach the fields of those from eclogites of the glaucophane-schist complexes.

The Isomorphous Content of Titanium in Garnets

In a whole series of cases, it has already been noted that, in the orange and red magnesian garnets from kimberlites, which as a rule, contain a low content of chromium and often occur in association with ilmenite, chemical analysis has revealed an amount of TiO_2, which in some cases reaches 1.0 - 1.5% (Sarsadskikh et al., 1960; Bobrievich et al., 1964; Sobolev, 1964a). Such garnets have been contrasted with the Cr-rich pyropes, colored crimson and lilac, and a tendency towards an increased amount of Ti and Cr in the garnets has been regarded as antipathetic (Khar'kiv and Makovskaya, 1970; Frantsesson, 1968).

During a study of the chromium-rich pyropes from peridotite xenoliths in the 'Udachnaya' kimberlite pipe, a significant amount of TiO_2 has been identified in individual cases, reaching 1.5%, which had not been recorded earlier in chromium-bearing garnets; such an amount (up to 1.0%) has also been found in the orange garnets from eclogite xenoliths and diamonds. The method of X-ray spectral quantitative micro-analysis which we have used, enables an unequivocal estimate to be made of the structural nature of the TiO_2 component, which during analysis of the powder by chemical means, could fall into earlier separate analyzed samples, owing to mechanical ingrowths of titanium-bearing minerals, especially ilmenite, rutile, and sphene. In this respect, comparative figures for the amount of TiO_2 in garnet, obtained from chemical analyses of the powder and with the aid of the microprobe for a particular sample, are respecitvely 1.47 and 0.05% (Knowles et al., 1969).

The pyropes, which we examined from the peridotites of the 'Udachnaya' pipe, described in greater detail in §1, are distinguished by increased n values, reaching 1.796. In addition to a significant

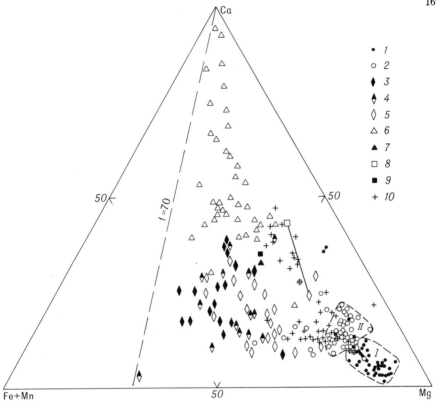

Fig. 23. Compositions of Ca-Mg-Fe garnets from kimberlite pipes. 1) chrome-pyropes from diamonds; 2) from peridotite xenoliths; 3) garnets from diamond-bearing eclogites; 4) Mg-Fe non-Cr garnets from diamonds; 5) from eclogite xenoliths; 6) from kyanite-eclogite xenoliths and grospydites; 7) from a diamond-bearing kyanite eclogite; 8) from a corundum eclogite; 9) from a diamond-bearing corundum eclogite; 10) from kimberlite concentrate; I) compositional field of magnesian garnets from diamonds; II) compositional field of most garnets from peridotite xenoliths.

amount of chromium, the titanium in these pyropes also exerts an influence on the increase in refractive index (Sobolev, 1949). The maximum amount of TiO_2 (from 0.82 to 1.5%) has been defined for five garnet samples. In the entire series of garnets studied from this pipe, compositions are also found with low TiO_2 content, down to 0.2%.

The discovery of a definite amount of TiO_2 in magnesian garnets, especially in the series of samples with variable amount of TiO_2, has enabled us to clarify the structural position of titanium in the pyrope–almandine garnets. Indirect estimates of the possible role of titanium in such garnets have so far been given on the basis of general considerations of the possibility of replacing silicon by titanium in silicates and by analogy with titanium–rich calcium garnets in alkaline rocks, for which a negative correlation has been clearly established in a number of cases between the amounts of silican and titanium (Sobolev, 1949; Sobolev, 1964a; Rickwood, 1968; Howie and Wooley, 1968).

However, even such a clear correlation between the amounts of silicon and titanium, in the opinion of some authors (Hartmann, 1969), is not proof of fourfold coordination of titanium in the calcium garnets, because in other schemes of possible replacement in the structural formulae of the garnets, such a correlation between silicon and titanium is also maintained during the transfer of part of the Fe^{3+} and Al into fourfold coordination. The partial possibility or replacement of silicon by titanium in silicates is also not denied by such a conclusion, but the predominant role is ascribed to titanium in a specifically sixfold coordination (Hartmann, 1969). At the same time, individual examples of clear correlation between titanium and Fe^{3+} are known with silicon remaining constant (Manning and Harris, 1970).

In estimating the role of titanium in the Mg-Fe garnets (when its content is low) with the aid of mathematical statistics, it is necessary to study the analyses used with extreme caution. In checking the association between the number of atoms of silicon, aluminum (plus chromium), and titanium in the structural formula of a garnet, we must use only those analyses in which amounts of these elements have been determined with maximum accuracy. Such an approach immediately excludes the possibility of using chemical analyses in view of the great likelihood, first, that mechanical precipitation of TiO_2 has occurred (Knowles et al., 1969) and, second, that SiO_2 has entered the material during grinding in an agate crucible at the time of preparation for analysis (Lebedev, 1959; Sobolev, 1964a). In this case, entry of microintergrowths is also not excluded.

The possibility of contamination with mechanical additives is almost completely eliminated only when using the method of local X-ray spectral microanalysis on the electron probe. This same method enables the detection of even a small amount of TiO_2 and the determination of its structural nature. It also excludes contamination by excess silica.

Thus, verification of the association of titanium in garnets is only necessary in analyses carried out by the above method. We have used information on garnets from pyrope peridotites, and also those from eclogite xenoliths, including the diamond-bearing types (Sobolev et al., 1973b). The quality of the analyses chosen for further treatment has been checked on the basis of the following criteria: 1) the total for the analysis, allowing for possible errors in the X-ray spectral analysis, must be within the range 98-101.5%; 2) the number of atoms of Si, which do not exceed 3000 (by a value of not more than 2 relative percent); the total of R^{3+}, which does not exceed 2000 (theoretical amount). Analyses deviating from this framework have not been considered satisfactory for further treatment.

Using such an approach, it has seemed possible to accept for further treatment, the analyses completed by Meyer and Boyd (1972), and also most of our own analyses. In addition, quite a substantial number of analyses of garnets have been published, which were carried out by the same method (Knowles et al., 1969). However, in all these analyses, the number of atoms of Si and Al are significantly in excess of our limits. It is clear that analytical errors in determining the light elements exceed 2% of their amount.

The 78 selected analyses of pyrope-almandine garnets are characterized by wide variations in the amount of TiO_2. Almost half of all the garnets, as seen in Figure 24, contain a very low amount of TiO_2, not exceeding 0.1 wt %. The compositions of these garnets, which on the basis of titanium content, form a separate maximum on the histogram, will in future be regarded as devoid of titanium. The titanium-bearing garnets include those with from 0.2 to 1.56% TiO_2 or from 0.012 to 0.089 atoms of titanium per 12 atoms of oxygen.

In order to check the position of titanium in the garnet structure, garphs have been constructed (see Fig. 24) and correlation coefficients have been calculated between the amount of Al + Cr, on the one hand, and Ti, on the other. The graphs have revealed a negative correlation between the amounts of Al + Cr and Ti, and the complete absence of correlation between Si and Ti. In the latter case, we have been limited only by the computation of the average number of atoms of Si in the 78 analyses of garnets used, which equals 3.019 ± 0.026 for the total selection.

A strong negative correlation has been revealed between the total amounts of Al + Cr and Ti in the overall selection of magnesium- and magnesium-iron, titanium-bearing garnets ($\underline{r}$ = 0.79 at 95% confidence level, with $\underline{N}$ = 46). The position of the regression line is determined by the equations

$$Al + Cr = 2005 - (3.22 \pm 0.46)Ti, \text{ and}$$
$$Ti = 716.8 - (0.36 \pm 0.05)(Al + Cr).$$

The establishment of this association makes it possible to assign titanium to the group of sixfold coordinated oxides and to determine the possible amount of Fe^{3+} by difference from the total sum of the atoms (2.00) in this group. Such a method has established a ratio of Fe^{3+} and Ti, which is apparently positive ($\underline{r}$ = 0.63) for this same selection. The position of the regression lines (see Fig. 24) is also determined by the equations

$$Ti = (0.36 \pm 0.09)Fe^{3+} + 1.26, \text{ and}$$
$$Fe^{3+} = (1.65 \pm 0.39)Ti + 35.$$

For the garnets, which contain less than 0.012 atoms of titanium, that is, non-titanium types, the average amount of Al + Cr has been calculated and this, for 40 analyses, appears to be 1.930 ± 0.028. This supports the hypothesis of the extremely constant amount of Fe^{3+} (andradite component) in garnets of the type examined, in any case, seldom exceeding 5 mol % (Sobolev, 1964a). In the case of an increased amount of titanium, the content of $\overline{Fe}^{3+}$ shows a clear trend towards increase, reaching, however, only 10% as a maximum.

Thus, the results of the investigations unequivocally indicate that the titanium additive in the pyrope-almandine garnets is to be treated as Ti-andradite.

When recalculating the garnet analyses, this component must be separated by combining titanium with an equivalent amount of calcium (equivalent to andradite) before the estimation of andradite. The maximum amount of the Ti-andradite component in the garnets so far studied reaches 6.5 mol % (see Table 4, Sample A-147).

In connection with the accurate identification of the structural position of titanium in pyrope-almandine garnets from kimberlites, it is necessary to dwell on the description of the highly-refractive acicular inclusions often found in garnets of this type, which resemble a sagenitic rutile network. These inclusions have a regular mutual

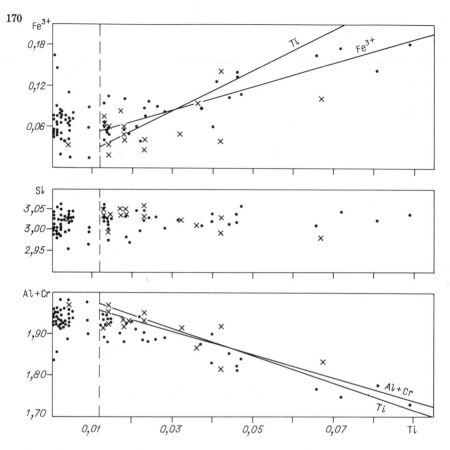

Fig. 24. Correlation between amounts of Ti, R^{3+}, and Si in magnesian chromium-bearing garnets (points) and Mg-Fe garnets (crosses) from kimberlite pipes.

orientation (Bobrievich et al., 1959) with the angle between the various directions close to 60°, and a more precise diagnosis has been difficult owing to the very small dimentions of the inclusions. A later study of these inclusions has shown that the maximum number of directions of orientation reached in individual samples is six, although in many cases, a significantly smaller number has been recorded visually. These data, and also the constancy of the angle between the directions, lying in the single plane (about 60°), suggest that the acicular inclusions have been oriented along the edges of the octahedron parallel to the faces of the rhombdodecahedron, and are most probably, exsolution textures.

Attempts to determine the composition of these inclusions with the aid of the microprobe, have established a ratio of Ca and Ti in an acicular inclusion measuring less than five microns, which is analogous to that in sphene. These data, in spite of the exceptionally small size of the inclusions, are quite accurate, since the intensity of the Ca and Ti -radiations has been identified in this case, and exceeds the 'background' itensity of the enclosing garnet by several times.

In another case, larger elongated inclusions have been examined, characterized by an analogous orientation in six directions. These are inclusions of ilmenite.

Thus, investigations of the composition and nature of the orientation of inclusions convincingly indicate the possibility of their formation as a result of dissociation of titanium-rich phases during lowering of temperature.

The Isomorphous Content of Sodium in Garnets

Although sodium is a widely distributed element, involved in eclogitic rocks, which are known both in the kimberlite pipes and in the metamorphic complexes, it has been concentrated only in the clinopyroxenes in these rocks, and the amount of Na_2O reaches 8-9% in them.

In the garnets, sodium cannot be concentrated in significant amounts, although an insignificant amount of it is evidently possible as a result of the probable presence of titanium or at particularly high pressures. A prerequisite for tackling the problem of investigating disseminated amounts of sodium in garnets has been the identification of a certain amount of Na_2O (up to 0.2%) in garnets enclosed in diamonds from Yakutia and the Urals.

The amount of Na_2O has been determined in 124 garnets with the aid of the microprobe, such garnets belonging to the eclogite and peridotite parageneses, under conditions ensuring a sensitivity level of 0.01% Na_2O (see Chapter I) (Sobolev and Lavrent'ev, 1971). In spite of the fact that the total amount of Na_2O in these garnets is low, the range of variation is quite wide (0.01-0.22%), which enables us to discern certain patterns.

We have separately considered the distribution features of sodium in the pyrope-almandine garnets with variable iron index and amount of calcium from eclogites and diamonds and in the chromium-bearing pyropes with low iron index and a normally low content of calcium. The garnets of the first type are associated with clinopyroxenes with an increased content of Na_2O (not less than 3% and up to 7.5%), and those of the second type are mainly associated with chrome-diopsides with an amount of $Na_2O = 1 - 2$% (with the exception of special parageneses with sodium- and chromium-rich pyroxenes, which may be termed chromium-rich eclogitic or peridotitic association). Although the garnets of the first recognized type are characterized by wide fluctuations in the amount of sodium, its regular pattern of distribution may be noted with individual, clearly distinguishable groups of rocks (Fig. 25). This applies primarily to the diamond-bearing eclogites, which are a unique group of rocks, characterized by the typomorphism of the pyroxene (Sobolev, 1968b). We have succeeded in examining garnets from a representative group of these xenoliths, comprising in all 16 samples (Table 39), of which 14 have been taken from the 'Mir' pipe.

As seen from the diagram (see Fig. 25), the garnets of these eclogites form a separate group with maximum amount of sodium, which along with other compositional features, is also characteristic of the yellow-orange garnets, found in the form of inclusions and in intergrowths with diamonds (see Table 46).

A completely different trend in the distribution of sodium is typical of the garnets from eclogites of the various metamorphic complexes. Its amount in these garnets is similarly low both in

172

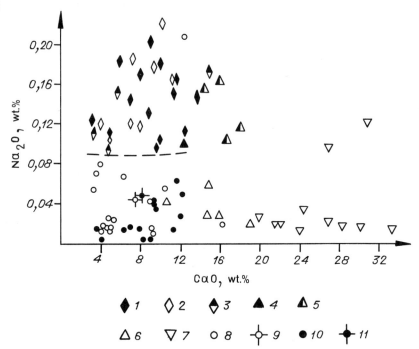

Fig. 25. Dependence of amounts of Na$_2$O (wt%) on those of CaO (wt%) in pyrope-almandine-grossular garnets from diamonds and eclogitic parageneses: 1) from diamond-bearing eclogites; 2) inclusions in diamonds; 3) intergrowths with diamonds; 4) from a diamond-bearing kyanite eclogite; 5) from kyanite eclogites of the 'Roberts Victor' pipe; 6) from kyanite eclogites of the 'Zagadochnaya' pipe; 7) from grospydites of the same pipe; 8) from xenoliths of eclogites; 9) from an eclogite xenolith with graphite; 10) from eclogites of metamorphic complexes; 11) from a quartz-garnet-jadeite rock, South Urals.

the eclogites of the gneissic complexes and in those associated with the glaucophane schists (0.01-0.06%). In more than half of the garnets examined from these rocks, the amount of Na$_2$O seemed to be almost at the limit of sensitivity of determination (0.01-0.02%).

It should be noted that isolated rare attempts to determine the amount of Na$_2$O in garnets of such type by chemical analytical methods (Bryhni et al., 1969) have led to definitely over-estimated results, owing to mechanical addition of omphacite in the analyzed material as a result of micro-inclusions in the garnet.

The recorded marked differences in the amount of sodium in garnets from eclogites of various types have been illustrated in Figure 26. Here, there is a clear recognition of histograms, showing the lowest amounts of sodium in the garnets from eclogites of the metamorphic complexes, and increased amounts, in those from diamond-bearing eclogites and inclusions in diamonds.

A varied picture of sodium distributed is typical of garnets from the eclogite and grospydite parageneses of xenoliths from the kimber-

TABLE 39. Amount of Na$_2$O and Some Features of the Composition of Pyrope-Almandine-Grossular Garnets from Diamonds and Various Eclogites

Sample No.	Samples	Na$_2$O, wt%	CaO, wt%	FeO/FeO++MgO (f), %	Na/Na+Ca (garnet), %	Na/Na+Ca (pyroxene), %	K_D^{Na}
1	MB-1	0,097	9,6	50	1,8	53	0,016
2	111060	0,101	12,3	35	1,7	50	0,017
3	M-49	0,105	9,8	49	1,9	54	0,017
4	84661	0,111	4,9	47	4,0	36	0,074
5	M-59	0,113	12,4	48	1,6	56	0,013
6	M-45	0,125	3,1	33	6,7	40	0,110
7	M-821	0,132	13,1	45	1,8	—	—
8	M-53	0,132	8,8	62	2,6	56	0,021
9	M-52	0,145	7,0	52	3,6	44	0,048
10	M-54	0,148	13,7	46	1,9	50	0,019
11	M-51	0,152	11,3	44	2,4	53	0,022
12	M-33	0,166	11,6	35	2,5	44	0,033
13	M-46	0,170	8,0	49	3,7	43	0,051
14	A-45	0,182	10,0	58	3,2	40	0,050
15	M-180	0,183	5,9	57	5,4	47	0,064
16	M-50	0,205	9,0	57	4,0	53	0,037
17	B-6	0,094	4,8	53	3,4	—	—
18	1^d	0,105	4,9	39	3,8	—	—
19	ME-131	0,111	3,3	57	5,7	—	—
20	B-19	0,118	7,9	55	2,6	—	—
21	8851	0,121	4,0	50	5,3	—	—
22	138	0,121	7,0	41	3,1	—	—
23	B-18	0,153	5,7	50	4,7	—	—
24	880	0,166	11,2	42	2,6	30	0,062
25	B-22	0,173	14,9	32	2,1	—	—
26	1011	0,178	9,3	55	3,3	—	—
27	850	0,186	7,2	42	4,5	—	—
28	15384	0,222	10,1	56	3,8	—	—
29	RV-1	0,103	16,6	49	1,1	—	—
30	110752	0,116	18,1	42	1,0	50	0,010
31	87375	0,155	14,4	54	1,4	—	—
32	RV-15	0,163	16,0	51	1,8	55	0,015
33	Z-49	0,020	18,5	42	0,2	42	0,003
34	Z-15	0,029	14,7	41	0,4	38	0,007
35	Z-52	0,029	15,9	32	0,3	35	0,006
36	Z-60	0,042	10,6	52	0,7	57	0,005
37	Z-24	0,059	14,9	55	0,7	48	0,008
38	Z-8	0,014	24,0	42	0,2	40	—
39	Z-128	0,017	33,3	54	0,2	—	—
40	Z-28	0,019	28,3	39	0,2	52	—
41	Z-33	0,019	30,1	38	0,2	53	—

№ п/п	Samples	Na$_2$O	CaO.	FeO/FeO+MgO (f), %	Na/Na+Ca (гранат), %	Na/Na+Ca (пироксен), %	K_D^{Na}
42	Z-43	0,021	22,0	38	0,2	34	—
43	Z-37	0,021	21,7	41	0,2	40	—
44	Z-34	0,024	26,9	41	0,2	46	—
45	Z-3	0,028	19,9	39	0,3	36	0,005
46	Z-6	0,035	24,4	54	0,3	49	0,003
47	Z-38	0,098	27,0	55	0.75	57	0,005
48	Z-1	0,123	31,0	62	0,7	65	0,004
49	O-160$_2$	0,010	9,2	20	—	26	—
50	M-114	0,013	4,0	16	—	—	—
51	O-309	0,014	4,9	20	—	23	—
52	O-270	0,015	5,0	29	—	27	—
53	O-379	0,015	4,9	32	—	14	—
54	O-166	0,016	9,1	36	—	28	—
55	O-145	0,020	4,2	22	0,8	23	0,027
56	O-160$_1$	0,020	16,2	24	0,2	25	0,006
57	NZ-2	0,024	5,3	30	—	18	—
58	OH-1	0,026	4,8	28	—	—	—
59	Z-18	0,043	9,0	42	0,9	44	0,012
60	M-100	0,044	7,5	23	1,0	29	0,024
61	M-60	0,055	3,3	28	3,2	32	0,070
62	U-1	0,056	10,5	44	0,9	50	0,009
63	Z-59	0,068	6,3	34	2,0	46	0,024
64	A-1	0,071	3,6	17	3,5	23	0,121
65	U-1/6	0,080	4,0	22	3,5	20	0,145
66	M-30	0,209	12,4	44	3,0	47	0,035
67	8312	<0,010	4,1	31	—	13	—
68	NR-3	<0,010	8,3	58	—	—	—
69	NR-4	<0,010	—	—	—	—	—
70	PVB-718	<0,010	8,9	53	—	—	—
71	344	0,015	—	—	—	—	—
72	85/2	0,015	11,2	89	—	49	—
73	MN-1	0,015	6,4	36	—	36	—
74	10333	0,016	3,5	40	—	—	—
75	303	0,017	12,0	—	—	62	—
76	K-44	0,028	12,0	75	0,4	37	0,007
77	SW-1	0,036	9,5	50	0,7	43	0,009
78	41	0,040	9,4	80	0,8	50	0,008
79	534	0,043	9,3	80	0,8	46	0,009
80	516-2	0,049	8,2	87	1,1	98	0,0002
81	302/143	0,050	12,2	85	0,7	48	0,008
82	170	0,063	11,6	85	1,0	50	0,010

Notes. Rock types for garnets investigated: 1-16) diamond-bearing eclogites: 1, 3, 5-16) diamond-bearing eclogites from 'Mir' pipe; 2) diamond-bearing kyanite eclogite from 'Roberts Victor' pipe (South Africa); 4) diamond-bearing eclogite from 'Newlands' pipe (South Africa); 17-28) inclusions in diamonds and garnet-diamond intergrowths: 17, 19, 23, 25) yellow-orange garnets from intergrowths with diamonds from 'Mir' pipe; 18, 21, 23, 24, 27, 28) yellow-orange garnets, enclosed in diamonds (Urals); 20, 26) yellow-orange garnets, enclosed in diamonds (Mir' pipe); 29-32) kyanite eclogites from 'Roberts Victor' pipe; 33-37) kyanite eclogites from 'Zagadochnayá' pipe (Yakutia); 38-48) grospydites from 'Zagadochnaya' pipe; 49-66) xenoliths of eclogites and websterites: 60, 61) websterites from 'Mir' pipe; 51-53, 55) websterites from 'Obnazhënnaya' pipe (Yakutia); 49) garnet-pyroxene zone from xenolith of complex composition from 'Obnazhënnaya' pipe; 56) garnet-pyroxene-corundum zone from same xenolith; 54) eclogite from 'Obnazhënnaya' pipe; 57) eclogite from Kakanui Breccia (New Zealand); 58) hypersthene eclogite from Oahu (Hawaiian Islands); 59, 63) eclogites from 'Zagadochnaya' pipe; 64, 66) eclogites from 'Mir' pipe; 62, 65) eclogites from 'Udachnaya' pipe (Yakutia); 60) graphite-bearing eclogite from 'Mir' pipe; 67-82) eclogites from metamorphic complexes; 67) bronzite pyropite, Sittampundi Complex, Madras (India); 68) Nordfjord, Norway; 69) kyanite eclogite, Nordfjord; 70) Swiss Alps; 71, 72, 75, 78) South Urals; 73) kyanite eclogite, Münchberg Massif (Western Germany); 74) Scotland; 76) Kazakhstan; 77) kyanite eclogite, Gorduntal, Switzerland; 79) Aktyuz, Tyan'-Shan'; 81, 82) California; 80) almandine-jadeite-quartz rock (South Urals).

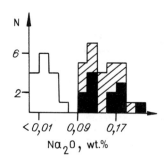

Fig. 26. Histogram of amounts of Na_2O (wt%) in garnets from diamonds (black), diamond-bearing eclogites (cross-hatched), and eclogites of the metamorphic complexes (white).

lite pipes (see Fig. 25), in which amounts have been defined that
vary over a wide range, typical of eclogites of different origin
and inclusions in diamonds.

A significantly lower average amount of sodium is characteristic
of the chrome-bearing pyropes of the peridotite associations and
inclusions in diamonds (Fig. 27). With the exception of Sample B-1,
which contains 0.13% Na_2O, all the garnets examined (36 samples)
contain from 0.01 up to 0.08% Na_2O. It is interesting to note
that the chrome-rich and calcium-poor garnets from the Yakutian
diamonds, for which we have assumed a paragenesis with olivine and
without clinopyroxene (Sovolev, Lavrent'ev et al., 1969), are
distinguished by the almost complete absence of sodium, like the
garnets of the serpentized diamond-bearing olivinites (V. S. Sobolev,
Nai et al., 1969).

The recognized differences in the amount of sodium in the garnets
from eclogites of various types cannot be associated with compositional features of the rocks, that is, with the amount of sodium in
them. The natural concentrator of sodium in eclogites is clinopyroxene, in which the amount of Na_2O varies from 3 often up to 7.5%.
Clinopyroxenes of such composition (omphacites) are typical both of
eclogites of the metamorphic complexes (Dobretsov et al., 1971) and
of diamond-bearing eclogites (Sobolev, 1968b), as well as the kyanite
eclogites and grospydites (Sobolev et al., 1968).

The data presented suggest that there is a possible relationship
between the entry of sodium into the garnets and pressure, defined
by the association with the diamond. Interest centers not on pressures of 12-15 kb (the lower limit of formation of eclogites), but
on significantly higher pressures, close to those necessary for the
formation of the diamond. Such a proposition also agrees closely
with the features of possible isomorphous entry of sodium into the
garnets.

Sodium may only enter the garnet structure in eightfold coordination, replacing divalent cations. In this respect, by analogy with
the pyroxenes of diopside-jadeite composition, the predominant
substitution, $Na \rightleftarrows Ca$, may be assumed. The fundamental possibility
of such substitution in the garnet structure has been demonstrated
experimentally in the case of a germanate, having the composition
(Ca, Na_2) $Ti_2Ge_3O_{12}$ (Geller, 1967). In this example, Na and Ca have
combined with Ti to form the hypothetical end members $CaTi_2Ge_3O_{12}$
and $Na_2Ti_2Ge_3O_{12}$.

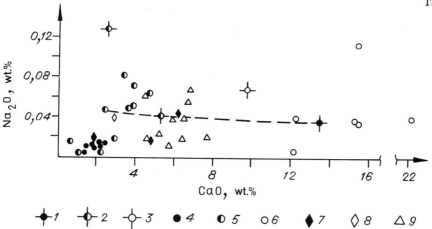

Fig. 27. Relationship between amount of Na_2O (wt%) and amount of CaO (wt%) in pyropes: 1) from diamond (Sample 57/9) in association with diopside and olivine; 2) from a garnet-diopside-diamond inter-growth (Sample B-1); 3) from a garnet-diopside-chromite inter-growth from kimberlite; 4) from diamonds; 5) from garnet-diamond intergrowths; 6) from kimberlites; 7) from diamond-bearing serpentinites; 8) from a dunite; 9) from peridotite xenoliths.

In the pyrope-almandine garnets, in which titanium, as a rule, is in small amount, the substitution may evidently be achieved mainly according to the pattern $NaSi \leftrightarrows CaAl$ for which conversion of part of Si^{IV} into Si^{VI} is necessary. In this respect, we have also not excluded the partial effect of titanium, but it may not be definite, since the garnets that contain a variable amount of TiO_2 (from 0.3 to 1.00%), are characterized by very similar figures for the amount of Na_2O, and in some cases the Na content is higher than Ti (see analysis 138, Table 28). Such conversion, according to Ringwood (1967), is possible only at a significantly increased pressure. In Ringwood's experiments (1967), up to pressures of 90 kb, 40% of pyrope and 60% of enstatite (stoichiometric ratio) crystallized from the original composition of 90% $MgSiO_3$.10% Al_2O_3. However, at increased pressures in the 90-110 kb range, a marked step-like increase in the amount of garnet from 40 to 80% occurred and then slowly to 95%, as the pressure was increased from 110 to 150 kb. These changes have been identified by a phase X-ray analysis and by increase in the a_0 value of the garnet. It is practically impossible by such a method, to identify the very initial phase of conversion of Si^{IV} into Si^{VI}. This can only be done with the aid of the X-ray microprobe.

For natural basalt compositions, analogous to those of the eclo-gites, the value of the pressure, necessary for complete garnetization of the rock, depends closely on the amount of Na_2O in the original composition (Ringwood, 1967), which is associated, as correctly assumed by Ringwood, with the stability of jadeite up to the highest pressures, because the greater the amount of sodium in the rock, the greater the amount of jadeite present in the experimental products. It has been established that, in a charge containing 2.7% Na_2O, only 65% garnet is formed at a pressure of 127 kb, and 70% at a pressure of 170 kb (Ringwood, 1967), as compared with a charge, containing

1% Na_2O (95% garnet at a pressure of 110 kb). This led Ringwood to suggest the impossibility of entry of alkalies into the garnet structure. However, in the above-noted work (Geller, 1967), and also in experiments with silicate compositions (Kushiro et al., 1967), the fundamental possibility of entry of alkalies into the garnet structure has been demonstrated. In the last work, it has been shown that in the absence of another phase, containing up to 5% of alkalies, K_2O may enter the garnet structure at pressures up to 96 kb[+]. In regard to sodium, the main difficulty is associated with the constant presence of a jadeite-bearing pyroxene as a natural phase, which concentrates sodium. In a later work (Ringwood and Lovering, 1970) accepted the possible entry of small amounts of Na_2O into garnet at pressures of the order of 100 kb, although in combination with titanium, and not with silicon. For the germanates with garnet structure, Na_2O even enters at atmospheric pressure (Geller, 1967).

Thus, until recently, direct experimental data on the entry of sodium into pyrope-almandine garnets at high pressures, have essentially been lacking, since the products of the experiments have not been analyzed for sodium content.

After identifying a trace of sodium in natural garnets (Sobolev and Lavrent'ev, 1971), sodium-calcium garnets, containing no titanium or aluminum, corresponding to the formula $Na_2CaSi_5O_{12}$, have been synthesized at pressures of 180 kb (Ringwood and Major, 1971), that is, the assumed nature of the isomorphism in garnets at especially high pressures has been proved experimentally.

As follows from the above results of analyzing garnets for sodium content, the garnets, associated with diamonds, are distinguished by a significantly greater amount of Na_2O than those from the eclogites of the metamorphic complexes. If we begin with an association between the amount of Na_2O (0.1-0.2%) and silica, then in the limiting case, then only about 1% of the Si present changes its coordination, that is, in other words, the amount of SiO_2 in the garnet may be increased by approximately 0.2-0.4%. Such increase in the amount of silica cannot be checked even with the microprobe, because instrumental error in the determination of Si is up to 2% of its amount.

This leads us to conclude that the initial phases of transfer of Si^{IV} into Si^{VI} may only be assessed from the amount of Na_2O in the garnet, and more precisely, from the distribution features of sodium in the garnets and the associated clinopyroxenes.

As an additional check of the possible association between the amount of sodium in the garnet and in the pyroxene, we have calculated the Na/(Na + Ca) ratios in these pairs (see Table 39), and also the coefficients of distribution of these ratios (K_D^{Na}), with the aid of widely employed methods, first used in mineralogy by Kretz (1961). As Kretz and many others have shown, the value of the coefficient of distribution depends on many factors, portion of which include the compositional features, and the conditions of formation. However, dependence on pressure in this particular case is evidently one of the principal factors.

[+]Later, these results were acknowledged to be in error (Erlank and Kushiro, 1970), and repeated determinations of the amount of K_2O in garnet demonstrated its almost complete absence, but a substantial addition of K_2O is present in the undiagnosed fine-grained phase.

The value of the coefficient of distribution (K_D^{Na}) has been calculated according to the formula (Kretz, 1961):

$$K_D^{Na} = \frac{x_1(1 - x_2)}{(1 - x_1)x_2}$$

where x_1 is the Na/(Na + Ca) ratio in the garnet, and x_2 is the same ratio in the clinopyroxene.

We have calculated the values of (K_D^{Na}) for garnet-pyroxene pairs with an amount of Na_2O in garnet not less than 0.03% in all for 44 pairs. With a lesser amount, which lies almost at the limit of sensitivity of the method, the error in the result increases sharply.

The largest values of (K_D^{Na}) are typical of garnet-pyroxene pairs from diamond-bearing eclogites. Sixteen pairs gave values of (K_D^{Na}) varying from 0.013 to 0.110 (average 0.039 $\pm$ 0.027). It is important to note that the garnet-omphacite association from a single crystal of Uralian diamond (Sobolev et al., 1971) is also characterized by a large value for (K_D^{Na}) = 0.06 (see Table 39), which is close to the average value for the pairs from the diamond-bearing eclogites. On the basis of this example, and of the established stable increased amount of sodium in the yellow and yellow-orange garnets, associated with the Uralian and Yakutian diamonds in the form of inclusions and in intergrowths (see Table 39), support is given to the view (Sobolev and Kuznetsova, 1966; Sobolev et al., 1971) that all garnets of this type in diamonds belong to the eclogite paragenesis. It is also evident that all the known analyzed garnets from African diamonds, similar in composition to those in the inclusions that we have examined, that is, characterized by a variable iron index and calcium content, must also be associated to the pattern discovered. This applies to Samples D 15, GL 16, and Sample 20f (Meyer and Boyd, 1972); in Sample D 15 and GL 16, 0.09 and 0.19% Na_2O have respectively been determined, and in the chrome-pyrope (Sample GL 25a), 0.02% Na_2O (Meyer and Boyd, 1972), these results being completely analogous to our data.

In the kimberlite pipes of South Africa, a considerable quantity of eclogites with diamonds has been found (Dawson, 1968; Gurney et al., 1969). We have examined material from two African samples (see Table 39). In order to obtain more complete and representative data, it is desirable to determine the amount of Na_2O in garnets of all known samples of diamond-bearing eclogites.

The significant variations in the values of (K_D^{Na}) for the pairs from diamond-bearing eclogites may indicate the presence of factors, which complicate the distribution of Na and Ca. Such factors may be: the influence of the garnet composition, and in the main, evidently, the variable values of pressure and certain fluctuations in the temperature of formation of the various diamond-bearing eclogites.

Of the 17 garnet samples from eclogites of metamorphic complexes of various regions of the globe investigated, only seven, that is, less than half, are characterized by some increase in the amount of Na_2O (0.03-0.06%) (see Table 39). Allowing for the compositions of the associated clinopyroxenes, the values of (K_D^{Na}), with fluctuations from 0.007-0.009, have been calculated for these pairs. It should be stressed that these values are maximal for the eclogites of the metamorphic complexes, since in the pairs, whose garnets contain about 0.01% Na_2O, the values will be practically an order lower. As an example of an exceptionally low value for (K_D^{Na}) with 0.05% Na_2O

present in garnet, we may cite the garnet-jadeite paragenesis from
a quartz-almandine-jadeite rock from the South Urals, associated
with eclogites of the glaucophaneschist complex. From the data in
Table 39, a (K_D^{Na}) value of 0.0002 has been calculated.
Of the 18 samples of garnets from xenoliths of eclogite composition
from kimberlite pipes, 10 are characterized by an amount of Na_2O
ranging from 0.01 to 0.03%. In our selection, we have given the
results of determining the amount of sodium in the garnets from the
'eclogitic' ejectamenta in breccias from Oahu (Hawaiian Islands) and
Kakanui (New Zealand) (see Table 39), in which low amounts of Na_2O
have also been recorded (0.02-0.03%).
When considering the features of the sodium content in the garnets
from xenoliths of eclogite composition (see Table 39), it is interesting to note that increased amounts of sodium are only
characteristic of individual xenoliths from these same kimberlite
pipes, where occurrences of diamond-bearing eclogites are known. This
applies to certain samples from the 'Mir' and 'Roberts Victor' pipes.
On the basis of sodium content, only isolated garnet compositions
from the xenoliths lie on the diagram (Fig. 26) in the compositional
field of the diamonds and diamond-bearing eclogites (or fall quite
closely). We must specially consider here the amount and distribution
of sodium in the garnet and clinopyroxene from the eclogite sample
in the 'Mir' pipe (see Table 39, Sample M-100). Judging by the
association, this rock was probably formed outside the field of
diamond stability. The (K_D^{Na}) value for the garnet-pyroxene pair
of this sample is below the average value for the diamond-bearing
eclogite type.
The differences revealed in the amounts of sodium in the garnets
from the kyanite eclogites fo the 'Roberts Victor' and 'Zagadochnaya'
pipes are extremely significant. In the former, two samples have been
found, which contain diamonds (Dawson, 1968; Switzer and Melson, 1969).
A small quantity of the garnet from one of the samples (111060)
was presented to us by courtesy of Dr. Switzer (U.S. National Museum).
The average amount of Na_2O in the garnets of the kyanite-bearing
xenoliths from the 'Roberts Victor' pipe is significantly greater
than in those of analogous composition from the 'Zagadochnaya' pipe.
The garnets from xenoliths of grospydites from this same pipe (11
samples) (see Table 39) are characterized by the same low average
amount of Na_2O. However, in two samples of grossular, an increased
amount of Na_2O (0.10 and 0.12%) has been found. When comparing
the data, an empirical relationship between the increased amount of
Na_2O in the pyrope-almandine-grossular garnets and their increased
iron index has been revealed (Fig. 28), and this applies not only to
the garnets similar in composition to grossular, but also to Sample
Z-60 from a kyanite eclogite. For comparison, we have given the values
of the Na/(Na + Ca) ratios for the garnets from eclogites in the
'Roberts Victor' pipe, which are significantly greater.
Thus, in addition to the established fact that the kyanite
eclogites of South Africa contain diamonds, the amount of sodium
in the garnets also indicates a higher pressure for the formation
of these rocks, as compared with the analogous parageneses of the
'Zagadochnaya' pipe. In this respect, the average value of (K_D^{Na}) of
0.010 for three pairs from the 'Roberts Victor' pipe exceeds by
several times the value for such pairs in the 'Zagadochnaya' pipe.
It has been shown earlier on (Sobolev et al., 1968) that the
amount of Na_2O in the pyroxenes from kyanite eclogites and associated

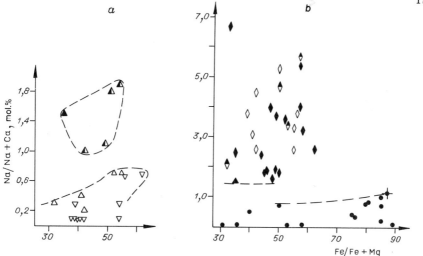

Fig. 28. Relationship between values of Na/(Na + Ca) ratio and iron index in garnets from kyanite eclogites and grospydites (a) and in garnets from eclogites and yellow-orange garnets from diamonds (b) (for symbols, see Fig. 25).

grospydites depends on the amount of calcium in the garnets with identical iron index in the latter, and the identical conditions of formation ($\underline{P}$ and $\underline{T}$) for the xenoliths of the 'Zagadochnaya' pipe have been emphasized. As exemplified by samples from the 'Roberts Victor' pipe, amongst which there are diamond-bearing xenoliths, there is support for the hypothesis (Sobolev, 1970) of still greater pressure during their formation, based on the increased amount of Na_2O in the pyroxene with garnets having a constant calcium content and iron index.

The picture of the distribution of sodium in garnets and pyroxenes is substantially less clear for the parageneses with chromium-bearing pyrope. Decreased amounts of Na_2O in the garnets are mainly associated with the low content of sodium in the pyroxenes (for many peridotite xenoliths); in the association with diamond, both minimal and maximal (up to 0.13%) concentrations of Na_2O have been identified.

In garnets associated with diamond, the almost complete absence of sodium or its very small amount, may be explained only by the features of the chemistry of the environment, that is, by the low content of sodium and the corresponding features of the paragenesis. In these cases, the predominant role is evidently played by the garnet-olivine association, also established from the compositional features of the garnet (Sobolev, Lavrent'ev et al., 1969). However, the garnets with low calcium content may be associated not only with olivine, but also may be stable in a paragenesis with a diopside-jadeite pyroxene (Sobolev, Zyuzin and Kuznetsova, 1966; Sobolev et al., 1968; Banno, 1967). Once again, such cases may be clearly distinguished by the amount of sodium in the garnets. An excellent illustration is the garnet-clinopyroxene-diamond association (Sample B-1), representing an intergrowth of these minerals (Sobolev et al., 1971a). The garnet of this paragenesis, being indistinguishable

180

TABLE 40. Amount of Na$_2$O and Some Features of the Chrome-Pyropes from Diamonds, Peridotite Xenoliths, and Kimberlite Concentrate

No.	Samples	Na$_2$O, wt.%	CaO, wt.%	Cr$_2$O$_3$, wt.%	Fe/Fe+Mg (f), %	Na/Na+Ca (garnet) %	no.	Samples	Na$_2$O wt%	CaO wt%	Cr$_2$O$_3$, Wt%	Fe/Fe+Mg (f), %	Na/Na+Ca (garnet), %
83	A-90a	0,017	4,8	8,7	18	—	104	ME-155	0,081	3,4	1,0	16	4,1
84	A-2a	0,020	1,9	8,2	14	—	105	B-1	0,128	2,5	3,1	14	8,6
85	A-4a	0,044	6,2	14,1	18	1,2	106	O-800	0,010				—
86	B-12	<0,010	2,2	10,8	12	—	107	Ud-3	0,012	5,7	5,1	18	—
87	B-3	<0,010	1,1	7,3	12	—	108	O-802	0,019	4,7	4,0	18	—
88	AO-434	<0,010	1,3	6,1	13	—	109	Ud-91	0,019	6,4	9,1	17	—
89	AO-449	<0,010	1,9	7,6	13	—	110	Ud-92	0,020	7,7	11,0	18	0,5
90	AO-458	<0,010	2,2	15,9	13	—	111	Ud-96	0,023	5,2	5,2	17	0,8
91	AO-432	<0,010	1,5	7,7	15	—	112	Ud-5	0,037	5,9	7,2	17	1,1
92	AO-429	0,011	1,8	10,0	15	—	113	Uv-126	0,039	6,5	7,4	17	1,1
93	931	0,012	2,2	15,6	13	—	114	Uv-196	0,040	2,9	12,7	14	2,5
94	AO-90	0,014	2,4	8,3	16	—	115	Uv-161	0,056	6,7	9,9	19	1,5
95	B-2	0,015	0,6	6,2	11	—	116	Ud-93	0,061	4,5	5,0	16	2,4
96	B-14	0,018	2,9	9,6	14	—	117	Ud-20	0,067	6,8	8,4	17	1,8
97	57/9	0,035	13,5	8,0	15	0,5	118	U-15	0,010	12,1	10,7	25	—
98	59/12	0,041	5,3	3,7	21	1,4	119	Z-53	0,036	15,5	5,2	30	0,4
99	B-4	0,47	2,4	6,8	17	3,4	120	Z-51	0,037	15,4	6,7	34	0,4
100	B-10	0,049	3,6	1,2	16	2,4	121	Z-47	0,039	12,3	6,0	25	0,6
101	B-11	0,050	3,7	2,6	12	2,4	122	Z-52	0,040	22,2	7,4	45	0,3
102	ME-137	0,064	4,7	5,5	18	2,3	123	Mz-2	0,069	9,8	9,6	25	1,3
103	B-20	0,071	3,9	3,3	16	3,2	124	Ms-59	0,113	15,4	10,0	29	1,3

Notes. Rock types for garnets examined (Sample numbers continue from Table 39): 83-85) diamond-bearing pyrope serpentinites, 'Aikhal' pipe, Yakutia; 86-105) chrome-pyropes, enclosed in diamonds and pyrope-almandine intergrowths; 88-92 and 94) inclusions in diamonds from Yakutia; 93) inclusions in diamonds (Urals); 97) pyrope, associated with olivine and chrome-diopside in the same diamond crystal (Yakutia); 86, 87, 95, 96, 99, 100-104) garnet-diamond intergrowths (Yakutia); 98, 105) garnet-clinopyroxene-diamond intergrowths (Yakutia); 106-117) peridotite xenoliths in Yakutian kimberlites: 106, 108) 'Obnazhënnaya' pipe; 107, 109-117) 'Udachnaya' pipe; 118-124) garnets from heavy fraction of kimberlite concentrate (Yakutia): 118) green garnet ('Udachnaya' pipe); 119, 120, 122) green garnets, associated with chrome-kyanite ('Zagadochnaya' pipe); 121) green garnet ('Zagadochnaya' pipe); 123) garnet-pyroxene-chromite intergrowth ('Mir' pipe); 124) green garnet ('Mir' pipe). For Samples 57/9, 59/12, B-1, Ud-5, and MZ-2, $K_D^{Na} = 0.041, 0.163, 0.087, 0.074$, and 0.044 respectively, and the Na/(Na + Ca) ratio in the pyroxenes is respectively 11, 8, 52, 13, and 23%.

from those of the pyrope olivinites and calcium-poor pyropes, enclosed in diamonds, on the basis of calcium content, is distinguished by an order in the content of sodium. The value of (K_D^{Na}) for this pair is of the same order as that for the pair from the diamond-bearing eclogites (0.08).

As an example of a significantly lower amount of sodium in garnet and pyroxene, we may cite Sample 57/9, which comprises the garnet-olivine-clinopyroxene-diamond association (Sobolev et al., 1970), with all the indicated minerals being identified as inclusions in the diamond. The amounts of sodium in the garnet and pyroxene here have been taken as the average of two determinations on each mineral, since both individual inclusions and also intergrowths of garnet and pyroxene were present in the diamond. The value of (K_D^{Na}), calculated from the data in Table 40, is 0.037, that is, it lies within the limits identified for the pairs from the diamond-bearing eclogites. A higher value (0.16) has been obtained for a garnet-pyroxene-diamond intergrowth (Sample 59/12, Table 47), which is characterized by a very low content of Na$_2$O in the pyroxene (0.87) and consequently, in the garnet. There is no doubt that the error in calculating (K_D^{Na}), as noted above, is much increased as the amount of sodium in garnet and pyroxene decreases.

TABLE 41. Average Amount of Na$_2$O (wt %) in Pyrope-Almandine and Magnesian Garnets from Diamonds, Peridotites, and Eclogites of Various Types

Rock types	N	$\bar{x}$	s	Limits of variation
Diamond-bearing eclogites	16	0,142	0,030	0,097—0,205
Yellow-orange garnets from Yakutian diamonds	12	0,146	0,037	0,094—0,220
Diamond-bearing eclogites and yellow-orange garnets from Yakutian diamonds . . .	28	0,143	0,036	0,094—0,220
Kyanite eclogites from 'Roberts Victor' pipe	5	0,128	0,025	0,103—0,163
Kyanite eclogites from 'Zagadochnaya' pipe	5	0,036	0,013	0,020—0,059
Grospydites from 'Zagadochnaya' pipe . .	11	0,038	0,035	0,014—0,123
Eclogites and garnet pyroxenites from Yakutian kimberlites	18	0,044	0,044	0,010—0,209
Eclogites of metamorphic complexes	16	0,027	0,017	0,010—0,063
Diamond-bearing serpentinites from 'Aikhal' pipe	3	0,027	0,012	0,017—0,044
Chrome-pyropes, enclosed in diamonds from Yakutia and the Urals	12	0,015	0,011	0,010—0,047
Pyrope-diamond intergrowths (Yakutia) . . .	5	0,063	0,012	0,049—0,081
Pyrope-clinopyroxene-diamond intergrowths (Yakutia)	2	0,085	—	0,041—0,128
Pyrope-olivine — chrome-diopside association, enclosed in diamond (Sample 57/9, Yakutia)	1	0,035	—	—
Pyrope peridotites from Yakutian kimberlites . .	12	0,034	0,018	0,010—0,067
Green garnets, rich in Ca and Cr, from Yakutian kimberlites	7	0,049	0,030	0,010—0,113

Notes. N = number of samples; $\bar{x}$ is the arithmetic average; and s is the standard deviation

Significant variations in the amount of sodium in the garnets (from 0.01 to 0.07% Na$_2$O) are typical of the peridotite xenoliths, which contain subcalcic clinopyroxene, from the 'Udachnaya' pipe, and which represent the highest-temperature (and deepest) of the known peridotite parageneses, not containing diamond (Sobolev, 1969).

On the basis of the results of determining Na$_2$O in garnets of different parageneses, reflected on the diagram (Fig. 27) and in Table 41, an approximate boundary has been plotted for the amount of Na$_2$O in garnets in paragenesis with olivine (below the boundary) and clinopyroxene (above the boundary) for the stability field of diamond. It is possible to assume a paragenesis with clinopyroxene for certain garnets, associated with diamond in intergrowths (Samples ME-137 and B-20). It is important, in our view, to emphasize the conclusions, based on amount of sodium in garnets, about the predominant distribution of the garnet-olivine association in the stability field of the diamond. Determination of the sodium content gives additional substantial information about the conditions of formation of the intergrowths of garnet-clinopyroxene composition (often with chromite), found in kimberlite concentrate.

Similar information may also be obtained on the conditions of crystallization of chromium- and calcium-rich garnets of pyrope-uvarovite composition (also containing the knorringite component), occasionally noted in the kimberlite concentrates. As investigations have shown (Sobolev, Lavrent'ev et al., 1969), there garnets belong partly to the paragenesis with clinopyroxene in the unusual chromium-bearing eclogites, and partly to the paragenesis with olivine.

Investigations of six samples of such garnets (see Table 40) have
shown that for only one sample, may we assume a specially deep-
seated character (MC-59), whereas the remainder are characterized
by low contents of Na_2O.

Thus, the investigations carried out have enabled us to discern
a series of patterns in the amount and distribution of Na_2O in the
garnets and clinopyroxenes.

In the garnets of the diamond-bearing eclogites and the yellow-
orange garnets from diamonds, the amount of Na_2O varies from 0.09 to
0.22%. On the basis of an analysis of the distribution of Na/(Na +
Ca) for inclusions of garnet and pyroxene from a single crystal of
Uralian diamond (Sample 880) and comparison with data on the diamond-
bearing eclogites, there is support for the hypothesis that all
the investigated yellow-orange garnets from diamonds belong to the
eclogite association. The features of the content of Na_2O in these
garnets indicate their possible association with the clinopyroxenes,
containing from 4 to 7% Na_2O. It should be emphasized that the
possible association with diopside (containing 1.4% Na_2O), identified
in the form of an inclusion in one of the African diamonds (Meyer
and Boyd, 1972) must evidently be of a minor nature, since the non-
chrome ferruginous garnet, associated with such a pyroxene, may
contain 0.03-0.05% Na_2O (depending on the amount of calcium in the
garnet).

The features of the distribution of sodium in the chrome-bearing
pyropes, which to belong to associations without clinopyroxene (garnet-
olivine), including both those with diamond, and with clinopyroxene
(peridotites and the peculiar chromium-bearing eclogites), enable
us to determine for the garnets enclosed in diamonds, the type of
association, starting from the amount of sodium in the garnet.
It has been emphasized that most of the garnets, associated with
diamond, belong to the garnet-olivine paragenesis, without clinopyroxene. In like manner, we may determine the nature of the conditions of formation of the rare green garnets from kimberlite
concentrates, rich in chromium and calcium, for which an association
has been demonstrated in a whole series of cases.

Because of the amount of Na_2O in the pyrope-almandine garnets
from the diamond-bearing eclogites and diamonds, and the values of
(K_D^{Na}) for the corresponding garnet-pyroxene pairs, are significantly
greater than in those from the eclogites of the metamorphic complexes,
we must check by experiment the relationship between the amount of
sodium in the garnets and the pressure. Such a check must be in
the form of a series of experiments with eclogite (basaltic) compositions with varying amounts of sodium at different pressures, mainly
in the range of 30-100 kb. The phases obtained (garnet and pyroxene)
must be analyzed with the aid of a microprobe for the amount of sodium
and calcium.

It is possible that such a method may more precisely define the
boundaries of the field of crystallization of the parageneses containing natural diamonds, and provide data on the distribution of sodium
and calcium in the garnets and pyroxenes dependent on pressure,
for that PT field in which phase transformations of the minerals
which play an important role in deep-seated mineral-formation, are
completely absent.

In spite of the exceptional variety of compositions of the
clinopyroxenes from deep-seated inclusions, it is possible to recognize among them in the overall complexity of composition, two types
of associations (ultramafic and basic (eclogitic)). The principal
distinguishing feature of the first type is the paragenesis with
olivine (and with enstatite), and also a trace of chromium and the
low iron index of the associated garnets. The last two features
are not clear, because there are pyroxenes with transitional compositions, characterized either by an increased amount of chromium (along
with a large amount of sodium) or by an increased amount of sodium
in association with magnesian garnets.

Clinopyroxenes of the Ultramafic Associations

The ultramafic associations are characterized by clinopyroxenes
with a variable amount of chromium and a comparatively low iron-index,
controlled by the paragenesis with magnesian garnets, olivine, and
orthopyroxene. An exceptionally interesting feature of the clinopyroxenes, associated with enstatite, is the presence of a solid
solution of orthopyroxene, defined by lowering of the $Ca/(Ca + Mg)$
ratio as the temperature is lowered. This feature of the peridotite
pyroxenes has been discussed in part in §2.

In the present section, in the histogram (Fig. 29), constructed
in the same way as that of Boyd (1970) and including our data (Sobolev,
1970), we have plotted the data on the $Ca/(Ca + Mg)$ ratio for all
the pyroxenes of the ultramafic associations. The compositions of more
than 100 clinopyroxenes from kimberlites have been represented, with
only 40% of the cases having a definitely known paragenesis with
enstatite (and pyrope). The remaining samples consist of chromium-
bearing magnesian pyroxenes from inclusions and intergrowths with
diamonds (often together with pyrope) and from garnet-pyroxene
intergrowths in kimberlite.

The nature of the distribution of the points on the histogram
emphasizes our earlier-stated proposition (Sobolev, 1969, 1970),
that the immiscibility gap (between $Ca/(Ca + Mg)$ = 44 and 36) in
the pyroxene series (which is characterized by the values of the
$Ca/(Ca + Mg)$ ratio from 50 down to 30), demonstrated by Boyd (1969,
1970) on the basis of a study of randomly selected numerous pyroxenes
from lherzolites and xenocrysts from kimberlite pipes, is only apparent
and depends on inadequate information alone.

The overall quantitative ratio of calcic and subcalcic pyroxenes
for the total selection of pyroxenes from kimberlites has not changed
substantially even with the availability of the new data. The calcic
pyroxenes, to which, according to Boyd (1970), pyroxenes with
$Ca/(Ca + Mg)$ = 50 - 44 may be assigned, markedly predominate over
the subcalcic types, to which we assign the pyroxenes with $Ca/(Ca + Mg) < 44$. Such a discontinuity, which applies mainly to diopside
xenocrysts from kimberlites, and to inclusions and intergrowths
with diamonds, may be explained in the following way: by the wide
distribution in kimberlites of comparatively low-temperature, that
is, less deep-seated pyroxenes, which is associated with the uneven
removal of material from the various levels of the mantle (Sobolev,
1971); by the displacement of the diopside-enstatite solvus curve
into the higher-temperature field during significant increase in

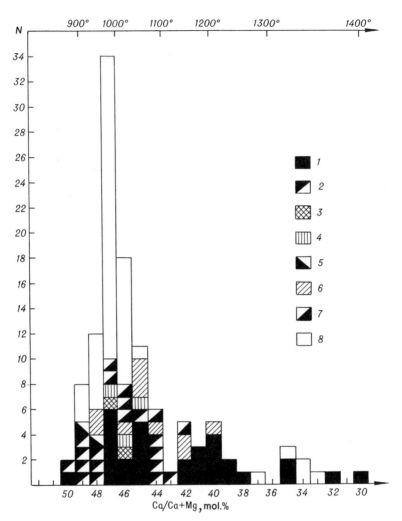

Fig. 29. Histogram of distribution of values of the Ca/(Ca + Mg) ratio in the clinopyroxenes of the ultramafic associations from kimberlites: 1) from pyrope peridotites; 2) pyrope websterites; 3) lherzolite paragenesis with diamond; 4) from diamonds; 5) from wehrlitic paragenesis with pyrope, enclosed in diamond; 6) from intergrowths with diamonds; 7) from garnet-pyroxene-chromite intergrowths from kimberlites; 8) xenocrysts from kimberlites; temperature values (°C) are indicated according to the points on the diopside (enstatite) solvus curve at P = 30 kb (Davis & Boyd, 1966).

pressure (associations with diamond); and by the presence of ultramafic parageneses, containing no enstatite, in the kimberlites.

The first opinion completely satisfactorily explains the dominant amount of peridotite xenoliths with calcic diopsides, selected at random, in the pipes, especially if we exclude from consideration the special type of high-temperature peridotites from the 'Udachnaya' pipe (see §2). However, the calcic nature of the diopsides, associated with enstatite in inclusions in diamond, may only be explained by the limited mutual solubility of the pyroxenes at especially high pressures (above 50 kb), which are still not available in experimental checks.

The reliable establishment of a new type of deep-seated garnet-bearing ultramafic paragenesis (wehrlitic) (Sobolev et al., 1970) indicates its definite role in the deep-seated rocks. In this paragenesis, the calcium index of the pyroxene does not depend on temperature. It is possible that such a paragenesis includes part of the inclusions and intergrowths with diamonds.

In spite of the fact that in a number of cases, we have to resort to different assumptions in order to interpret the temperatures of equilibrium of pyroxenes of the ultramafic parageneses (see §2), wide fluctuations in temperatures, determined with the aid of the two-pyroxene geothermometer, are an established fact. Such fluctuations (from 900 to 1400°C) are typical of pyrope peridotites from kimberlites, which contain the two-pyroxene association.

The clinopyroxenes of various ultramafic associations (including those transitional to the eclogites) are distinguished by a number of compositional features (Figs. 30, 31). Accumulated data at present enable to determine a significantly more detailed separation of the paragenetic types of these pyroxenes (Table 42) than has been done in other works (Dobretsov et al., 1971).

We have recognized new types of pyroxenes, associated with diamond (in inclusions and intergrowths). They probably include several associations, and a common feature of the pyroxenes is the increased amount of sodium and chromium (and wide fluctuations in their amounts), and the low iron-index.

Among the pyroxenes of the pyrope peridotites, including the websterites, four types have been recognized on the basis of interpretation of 55 analyses (see Table 42). Significant interest centers on the type of subcalcic pyroxenes of especially deep-seated pyrope peridotites from the 'Udachnaya' pipe (Type 2), established for the first time. Whereas earlier occurrences of subcalcic pyroxenes were extremely rare and comprised about 10% of the pyroxenes from kimberlites, selected at random (Boyd, 1969), a stable type of deep-seated peridotite paragenesis has been identified for the first time in the 'Udachnaya' pipe (see §2).

Type 3 comprises pyroxenes from lherzolites of various pipes in Yakutia and Africa, and is distinguished by the higher value of the Ca/(Ca + Mg) ratio. The calcium index of the pyroxenes from peridotites, known in of metamorphic terrane, is still higher (Type 4).

All three recognized types of pyroxene from pyrope lherzolites are characterized by a similar amount of jadeite, equal to 10-13%. The amount of Al + Cr + Ti is almost equal to that of sodium with a small excess, which indicates the insignificant role of Al^{IV} in their composition, even allowing for the role of Fe^{3+}. In previous considerations (Dobretsov et al., 1970, 1971) and in Types 3 and 4 recognized by us, the role of Al^{IV} in the pyroxenes of the pyrope peridotites, as a rule, has been underestimated in the average compositions, which has led to under-compensation of sodium and is evidently associated with the underestimate of silica in the chemical analyses. The most reliable data based on the Al^{IV}/Al^{VI} ratio may, in our view, be regarded as those for pyroxenes of Types 1 and 2 (see Table 2), analyses of which have been carried out with the aid

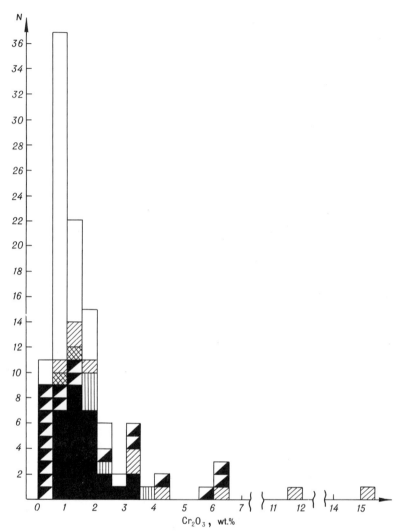

Fig. 30. Histogram of distribution of Cr_2O_3 in clinopyroxenes from ultramafic associations in kimberlites (for symbols, see Fig. 29).

of the microprobe. The overall low content of iron in these pyroxenes and comparison with chemical analyses, where iron has been separated, suggest that the role of Fe^{3+} is not significant.

The pyroxenes of the websterites have been separated from those of the lherzolites by the larger amount of jadeite (19%), and also by the increased amount of Al, indicating the more significant role of Al^{IV} in these pyroxenes, as compared with those of the lherzolites. We emphasize that the compositions of the associated garnets, in their calcium and iron indices, do not reveal significant differences from those of the lherzolites.

The general conclusion, which may be reached on the basis of compositional features of the pyroxenes of the ultramafic associations concerns the more substantial role of sodium in their composition,

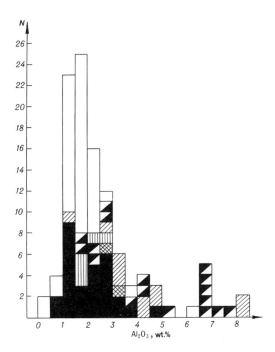

Fig. 31. Histogram of the distribution of Al_2O_3 in clinopyroxenes from ultramfaic associations in kimberlites (for symbols, see Fig. 29).

the average amount of which for 55 pyroxenes from pyrope peridotites (13% Na-component) is almost twice that established earlier on. The importance of sodium in the composition of the pyroxenes of the ultramafic associations may also be assessed by comparing the total amount of Al + Cr in the various pyroxenes, which also takes into account the compositional features of those numerous samples (mainly xenocrysts from kimberlite), for which there are partial analyses without sodium determinations (Fig. 32). As may be judged from the composition of the pyroxenes from the lherzolites, the total of R^{3+} is practically equivalent to the amount of sodium (Fig. 33). Another feature, which applies exclusively only to the pyroxenes of the lherzolites from the pipes, is the wide variation in the calcium index of the pyroxenes whereas that of the associated garnets remains constant.

Clinopyroxenes of the Eclogites and Grospydites

The principal features of the pyroxenes of the eclogites are the increased amount of sodium and the marked predominance of Al^{VI} over Al^{IV}. The latter feature is a clear distinguishing characteristic of the pyroxenes of typical eclogites (including those from kimberlites and from metamorphic complexes) from various eclogitic rocks (White, 1964).

The compositional features of the pyroxenes from xenoliths of eclogites and grospydites have been considered on the basis of data from 66 analyses, most of which are original. During the comparison

188

TABLE 42. Average Amounts (above sign) and Average Standard Deviations (below sign) of Cations in

	Paragenetic types	N	Al^{IV}	Al^{VI}	Ti	Cr	Fe^{3+}
Peridotites	1. Inclusions and intergrowths with diamonds	17	7/9	131/73	5/4	105/116	
Peridotites	2. Porphyritic pyrope lherzolites of the 'Udachnaya' pipe	14	13/19	49/11	6/3	42/14	—
Peridotites	3. Pyrope lherzolites from kimberlites	12	44/16	64/19	6/4	42/24	—
Peridotites	4. Pyrope peridotites of metamorphic complexes	15	52/21	53/31	4/4	22/13	22/15
Eclogites	5. Websterites (enstatite eclogites from kimberlites)	14	61/29	171/77	11/7	15/10	34/22
Eclogites	6. Eclogites from kimberlites	20	51/30	311/88	12/4	3/2	58/18
Eclogites	6a. Eclogites from kimberlites	8	33/21	198/72	11/2	6/2	67/17
Eclogites	7. Inclusions in diamonds	4	23	292	8	2	24
Eclogites	8. Diamond-bearing eclogites	16	7/8	455/92	12/4	1/1	24/25
Eclogites	9. Grospydites and kyanite eclogites from 'Zagadochnaya' pipe	20	73/35	583**/129	5/3	—	—
Eclogites	9a. Kyanite eclogites	5	89/45	518**/62	5/2	—	—
Eclogites	9b. Grospydites	8	83/23	519**/61	5/4	—	—
Eclogites	9c. Grospydites	6	53/26	740**/113	5/3	—	—
Eclogites	10. Kyanite eclogites from 'Roberts Victor' pipe	3	90	646	9	1	—
Eclogites	11. Corundum eclogite	1	256	483	3	1	19
Eclogites	12. Diamond-bearing corundum eclogite	1	27	751	12		—

Note. $^+$ All iron in form of Fe^{2+}; $^{++}$ together with Cr^{3+}. For Type 2, the average data on coexisting garnets

Clinopyroxenes from Deep-Stated Inclusions (based on 6000 atoms of oxygen) and Some Features of the Composition of the Associated Garnets

Fe^{2+}	Mn	Mg	Ca	Na	f'	f	Ca/Ca+Mg	Ca/Ca+Mg+Fe	Garnets Ca comp.	f	Cr comp.
58* / 13	3 / 2	784 / 168	657 / 167	261 / 171	—	7,2 / 2,3	45,6 / 2,6	43,5 / 2,8	—	—	—
89* / 17	1 / 1	1017 / 24	697 / 44	105 / 13	—	8,1 / 1,3	40,6 / 2,3	38,7 / 2,2	16,2 / 2,6	17,3 / 0,4	22,5 / 6,3
91* / 31	3 / 1	954 / 107	730 / 106	134 / 42	—	8,9 / 2,3	43,3 / 6,0	41,2 / 6,1	12,8 / 1,8	19,7 / 3,2	8,0 / 2,2
59 / 20	2 / 1	919 / 58	838 / 65	104 / 37	6,0 / 2,1	8,1 / 2,7	47,7 / 1,9	46,1 / 1,9	12,6 / 1,8	25,4 / 10,2	5,2 / 5,2
45 / 24	1 / 1	818 / 90	720 / 75	188 / 60	5,2 / 3,0	8,2 / 3,2	46,8 / 2,5	45,5 / 2,6	10,9 / 1,4	22,5 / 7,7	—
80 / 43	2 / 2	629 / 113	569 / 76	325 / 82	10,9 / 4,8	18,0 / 4,3	47,6 / 1,3	44,5	19,1 / 7,0	39,5 / 7,7	—
97 / 45	4 / 2	753 / 97	610 / 62	251 / 60	11,7 / 5,4	18,5 / 5,1	44,8 / 2,1	41,8 / 2,3	10,8 / 1,4	32,4 / 8,1	—
170	2	594	616	302	22,6	25,1	50,9	44,6	30,9	45,9	—
104 / 30	1 / 1	455 / 89	463 / 50	486 / 78	18,4 / 3,5	21,5 / 5,4	50,7 / 2,4	45,6 / 2,2	24,7 / 8,3	49,4 / 7,3	—
56* / 19	0	393 / 132	477 / 104	427 / 97	—	13,5 / 4,8	56,2 / 6,6	51,5 / 6,5	62,3 / 17,2	44,2 / 7,8	—
72* / 16	—	466 / 80	499 / 67	398 / 74	—	13,9 / 4,1	51,9 / 1,1	48,1 / 1,2	40,5 / 6,4	43,0 / 9,2	—
61* / 6	1 / 1	454 / 59	540 / 45	362 / 41	—	11,9 / 1,6	54,5 / 1,7	53,1 / 1,9	61,0 / 6,4	41,9 / 4,7	—
36* / 12	—	237 / 110	365 / 94	539 / 84	—	15,2 / 6,4	62,8 / 7,3	59,4 / 7,3	82,3 / 6,2	48,7 / 8,0	—
52	1	333	427	504	—	13,7	55,9	55,2	41,6	40,4	—
22	1	518	695	230	4,1	7,3	57,3	56,3	42,3	23,5	—
39*	0	218	249	712	—	15,2	53,3	49,2	35,2	35,4	—

are presented for 12 samples, and for Type 9, for 19 samples.

190

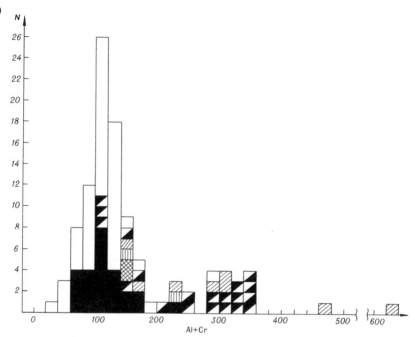

Fig. 32. Histogram of the distribution of Al + Cr (based on 6000 atoms of oxygen) in clinopyroxenes from kimberlites (for symbols, see Fig. 29).

of the compositions of the pyroxenes in all cases, data were also used on the composition of the associated garnets. This has enabled us almost to exclude the indeterminacy of the parageneses and to assess the features of the joint change in the compositions of the associated garnets and pyroxenes.

For 20 pyroxenes from eclogites of Yakutia and Africa, we obtained data, coinciding with those of Dobretsov et al. (1971), but distinguished from the earlier generalizations (Dobretsov et al., 1970). It should be noted that the standard deviation in the amount of sodium, according to the new data, is almost twice lower compared with earlier data. The amount of sodium in the pyroxenes from eclogites is significantly greater than in those from websterites. Data on the associated garnets have made it possible to recognize the link between ultramafic associations of the lherzolite-websterite type and eclogites with soda-rich pyroxenes. From the available pairs of minerals, we have selected only the pyroxenes, associated with garnets with low calcium index, similar to that of garnets from peridotites. An intermediate amount of sodium has been obtained in the pyroxene in this selection (see Table 42, Type 6).

The pyroxenes from diamond-bearing eclogites (see Table 42, Type 8) are characterized by maximum amounts of sodium. This applies not only to the entire selection of eclogites, the garnets of which are quite rich on average in calcium, but also to individual samples with garnets of low-calcium index.

At the same time, wide variations in the amount of sodium are typical of five pyroxenes from diamonds (2.59 ± 1.27). In comparing

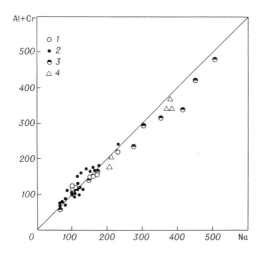

Fig. 33. Association between amounts of Na and those of Al + Cr (based on 6000 atoms of oxygen) in the clinopyroxenes from ultramafic parageneses: 1) inclusions in diamonds; 2) lherzolites from kimberlites; 3) intergrowths with diamonds; 4) intergrowths in kimberlite.

them with the compositions of pyroxenes from diamond-bearing and even from normal eclogites, we may conclude that pyroxenes of such composition as Sample M (Meyer and Boyd, 1972), are a great rarity (see Table 31). If we rely on data on compositions of those pyroxenes from diamonds, for which associated garnets are reliably known, that is, typical eclogitic associations, then the average amount of sodium appears to be significantly greater (3.10) (about 31% jadeite component).

From the viewpoint of emphasizing the role of sodium in accordance with increase in pressure, special interest centers round the eclogites, containing monomineralic parageneses, especially kyanite and corumdum eclogites. The average composition of pyroxenes from grospydites (see Table 42, Type 9) supports the conclusions (Sobolev, 1968b; Sobolev et al., 1968) that sodium plays a significant role in the pyroxenes of the deep-seated eclogites.

For the pyroxenes from grospydites, an excess of Al^{VI} has also been identified, which is positively correlated with the calcium index of the associated garnets, and which is especially clearly manifested in the pyroxenes, associated with grossulars (more than 75% calcium component). The compositions of these pyroxenes have been separated into a special type (see Table 42, Type 9c). It is evident that in this case, we may assume both a solid solution of 'grossular' and 'kyanite'. Apparently, the solid solution of grossular is a special equivalent of Tschermak's component in the pyroxenes of the grospydites.

In order to illustrate the role of sodium in the pyroxenes of the deep-seated eclogites, there is great interest in a comparison of the compositions of the pyroxenes from the kyanite eclogites of the 'Zagadochnaya' pipe and the equivalent eclogites of the 'Roberts Victor' pipe, including the diamond-bearing type. Previously (see §6 and §8), we emphasized the difference in the conditions of formation of the kyanite eclogites of these pipes, identified on the basis of a series of independent features. The present comparison of the compositions of the pyroxenes from the kyanite eclogites, associated with garnets of identical calcium- and iron-indices, demonstrates the significant differences in the amount of sodium, which may, most probably, explain the increased role of Al^{VI} and, consequently, sodium during increase in pressure.

There is even greater interest in a comparison of the compositions of pyroxenes from corundum eclogites, formed at different pressures (see Table 42, Types 11 and 12). The composition of the pyroxene from the 'Obnazhënnaya' pipe is similar in many ways to that studied experimentally by Boyd (1970) in the appropriate system without iron and sodium. This is explained by the relative closeness in the composition of the corundum eclogite to that used in the experiments. An increase in the amount of iron and sodium in the system, determined by the compositions of garnet and clinopyroxene, lowers the overall content of Al^{IV} in the pyroxene owing to an increase in the amount of jadeite. However, as is seen from the example cited, with insignificant amounts of additional components in the pyroxene composition, even that forming under conditions of high pressure, Al^{IV} plays a significant role (see Table 42, Type 11). In the analogous association, formed under conditions of even greater pressure (the diamond-bearing paragenesis), the amount of Al^{IV} becomes negligible, and all the aluminum passes into Al^{VI}. It must be emphasized that the composition of the pyroxene from the diamond-bearing corundum eclogite (see Table 42, Type 12), like that of the pyroxenes from certain grospydites, falls once again, on the basis of amount of sodium, into the middle of the diopside-jadeite immiscibility range (Dobretsov et al., 1971a; Ginzburg, 1970), which again emphasizes the necessity to link the features of isomorphism in the pyroxenes with the actual conditions of their formation, primarily pressure and temperature. The existence of natural clinopyroxenes, containing 60-80% of jadeite, has also been established for the pyroxenes from unusual eclogite xenoliths from the kimberlite pipes in Arizona, USA (Watson and Morton, 1969).

We shall dwell on some more general features of the composition of the deep-seated clinopyroxenes, most of which have been recognized for the first time and were not accounted for previously in the classification of the pyroxenes. Such features primarily include the presence of all the intermediate compositions in the diopside-jadeite series. It must also be emphasized that the presence of a wide series of solid solutions of diopside-clinoenstatite composition has been demonstrated in the naturel pyroxenes. This series is continuous in the range of values of the Ca/(Ca + Mg) ratio from 50 to 30, which corresponds to an amount of up to 40% of clinoenstatite component (see Fig. 29).

The new data on the composition of pyroxenes, associated with diamond, and from deep-seated inclusions, has demonstrated the significant role of the ureyite component ($NaCrSi_2O_6$), the amount of which, according to the results of all previous investigations, did not exceed 10 mol %. We have found a whole series of compositions, containing up to 20% of this component, and in two pyroxenes, even 30 and 46%. These data are a direct proof of isomorphism in the series of jadeite-ureyite compositions, which is shown in Figure 34.

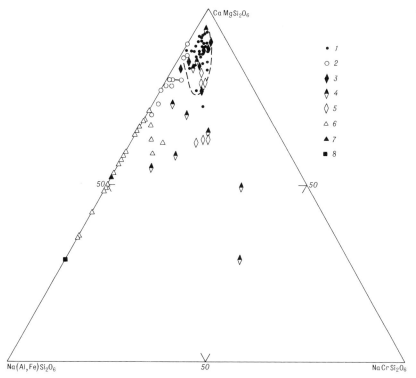

Fig. 34. Compositional features of iron-poor clinopyroxenes with
variable content of Na and Cr from kimberlite pipes: 1) from
lherzolites; 2) from websterites; 3) from inclusions in diamonds;
4) from intergrowths in diamonds; 5) from garnet-pyroxene-chromite
intergrowths in kimberlites; 6) from grospydites and kyanite
eclogites; 7) from a diamond-bearing kyanite eclogite; 8) from
a diamond-bearing corundum eclogite. Compositional field of pyroxenes from lherzolites surrounded by dashed line.

The Isomorphism of Potassium in the Clinopyroxenes

During investigations of probable material of the upper mantle,
such as the deep-seated xenoliths in kimberlites and inclusions in
diamonds, there has been great interest in an attempt to clarify
the role of potassium in the upper mantle. The potassium-rich
material, phlogopite, has a limited distribution in the deep-seated
rocks. The recently discovered potassium-rich amphibole, richterite
(found in certain phlogopite-bearing xenoliths (Erlank and Finger,
1970)), is also extremely rare and may not be a concentrator of
potassium in the mantle.

Attempts to determine K_2O with the aid of the microprobe in such
widely distributed minerals of the deep-seated rocks as olivine,
garnet, and enstatite, have demonstrated its very low concentrations
(less than 0.004%), which have essentially given a negative result
(Erlank and Kushiro, 1970). In the clinopyroxenes from lherzolites,
an amount of K_2O of the order of 0.01% has been identified, and in
two chrome-diopsides, it reached 0.04% (Erlank, 1969; Erlank and
Kushiro, 1970).

194
A substantially larger amount of K_2O has been identified in
certain omphacites from South African eclogites (0.10-0.17%), although
these figure appeartain to most-altered pyroxenes, and the freshest
samples contain not less than 0.002% K_2O. These results do not enable
the authors (Erlank and Kushiro, 1970) to judge the nature of the
potassium content, who have concluded that micro-ingrowths of
amphibole are possible in certain omphacites, the more so since
the experimental check of the possibility of entry of potassium
into the pyroxene structure at pressures up to 32 kb, carried out
by them, has shown a negative result. We have carried out special
investigations into the amount of K_2O in clinopyroxenes from deep-
seated xenoliths and inclusions in the diamonds of Yakutia, with
the principal requirement for the material analyzed, being the
freshness of the clinopyroxene (Sobolev et al., 1971). The amount
of K_2O (in the form of K) has been determined in 50 clinopyroxenes,
including 27 chrome-diopsides and 23 omphacites, with the aid of
the microprobe (Fig. 35). The use of a PET crystal-analyzer has
enabled us to determine the amount of potassium with high sensitivity
(limit of detection, 0.008% K_2O). As a standard, we have used an
orthoclase (Sample Or-1), employed for similar purposes in the
Geophysical Laboratory in the USA and kindly supplied to us by Dr.
F. R. Boyd (see Chapter I). The results of determining the amount
of K_2O are set out in Table 43. A low K_2O content is characteristic
of six clinopyroxenes from peridotites in the 'Mir' and 'Obnazhennaya'
pipes, and four samples contained less than 0.008% K_2O.

A very interesting result was obtained during the investigation
of 12 clinopyroxenes from especially deep-seated peridotites in the
'Udachnaya' pipe. The amount of K_2O in these pyroxenes varies from
0.025 to 0.086%, and the average amount in 0.046%. The stable nature
of the K_2O in the fresh samples is particularly noteworthy.

Even wider limits of variation have been identified in four
chrome-diopsides, enclosed in diamonds (from 0.031 to 0.152%). It
is interesting that the amount of K_2O in fresh unaltered clinopyroxene
(inclusion in diamond, Sample 57/9) is almost four times greater
than the maximum concentration (0.04%) determined in the South
African chrome-diopsides. Amounts, exceeding 0.04% K_2O have been
determined in seven samples of Yakutian chrome-diopsides, both from
peridotites and diamonds.

The chromium-bearing pyroxenes from intergrowths with diamonds
are characterized by comparatively wide variations in the amount
of K_2O (less than 0.008 to 0.058%).

During the investigation of omphacites, we paid particular attention to analysis of pyroxenes from diamond-bearing eclogites, using
material entirely from our own collection of these unique eclogites
(15 samples). Six pyroxenes taken from eclogites without diamond
and from grospydites, have revealed comparatively low quantities of
K_2O (less than 0.008 up to 0.053%). The omphacites from diamond-
bearing eclogites are characterized by a stable amount of K_2O, and
in all cases, only unaltered pyroxenes have been used. The limits
of variation in K_2O content for the 15 samples are 0.046-0.154%,
with 0.089% on average. In spite of the fact that material of the
greatest purity was used by us in all determinations, Sample M-54,
which is the most rich in K_2O (0.154%), was subjected to a careful
check for homogeneity and showed excellent reproducibility.

The most unexpected results were those obtained during examination
of three omphacite inclusions in diamonds. In one sample (7a)

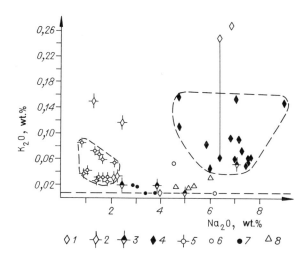

Fig. 35. Relationships between amounts of K_2O and Na_2O in clinopyroxenes from kimberlite pipes: 1) from diamonds (omphacites);
2) from diamonds (chrome-diopsides); 3) from intergrowths with
diamonds (chrome-bearing); 4) from diamond-bearing eclogites;
5) from pyrope peridotites in the 'Udachnaya' pipe; 6) from xenoliths of pyrope peridotites; 7) from xenoliths of eclogites; 8)
from xenoliths of grospydites. Compositional fields of pyroxenes
from diamond-bearing eclogites and peridotites in the 'Udachnaya'
pipe surrounded by broken lines; horizontal broken line marks limiting sensitivity for determining K_2O; compositional points for clinopyroxene from diamond and enclosing diamond-bearing eclogite (M-46)
joined by continuous line.

from a Uralian diamond, potassium was practically absent (less than
0.008% K_2O). The other two omphacites from diamonds in the 'Mir'
pipe (Samples AV-6 and M-46a) contain respectively 0.271 and 0.30%
K_2O. Such an amount is the largest recorded, and it is all the more
interesting in that it has been found in unaltered pyroxenes,
distinguished by homogeneity of composition on a micron level.
    Thus, as a result of the investigation of the amount of potassium
in unaltered pyroxenes, we may reach a conclusion about the structural nature of the isomorphous K_2O content. The stable amount of
K_2O in pyroxenes from especially deep-seated peridotites in the
'Udachnaya' pipe, and also from diamond-bearing eclogites and certain
inclusions in diamonds, suggests that clinopyroxene is a concentrator
of potassium in the upper mantle, but under conditions of particularly
high pressure.
    The greatest interest in understanding the behavior of potassium
under upper-mantle conditions centers around the results of investigating its amount of pyroxene, enclosed in diamond from a diamond-
bearing eclogite (M-46) and in pyroxene from the same eclogite
itself (respectively 0.30 and 0.080%) (V. S. Sobolev, Sobolev and
Lavrent'ev, 1972). The fourfold difference in these amounts, in
pyroxenes of general similarity, suggests that potassium has been
extracted from the pyroxene of the eclogite itself, which took place
after the formation of diamond, in which pyroxene of original composition was preserved (see §26).

TABLE 43.   Amount of K_2O in Clinopyroxenes from Kimberlites

No.	Samples	K_2O	No.	Samples	K_2O	№	Samples	K_2O
1	M-33	0,046	19	M-30	0,008	36	Ud-111	0,086
2	M-59	0,049	20	A-45	0,053	37	M-40	<0,008
3	M-180	0,061	21	Z-8	0,016	38	M-67	0,008
4	M-46	0,080	22	Z-43	0,017	39	O-172	<0,008
5	M-53	0,063	23	Z-54	0,018	40	O-802	<0,008
6	A-45/2	0,073	24	Z-34	0,031	41	O-466	0,018
7	M-51	0,073	25	Ud-94	0,025	42	O-145	0,019
8	M-50	0,080	26	Ud-101	0,028	43	B-9	0,008
9	M-52	0,083	27	Ud-3	0,031	44	B-24	0,008
10	MB-1	0,090	28	Ud-40	0,031	45	B-23	0,020
11	M-49	0,093	29	Ud-98	0,033	46	29/9	0,031
12	M-45	0,110	30	Ud-92	0,037	47	AV-14	0,033
13	F-1508	0,147	31	Ud-5	0,039	48	B-8	0,039
14	M-821	0,147	32	Ud-93	0,054	49	B-7	0,058
15	M-54	0,154	33	Ud-18	0,059	50	AV-34	0,117
16	7a	<0,008	34	Uc-161	0,069	51	57/9	0,152
17	M-46a	0,300	35	209/576	0,071			
18	AV-6	0,271						

Notes. Rock types for clinopyroxenes examined: 1-15) diamond-bearing eclogites: 1-12, 14, 15) 'Mir' pipe; 13) 'Newlands' pipe (South Africa); 16-18) inclusions in diamonds: 16) Urals; 17, 18) 'Mir' pipe; 19-20) eclogites from 'Mir' pipe; 21-24) grospydites from 'Zagadochnaya' pipe; 25-36) pyrope peridotites from 'Udachnaya' pipe; 37-42) peridotite xenoliths (Yakutia): 37, 38) 'Mir' pipe; 39-42) 'Obnazhënnaya' pipe; 43-51) chrome-diopsides, associated with diamonds (Yakutia): 46, 47, 50, 51) inclusions in diamonds; 43-45, 48, 49) intergrowths with diamonds.

The significant unevenness in the amount of potassium, even in homogeneous pyroxenes enclosed in diamonds, is also emphasized by comparison with the pyroxenes from South African diamonds (Meyer and Boyd, 1972). Such unevenness indicates variation in the amount of potassium even in the deeper portions of the upper mantle, in the stability field of the diamond. Since the content of K_2O of the order of 0.3% in the clinopyroxenes is already an established fact, and the extraction of potassium from pyroxene under deep-seated conditions has also been established, we may assume that the source of potassium in the upper parts of the mantle, in the field of predominant development of phlogopite, may be deep-seated pyroxenes, and not only omphacites from the quantitatively minor deep-seated rocks (eclogites), but also the chrome-diopsides, in which amounts of K_2O up to 0.15% have already been identified.

If we consider the pyroxenes which we have studied, separately in respect of types of rocks and depth facies, we may note patterns of increase in the upper limit of K_2O content with increase in depth of the associations. Thus, for the grospydites from the 'Zagadochnaya' pipe, such limit is 0.031%, and normal peridotites with calcic diopside contain up to 0.02% K_2O. In the peridotites from the 'Udachnaya' pipe, this amount increases significantly to 0.086%. In the pyroxenes from the facies of diamond-bearing eclogites and pyrope peridotites, an even greater amount of K_2O (up to 0.15 and even 0.30%) has been identified.

Consequently, a regular increase in the upper limit of the amount of K_2O has been recorded in pyroxenes with increase in pressure, that is, increase in pressure must favor the entry of potassium into the clinopyroxenes (Sobolev, Lavrent'ev and Pospelova, 1971). In the light of these patterns, it is possible in a different way to interpret the experimental data (Erlank and Kushiro, 1970) based on the study of the amount of potassium in synthetic pyroxenes, obtained at pressures of up to 32 kb and temperatures up to 1150°C. The low content of K_2O (up to 0.015%), determined with the aid of the microprobe in the synthetic pyroxenes, indicates not so much the impossibility of entry of potassium into the pyroxene structure, but the necessity to increase pressure (to more than 40 kb) in order to obtain a positive result. There is no doubt that experiments to examine the possibility of the entry of potassium into the pyroxene structure at pressures of the order to 50-150 kb would be of extremely great interest. Such an experimental approach could provide the final answer to the problem of the possible limit of isomorphous entry of potassium into pyroxene under the conditions of significant depths in the upper mantle. The first work of such a kind, carried out in the 40-100 kb interval (Shimizu, 1971) has experimentally supported the hypothesis of increased solubility of K_2O in pyroxene with increase in pressure.

§14. ENSTATITE

The principal feature of the enstatites, associated with pyroxene, is the decreased amount of aluminum (along with chromium which replaces it) and the decreased iron index. The amount of CaO is also of diagnostic value, being a solid solution of diopside in enstatite, and increasing in the enstatites from high-temperature parageneses. Since the amount of Al_2O_3 does not usually exceed 1.0-1.5%, and the amount of CaO in the enstatites from the most widely distributed peridotites usually does not exceed 0.5%, it is absolutely necessary to step up the quality requirements in the analyses of enstatites, mainly to prevent possible mechanical contamination from ingrowths. In this respect, two possible cases of distorted composition must be kept in mind. The first is the amount of clinopyroxene in the dissociation textures, which is partly removed during the process of preparing the material for analysis, and which partly remains in the form of especially fine ingrowths. In this case, the analytical data do not truly reflect either the features of high-temperature crystallization (before dissociation), or the conditions of dissociation, and they have a seemingly intermediate character. The second case is that of mechanical addition of clinopyroxene in such samples, in which solid-solution dissociation textures are absent, which applies mainly to the pyrope peridotites, known in metamorphic terranes.

Thus, when selecting analyses of orthopyroxenes for statistical components it is necessary to allow for the possibility of mechanical addition (mainly by clinopyroxene, and possibly pyrope also), which will be reflected in the result in the distorted amount of Al and Ca in the form of over-estimation. The most objective information on the composition of enstatites may be obtained by the use of quantitative X-ray spectral microanalyses.

We have employed about 20 new analyses of enstatites, completed with the aid of the microprobe, and several chemical analyses available in the literature. The compositional features of enstatites from different parageneses (inclusions in diamonds and various pyrope peridotites), based on the content of Al, Cr, and Ca, are shown in Figure 36.

A comparison of the data in Figure 36 with those for enstatites from pyrope peridotites and kimberlites (Dobretsov et al., 1970,

198

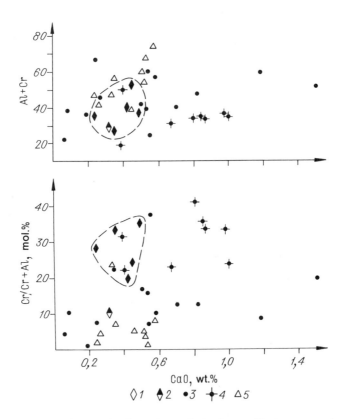

Fig. 36. Content of CaO (wt%), Al, and Cr (atom %) in enstatites from kimberlites: 1) from diamonds (black rhombs); 2) from diamonds (lherzolite paragenesis); 3) from peridotite xenoliths; 4) from peridotite xenoliths of the 'Udachnaya' pipe; 5) from pyrope peridotites of the metamorphic complexes. The dashed line indicated the compositional field of the majority of enstatites from diamonds.

1971) demonstrates that many analyses, used for statistical treatment, have been contaminated with mechanical additives. This shows up particularly in the comparatively large amount of Al + Cr cations (based on 6000 atoms of oxygen) in previously calculated average compositions of enstatites from garnet peridotites and kimberlites: respectively 57 and 61, and 58 and 78, according to Dobretsov et al. (1970, 1971). The new materials demonstrate that the amounts of Al + Cr cations on average do not exceed 40 for the pyrope peridotites from the kimberlites and 55, for the pyrope peridotites, known within metamorphic terranes, and also indicate the impossibility of separately considering the compositions of the enstatites of these rocks, which are apparently very similar in composition.

In regard to the amount of Al_2O_3 in the orthopyroxene, the original experimental determination on a pyrope showed its high solubility, of the order of 14–19% Al_2O_3, from which the conclusion was reached that further increase in pressure must even further increase this limit. When considering the results, it was noted that such a trend cannot be maintained, since half of the aluminum in the enstatite occurs in fourfold coordination (V. S. Sobolev, 1963a). Whereas the formation of aluminous enstatite takes place at the expense of corierite, that is, up to pressures of about 20 kb, in a system with an excess of silica, this must lead to a gain factor in volume, and the pressure will contribute to the solubility of Al_2O_3 in enstatite. However, in the stability field of pyrope, where all the aluminum occurs in sixfold coordination, the volume effect will be inverse, and further increase in pressure must lead to decrease in the solubility of Al_2O_3. These theoretical considerations were completely supported by subsequent experiments (Boyd and England, 1964) and an independent investigation of natural enstatites from pyrope peridotites (O'Hara and Mercy, 1963; Sobolev, 1964c; Banno et al., 1963).

It is clearly evident in Figure 17 that with further increase in pressure, the content of Al_2O_3 in enstatite, in equilibrium with pyrope, is diminished in the isothermal section (Boyd, 1970). This trend is also emphasized by the available factual material. In fact, a comparison of the total content of Al + Cr in the enstatites of the pyrope peridotites, known within metamorphic terranes, with those in an overall selection of analogous peridotites from kimberlites, and especially with those from the very deep-seated peridotites f om the 'Udachnaya' pipe (see Fig. 36), demonstrates a decrease in the content of these cations. However, with further increase in pressure, the amount of Al + Cr in enstatite will evidently be decreased even more slowly, since the enstatites from diamonds are not distinguished on the basis of this content from those from the peridotites of the 'Udachnaya' pipe (see Table 7).

The experimentally established importance of the amount of Ca in the composition of enstatites from two-pyroxene parageneses, as an indicator of equilibrium temperature (Davis and Boyd, 1966; Boyd, 1970) is strongly emphasized by the data on natural parageneses. In analyses, carried out on the microprobe, the clearest correlation has been observed between the amount of Ca in enstatite and the calcium-index of the associated clinopyroxene. Chemical analyses only provide a sufficiently clear picture for the particularly high-temperature parageneses (e.g. Sample E-3, Nixon et al. (1963)), where dissociation textures are absent.

We shall dwell briefly on the results of a study of the amount of chromium in enstatites. The role of chromium in coexisting minerals of deep-seated rocks is described in detail below (see §22). We shall note here, in respect of enstatite, that its chromium-index $(Cr/(Cr + Al))$ is positively correlated with that of the associated minerals, especially pyrope. The highest chromium-index is characteristic of the enstatites, associated with the most chromium-rich pyropes. In the case of unknown paragenesis, for instance, the isolated inclusions of enstatite in diamonds (see §9), we may reach the conclusion as to the high chromium-index of the pyropes, possible in paragenesis with five individual inclusions of enstatite available to us (see Table 31).

In conclusion, we once again amphasize that the basic information on the conditions of formation of magnesian enstatites from deep-seated parageneses is embraced in the features of the amounts of such oxides as CaO. Therefore, exceptionally careful selection of material is necessary, both for analyses, and for subsequent comparisons.

Demonstration of the overall increase in the role of aluminum
in sixfold coordination in clinopyroxenes with increase in pressure
(V. S. Sobolev, 1947, 1965), and also of the entry of sodium even
into garnets at especially high pressures (Sobolev and Lavrent'ev,
1971) has posed the problem as to the possible entry of sodium also
into the enstatites of the pyrope peridotites. A check has been
carried out, using the microprobe, on a series of enstatites,
associated with clinopyroxenes, characterized by wide fluctuations in
the amounts of Na_2O (from 1.0 to 4.3%) (Sobolev et al., 1971). Determinations were carried out at a sensitivity level of 0.01% Na_2O,
as for the garnets (see Chapter I). The range of identified fluctuations in the amount of Na_2O in the enstatites seems quite wide:
0.01-0.25% (Table 44).

All the enstatites examined (18 samples) may be roughly separated
into two groups, based on the content of Na_2O. The first group,
contains from 0.09 to 0.10%, and the second group is in the higher
range (0.13-0.25%).

We have also compared the Na_{en}/Na_{di} ratio in associated pairs
of pyroxenes. The values of this ratio vary widely (0.01-0.18),
but they may also be separated into two groups: 0.01-0.05, and
0.06-0.18. The only enstatite which we have examined from the Czech
Massif belongs to the first group along with some xenoliths from
kimberlites.

The data obtained suggest an increase in the coefficient of
distribution of sodium in the associated pyroxenes as pressure
increases. In this case, it is noteworthy that two enstatites
from inclusions in diamond, on the basis of Na_2O content, fell into
different groups. A low Na_2O content (0.07%) is characteristic of
the enstatite from Sample AV-42 which, on the basis of a number of
features, has been assigned to the harzburgite-dunite paragenesis
(without clinopyroxene), and a high content is typical of the
enstatite from the lherzolite paragenesis. In this case, the low
Na_2O content or its absence may be regarded, by analogy with the
chrome-pyropes from diamonds (see §9), as a feature of the absence
of clinopyroxene from the paragenesis. It is also evident that the
four samples of enstatites from diamonds of Africa (see Table 31),
which are very similar in compositional features to Sample AV-42
(see Fig. 36), must also contain a low Na_2O content. Unfortunately,
sodium has not been determined in these enstatites.

The investigations have not only identified an additional feature
of the deep-seated nature of the xenoliths in kimberlites, but have
also made it possible to use the content of sodium in enstatites,
associated with diamonds, as a diagnostic feature for identifying
the nature of the paragenesis (with or without clinopyroxene) of
individual enstatite inclusions.

The uniformity in the content of the relatively high concentrations
of sodium in the enstatites conclusively indicates the structural
nature of this material. The investigations have enabled us for the
first time to determine accurately the amount of Na_2O in enstatite,
which reaches 0.25%, which has also been emphasized for the samples
from South Africa (Boyd, 1973). The entry of such amounts of sodium
into enstatite is most likely, in combination with trivalent cations
(Al, Cr, and Fe), and also with titanium. This hypothetical solid

No.	Samples	Na$_2$O$_{(en)}$	Na$_2$O$_{(di)}$	Na$_2$O$_{(en)}$/Na$_2$O$_{(di)}$	№	Samples	Na$_2$O	Na$_2$O	Na$_2$O /Na$_2$O
1	Ud -111	0,042	0,84	0,050	11	AV -34	0,155	2,44	0,064
2	Ud -3	0,065	1,79	0,036	12	M-114	0,045	1,37	0,033
3	Ud -92	0,127	0,91	0,140	13	O-379	0,056	1,89	0,030
4	Ud -101	0,181	—	—	14	M-60	0,060	4,26	0,014
5	Ud -5	0,177	0,98	0,181	15	O-172	0,042	3,90	0,011
6	Ud -4	0,146	—	—	16	O-145	0,049	2,87	0,017
7	Ud -40	0,201	1,68	0,120	17	O-802	0,099	3,39	0,029
8	Ud -18	0,165	1,66	0,100	18	T7/272	0,033	0,96	0,034
9	Ув -16$_1$	0,254	1,50	0,169					
10	AV -42$_1$	0,073	—	—					

Notes. Rock types for enstatites examined: 1-9) pyrope peridotites from 'Udachnaya' pipe; 10, 11) inclusions in diamonds from 'Mir' pipe; 12-17) pyrope peridotites from Yakutian kimberlites: 12, 14) 'Mir' pipe; 13, 15-17) 'Obnazhënnaya' pipe; 18) pyrope peridotite from Czech Massif

solution, which may provisionally be termed 'jadeitic', we shall represent in the form of NaR^{3+}.

§15. OLIVINE

Amongst all the silicates, which make up the deep-seated pyrope-bearing rocks, olivine is distinguished by its most simple composition. As compared with the pyroxenes and garnets, in which the possibility of wide-ranging isomorphous substitutions of cations in eightfold and sixfold coordination has been established, the minerals of the olivine group represent two-component isomorphous series with complete miscibility in the forsterite-fayalite range. The value of the iron index (and inversely, the content of forsterite) is the principal and, perhaps, the only characteristic of the composition of the olivines under consideration, and in the parageneses of the pyrope-bearing peridotites, this value fluctuates from 5 to 15%. Additional information is also provided by determinations of the amount of nickel, which is usually positively correlated with the amount of forsterite (Ukhanov, 1968; Simkin and Smith, 1970). The maximum amount of NiO identified in the most magnesian olivines from diamonds is 0.43% (Meyer and Boyd, 1972).

Special investigations into the composition of the magnesian olivines from various ultramafic rocks (Il'vitsky and Kolbantsev, 1968) have shown insignificant differences and, consequently, the impossibility of separating them on the basis of iron index. The tendency towards decrease in iron index has been noted only in olivines from diamonds (see §9).

Isomorphism of Chromium in Olivines

All the known isomorphous additives enter olivine in the group of divalent cations, because the nature of the olivine structure excludes the possibility of the entry of trivalent cations. Nevertheless, during detailed investigations of olivines, enclosed in diamonds (Meyer and Boyd, 1972), an apparently completely anomalous amount of chromium, usually known in natural minerals in the form of Cr^{3+}, has been accurately identified.

When discussing the causes for the appearance of an anomalous

amount of chromium in olivine (up to 0.1% Cr_2O_3), it has been correctly noted that such an addition cannot be associated simply with
the high content of chromium in the melt, since the olivines,
associated with chromitites, contain less than 0.01% Cr_2O_3 (Meyer
and Boyd, 1972). An unequivocal assessment of the reasons for the
entry of chromium into olivine has been given only after finding
larger amounts of chromium (up to 0.21% Cr_2O_3) in olivines from lunar
basalts, collected by 'Apollo-11' (Haggerty et al., 1970). A similar
and even larger amount of Cr_2O_3 (up to 0.28%) has been found in many
phenocrysts of olivine from various samples and small fragments of
lunar basalts (Brown et al., 1970; Agrell et al., 1970). A detailed
study of olivine from the lunar basalts has shown that chromium
most probably enters in the form of Cr^{2+}, which may depend both for
the lunar rocks and for inclusions in diamonds, on the exceptionally
low activity of oxygen. Further investigations of olivine from
lunar basalts have revealed even larger amounts of chromium (up to
0.4% Cr_2O_3) in large olivine phenocrysts with an iron index of 30%
(Boyd et al., 1971).

Although the presence of Cr^{2+} in natural olivines was established
quite recently, the possibility of entry of Cr^{2+} into the olivine
structure had been demonstrated experimentally as long ago as 1964
in experiments carried out in an atmosphere of hydrogen in the
system Cr_2O_3 - Cr - SiO_2 (Tsvetkov et al., 1964), with the final
product being an orthosilicate, Cr_2SiO_4, analogous to olivine. The
principal reason for the rarity of such an additive in natural
minerals is the unusual nature of the conditions (the very low
activity of oxygen).

An attempt to find chromium in magnesian olivines of several
terrestrial ultramafic rocks has essentially yielded a negative
result, having revealed only about 0.01% Cr_2O_3 (Meyer and Boyd,
1972), and the object of determining chromium was not even achieved
during the special investigations of amounts in olivines (Simkin
and Smith, 1970).

Arising from the positive results of determining the amounts in
olivines from diamonds (Meyer and Boyd, 1972), we proposed the
more significant role of chromium in olivines from the particularly
deep-seated parageneses (diamonds, the kimberlites themselves, and
peridotite xenoliths). At the same time, we made an attempt to
determine the amount of chromium in olivines from the pyrope peridotites of the Czech Massif, the lherzolitic inclusions in
basaltoids, and the Alpine-type ultramafic rocks and meimechites,
totalling in all 69 samples (Sobolev, Lavrent'ev and Pospelova,
1971). In parallel with the determination of chromium in all the
olivines, we determined the amount of iron (in the form of FeO),
which is entirely adequate for the complete characterization of
olivine. The results of 79 determinations, including 10 for
chromium in olivines from diamonds in Africa and Venezuela (Meyer
and Boyd, 1972), are presented in Table 45, along with data on the
amount of iron, the iron index, and forsterite, calculated on the
basis of the general features of olivine compositions.

The results of determining the amounts of Cr_2O_3 and FeO in the
olivines are also presented in Figure 37. The comparatively wide
variations in the amount of chromium are typical of the olivines from
diamonds: 0.017-0.08% Cr_2O_3, and the new data do not show any increases
in the earlier maxima established (0.08%). Great interest also
centers around the wide variations in the amount of Cr_2O_3 in individ-

TABLE 45. Amount of Cr_2O_3 and Compositional Features of Magnesian Olivines from Various Rocks

№	Samples	Cr_2O_3	FeO	f	fo	№	Samples	Cr_2O_3	FeO	f	fo
1	AV-34	0,017	7,01	7,3	92,5	41	O-802	0,026	8,76	9,1	90,9
2	AV-14	0,017	6,30	6,5	93,4	42	Ud-3	<0,007	7,13	7,4	92,6
3	G16a	0,02	7,10	7,0	93,0	43	Ud-11	<0,007	7,59	8,0	92,0
4	38/8	0,03	6,15	6,1	93,9	44	Uv-128	0,027	9,30	9,7	90,3
5	23в	0,04	7,60	7,6	92,2	45	Ud-101a	0,028	6,88	7,2	92,8
6	57/9	0,047	6,00	6,0	93,1	46	Ud-92	0,044	8,10	8,6	91,4
7	G10b	0,05	6,91	6,9	93,0	47	Ud-101b	0,049	8,73	9,1	90,9
8	41/33	0,052	7,39	7,6	92,4	48	Ud-5	0,056	8,79	9,2	90,8
9	57/33	0,056	7,45	7,7	92,3	49	Uv-196	0,063	6,87	7,2	92,8
10	10c	0,06	7,09	7,2	92,6	50	Ud-4	0,070	7,95	8,3	91,7
11	13c	0,06	7,11	7,3	92,5	51	Uv-126	0,070	8,55	8,9	91,1
12	GL24c	0,07	6,65	6,6	93,4	52	Uv-161	0,079	9,28	9,8	90,2
13	GL2qb	0,07	7,14	7,2	92,8	53	M-300	<0,007	7,81	8,1	91,9
14	G17a	0,07	8,34	8,5	91,6	54	S-104	<0,007	6,71	6,9	93,1
15	11a	0,08	5,52	5,5	94,3	55	S-434	0,007	6,52	6,7	93,3
16	GL47a	0,08	6,21	6,0	93,8	56	S-455	0,009	6,91	7,2	92,8
17	Un-11	<0,007	7,50	7,8	92,2	57	S-452	0,017	6,79	7,1	92,9
18	Un-2	0,007	13,1	13,9	86,1	58	A-147	0,026	13,2	14,0	86,0
19	Un-7	0,017	7,53	7,8	92,2	59	Sk1-1	<0,007	9,26	9,7	90,3
20	Un-15	0,018	7,04	7,3	92,7	60	T7/272	<0,007	8,28	8,7	91,3
21	Un-13	0,021	8,73	9,1	90,9	61	T7/317	<0,007	8,45	8,9	91,1
22	Un-12	0,022	6,98	7,2	92,8	62	AV-10	<0,007	9,21	9,6	90,4
23	Un-14	0,032	6,52	6,8	93,2	63	HW-8	<0,007	9,19	9,6	90,4
24	Un-10	0,033	7,29	7,6	92,4	64	AV-3	0,008	9,11	9,5	90,5
25	Un-5	0,034	7,02	7,2	92,8	65	Sm-1	0,011	8,77	9,2	90,8
26	Un-8	0,035	6,98	7,2	92,8	66	HW-6	0,021	15,1	16,0	84,0
27	Ush-2	0,037				67	HW-7	0,028	13,4	14,2	85,8
28	Un-6	0,041	6,99	7,2	92,8	68	309	<0,007	5,49	5,7	94,3
29	Un-9	0,043	6,71	6,9	93,1	69	308a	<0,007	7,54	7,9	92,1
30	Un-1	0,059	6,69	6,9	93,1	70	353b	<0,007	7,54	7,9	92,5
31	Un-3	0,059	9,02	9,4	90,6	71	104	<0,007	8,80	9,2	90,8
32	Ob-5	<0,007	6,78	7,1	92,9	72	147a	<0,007	9,24	9,6	90,4
33	Ob-4	<0,007	7,93	8,3	91,7	73	326b	<0,007	7,97	8,3	91,7
34	Ob-1	<0,007	7,44	7,7	92,3	74	193a	0,010	7,21	7,5	92,5
35	Ob-2	<0,007	7,30	7,7	92,3	75	4V-5	0,009	10,2	10,8	89,2
36	Ob-6	<0,007	7,40	7,7	92,3	76	56-1V	0,016	14,6	15,5	84,5
37	Ob-3	<0,007	7,16	7,5	92,5	77	120-V	0,069	10,5	11,1	88,9
38	Ob-7	0,009	7,81	8,2	91,8	78	MT-1	0,085	8,64	9,0	91,0
39	O-606	<0,007	7,35	7,7	92,3	79	113R	0,110	8,64	9,0	91,0
40	O-800	<0,007	8,41	8,7	91,3						

Notes: 1-16) inclusions in diamonds: 1, 2, 4) 'Mir' pipe; 6) 'Udachnaya' pipe; 8, 9) Yakutia; 5, 10, 11, 15) Africa; 3, 7, 14) Ghana; 12, 13, 16) Venezuela; 17-31) xenocrysts from kimberlite ('Udachnaya' pipe); 32-38) xenocrysts from kimberlite ('Obnazhënnaya' pipe); 39-58) xenoliths of pyrope peridotites from kimberlite pipes: 39-41) 'Obnazhënnaya' pipe; 42-52) 'Udachnaya' pipe; 53, 58) 'Mir' Pipe; 54-57) 'Bultfontein' pipe (South Africa); 59-61) pyrope peridotites of Czech Massif; 62-67) peridotite xenoliths in basaltoids: 62, 64) Avacha Volcano (Kamchatka); 63, 66, 67) Hawaiian Islands; 65) Smrchi (Czechoslovakia); 68-74) Alpine-type ultramafic rocks of the Anadyr-Koryak fold system; 75, 76) dunites of the Gula intrusive (northwest of Siberian Platform); 77-79) meimechites (northwest of Siberian Platform).

ual olivine grains from the Yakutian kimberlite pipes. In 22 samples, we have identified from less than 0.007 up to 0.059%. Still wider variations are typical of the olivines from xenoliths of peridotites: from less than 0.007 up to 0.079%. The olivines from ultramafic xenoliths in basaltoid rocks contain little chromium (up to 0.28% Cr_2O_3), and those from Alpine-type ultramafic rocks of the Anadyr-Koryak fold system are almost devoid of this material.

An important diagnostic feature is the identification of increased amounts of Cr_2O_3 in the olivines from meimechites. Three samples

204

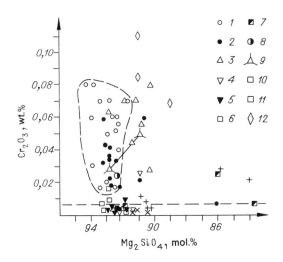

Fig. 37. Amount of Cr_2O_3 and forsterite component in magnesian olivines: 1) from diamonds; 2) kimberlites of the 'Udachnaya' pipe:
3) pyrope peridotites of the 'Udachnaya' pipe; 4) pyrope peridotites
of the 'Obnazhënnaya' pipe; 5) kimberlite of the 'Obnazhënnaya'
pipe; 6) pyrope peridotites of South Africa; 7) pyrope peridotites
of the 'Mir' pipe; 8) kimberlite of the 'Dal'nayaya' pipe (groundmass and phenocrysts); 10) lherzolite xenoliths in alkaline basaltoids; 11) Alpine-type ultramafic rocks; 12) meimechites.

studied contained from 0.069 up to 0.11%, that is, in these olivines
from meimechites, the highest maximum amount of chromium amongst
olivines of terrestrial origin has been identified. This feature
directly supports the view that these rocks are of a particularly
deep-seated nature (Sheinmann, 1968).

On the whole, without taking account of the earlier accurately
identified olivines with chromium present (from diamonds), this
investigation has enabled us to find an amount of Cr_2O_3 (more than
0.02%) in 27 samples, and more than 0.05% in 10 samples of olivines
from various rocks.

During a study of olivine grains from kimberlite in the 'Udachnaya'
and 'Obnazhënnaya' pipe, we succeeded in establishing a definite
difference in their chromium content, along with an almost identical
iron index. The 'Udachnaya' olivines contain on average 0.038%
Cr_2O_3, whereas chromium is practically absent in the 'Obnazhënnaya'
olivines. The same trend is also typical for the olivines from the
xenoliths of peridotites from these pipes.

Such clear differences, obtained from representative material,
collected at random (for the grains from the kimberlite) are of
significant interest, especially in connection with the fact that the
'Udachnaya' is diamond-bearing, and the 'Obnazhënnaya' is completely
devoid of diamonds, which may be primarily associated with their
different depth situation. The presence of Cr_2O_3 in the olivines
from the peridotites in the 'Udachnaya' pipe ($\bar{x}$ = 0.045% for 14
samples) is additional evidence for their deep-seated nature as compared
with other analogous rocks.

Arising from the hypothesis that the activity of oxygen decreases
with depth, we should expect very low concentrations of chromium

in the olivines from shallower depths as compared with the xenoliths
of rocks from the pipes (Alpine-type ultramafics and inclusions in
basaltoids), which has also been emphasized by the analyses.

§16. KYANITE

Kyanite has very rarely been found in the xenoliths of the kimberlite pipes, which is, in all probability, the result of the rareness
of eclogite compositions, rich in CaO and Al_2O_3, simultaneously
with low iron index (Sobolev et al., 1968; Godovikov, 1968).
A comparison of the compositional features and physical properties
of the kyanites from xenoliths of eclogites and grospydites from
Yakutia and South Africa (Sobolev et al., 1968; Mathias et al.,
1970; Gurney et al., 1971), and also those from eclogites of the
metamorphic complexes (Matthes et al., 1970) and kyanites from other
metamorphic rocks (Chinner et al., 1969; Albee and Chodos, 1969),
demonstrates their complete similarity. This is explained by the
practical absence of trace-elements in the kyanites of various rocks,
which depends mainly on the chemical composition of the rocks.
The chrome-bearing kyanites of the 'Zagadochnaya' pipe. Against
the background of exceptional uniformity in the kyanite compositions,
a marked anomaly showed up in the discovery of an amount of Cr_2O_3
up to 1.8% in the green kyanites of the Chainit corundum deposit
in Yakutia (Ozerov and Bykhover, 1936). Even more unexpected was
the discovery of a whole series of kyanites with variable Cr_2O_3
content (Sobolev et al., 1968), almost up to 17% (see Table 19),
in the 'Zagadochnaya' pipe. These kyanites form an almost continuous
component (Cr_2OSiO_4) of up to 20 mol %.
X-ray structural data (Sobolev, Kuznetsova and Zyuzin, 1966;
Sobolev et al., 1968) for a chrome-bearing kyanite with up to 15%
Cr_2O_3, demonstrate an increase in the values of the a_0 and b_0
parameters, as compared with the normal natural kyanites: a_0 =
7.14 and 7.11 Å, and b_0 = 7.88 and 7.84 Å. The structural nature
of the Cr_2O_3 additive is also indicated by the positive correlation
between   for the kyanites and the amount of Cr_2O_3 up to 1.794 in
Sample Z-51 (see Table 19). An unevenness in the coloring of
kyanite individuals has been established: from colorless to blue
and blues of differing intensity in the one grain. This is associated with the uneven distribution of Cr_2O_3, which has been illustrated
in the example of a kyanite grain from Sample Z-13, containing in
different portions, 5.55 and 8.41% Cr_2O_3 (see Table 19).
The amount of Fe_2O_3 is similar in all the samples examined (see
Table 19), and it is most probable that the increase in the intensity
of the blue colors (with a greenish tint) and the pleochroism, is
associated only with increase in the amount of Cr_2O_3. Experimental
data on the stability of the kyanite containing Cr_2O_3 (Seifert and
Langer, 1970), have supported our hypothesis (Sovolev et al., 1968)
of limited miscibility in the system Al_2OSiO_4 - Cr_2OSiO_4, in
dependence on pressure (see §6).

§17. THE MINERALS OF THE Al_2O_3 - Cr_2O_3 ISOMORPHOUS SERIES

The natural corundums, like the kyanites, contain a very small
amount of other oxides, especially Cr_2O_3 (Kolesnik and Guletskaya,
1968). Thus, the largest amount of Cr_2O_3 in natural corundums
(1.96%) has been found in only one case in the chainit corundum

TABLE 46. Results of X-Ray Structural Analysis of Chromium-Bearing Corundum (URS-50-IM Diffractometer, Cu-anticathode)

Line number	hkil	Cr₂O₃ (ASTM—6—0504)		Cr₂O₃		Chrome-corundum			Corundum O-160		λ-Al₂O₃ (corundum) (after Mirkin, 1961)	
		I	d, Å	I	d, Å	I	d, Å	d, Å	J	d, Å	I	d, Å
1	2	3	4	5	6	7	8	9	10	11	12	13
1	01$\bar{1}$2	74	3,633	65	3.63	70 70	3,59* 3,53**	3,59* 3,53**	60	3,48	72	3,479
2	10$\bar{1}$4	100	2,666	100	2,66	85 85	2,63* 2,59**	2,63* 2,59**	90	2.55	92	2,552
3	11$\bar{2}$0	96	2,480	80	2,47	55 55	2,45* 2,42**	2,45* 2,41**	50	2.38	41	2,379
4	0006	12	2,264	10	2,27	— —	— —	— —	—	—	—	—
5	11$\bar{2}$3	38	2,176	35	2,18	55 85	2,15* 2.12**	2,15* 2.12**	100	2.09	103	2,085
6	02$\bar{2}$4	39	1,815	35	1.815	40 55	1,796* 1,765**	1,794* 1,764**	50	1.741	41	1,740
7	11$\bar{2}$6	90	1,672	75	1,672	70 100	1,652* 1,624**	1,651** 1,624**	90	1.602	83	1,601
8	12$\bar{3}$2	13	1,579	10	1,580	15 15	1,568* 1,537**	1,562* 1,536**	10	1,513	2	1,510
9	21$\bar{3}$4	25	1,465	25	1.465	30 55	1,448* 1,424**	1,448* 1,424**	30	1,406	38	1,404
10	30$\bar{3}$0	10	1,4314	35	1,432	40 55	1,415* 1.394**	1,416* 1,393**	60	1.374	42	1,374
11	101$\bar{1}$0	20	1,2961	20	1,297	30 30	1,277* 1,256**	1,277* 1,256**	10	1.240	16	1,239
12	22$\bar{4}$0	17	1,2398	10	1,241	— —	— —	— —	10	1,191	6	1,1898
13	30$\bar{3}$6	7	1,2101	5	1.211	— —	— —	— —	—	—	—	—
14	02$\bar{2}$10	10	1,1488	5	1,150	— —	— —	— —	—	—	—	—
15	22$\bar{4}$6	17	1,0874	15	1,088	— —	— —	— —	10	1.043	13	1,0426
16	21$\bar{3}$10	16	1,0422	15	1,042	15 15	1,029* 1,011**	1,028* 1.011**	—	—	—	—
17	32$\bar{5}$4	13	0,9462	5	0,9475	— —	— —	— —	10	0.9084	12	0,9076
18	13$\bar{4}$10	14	0,8957	10	0,8964	15 15	0,8845* 0,8697**	0,8846* 0,8697**	20	0.8589	12	0,8580
19	41$\bar{5}$6	23	0,8658	15	0,8667	15 15	0,8563* 0,8421**	0,8563* 0,8422**	20	0.8312	22	0,8303
20	101$\bar{1}$6	11	0,8331	10	0,8343	— —	— —	— —	—	—	—	—

(table 46)   207

1	2	3	4	5	6	7	8	9	10	11	12	13
				$a_H = 4,963\text{Å} \pm 0,003\text{Å}$ $c_H = 13,60\text{Å} \pm 0,01\text{Å}$ $\dfrac{c_H}{a_H} = 2,739;$ $a_r = 5,36\text{Å} \pm 0,01\text{Å}$ $\alpha = 55°08' \pm 8'$		$a_H = 4,907\text{Å} \pm 0,003\text{Å}$ $c_H = 13,39\text{Å} \pm 0,01\text{Å}$ $\dfrac{c_H}{a_H} = 2,728;$ $a_r = 5,29\text{Å} \pm 0,01\text{Å}$ $\alpha = 55°20' \pm 8'$ $a_H^{**} = 4,826\text{Å} \pm 0,003\text{Å}$ $c_H = 13,16\text{Å} \pm 0,01\text{Å}$ $\dfrac{c_H}{a_H} = 2,726;$ $a_r = 5,20\text{Å} \pm 0,01\text{Å}$ $\alpha = 55°20' \pm 8'$		$a_H = 4,762\text{Å} \pm 0,003\text{Å}$ $c_H = 13,00\text{Å} \pm 0,01\text{Å}$ $\dfrac{c_H}{a_H} = 2,730;$ $a_r = 5,13\text{Å} \pm 0,01\text{Å}$ $\alpha = 55°16' \pm 8'$		

$^+$ Phase I with larger amount of Cr_2O_3.
$^{++}$ Phase II with smaller amount of Cr_2O_3.

deposit (Belyankin et al., 1941). At the same time, in the series of compositions, isostructural with corundum, besides Fe_2O_3 (hematite), which does not form solid solutions with corundum, the mineral eskolaite (Cr_2O_3) is known (Kouvo and Vuorelainen, 1958), identified only in Finland under specific geochemical conditions of enrichment in Cr_2O_3. It must be emphasized, however, that about 4% V_2O_3 has been found in the eskolaite and Al_2O_3 is completely absent. This case may be regarded as somewhat analogous to that of the minerals of the garnet group, where a pure chrome-garnet (uvarovite) occurs as a later mineral in chromite deposits, and the Al-garnets, containing a variable amount of Cr_2O_3, are formed not only with the appropriate chemical composition of the rocks, but also at increased pressures.

A continuous series of solid solutions, Al_2O_3 - Cr_2O_3, has been obtained experimentally at high temperatures. The experimental data have been collated in a reference work (Toropov et al., 1965). In addition, there are results from a hydrothermal synthesis of the continuous series, Al_2O_3 - Cr_2O_3, at $T = 800°C$ and $P = 2000$ atm (Neuhaus et al., 1962). Dissociation textures have been identified during these investigations in a compound of intermediate composition with lowering of temperature, forming laminar ingrowths.

In the synthesized Al_2O_3 - Cr_2O_3 series, a positive correlation has been identified between the values of the refractive index and the amount of Cr_2O_3 (Grum-Grzhimailo, 1946, 1953; Grum-Grzhimailo and Utkina, 1953), and also the analogous dependence of the a_0 value on the amount of Cr_2O_3 (Toropov et al., 1965; Steinwehr, 1967). The alexandrite effect was identified during an examination of an Al_2O_3 - Cr_2O_3 mixture (Grum-Grzhimailo, 1946).

During a study of the grospydites and mineral intergrowths, containing chrome-kyanite, we have found laminar crystals of red and crimson color; the grains, containing one phase of reddish-brown color and another, opaque (Sample Z-51), have been found as ingrowths in chrome-kyanite (Sobolev, Lavrent'ev and Zyuzin, 1976).

α X-ray investigations of two-phase ingrowths from Sample Z-51 have

been carried out by N. I. Zyuzin on a URS-50-IM diffractometer
(Cu$\underline{K}$ α radiation) (Table 46). Pure Al_2O_3 and Cr_2O_3 were used as
standards. The X-ray patterns are shown in Table 38. The para-
meters of the unit cells have been calculated on the (416) and
(1.3.10) reflections.

Results of optical investigations have supported the assignment
of these compositions to corundums, enriched in Cr_2O_3, and the
dependence of the γ values on the amount of Cr_2O_3, identified on
synthetic corundums (Grum-Grzhimailo, 1946; Grum-Grzhimailo and
Utkina, 1953).

An analysis with the aid of the microprobe has shown that the
samples belong a series of Al_2O_3 - Cr_2O_3 solid solutions (Table 47).

During the interpretation of the data obtained, special attention
was focussed on the composition of the minerals of the Al_2O_3 - Cr_2O_3
series in Sample Z-51, where they are present as ingrowths in the
chrome-diopside. Results of an examination in X-ray Al$\underline{K}$ and Cr$\underline{K}$
radiation and in absorbed electrons on a scale for the side of the
square, equal to 200 microns (Photo 19), have emphasized that Cr_2O_3-
rich phase is present in the form of laminar ingrowths. Investigation of the composition of the two phases immediately on the contact
(see Table 46, Z-51/1 and Z-51/2) emphasizes that the amount of
Cr_2O_3 in them is markedly variable. The coexistence of two phases
may serve as an indication of a break in miscibility, probably
dependent on dissociation of an intermediate compound during lowering of temperature, similar to that obtained experimentally (Neuhaus
et al., 1962). The phase with decreased content of Cr_2O_3 has a
complex, quite variable composition, which is indicated by fluctuations in the values, and also variations in the composition of
the analyzed sectors.

In Sample Z-51, a grain has been found with a very large content
of Cr_2O_3, and uniform. Samples have also been studied with a
relatively low amount of Cr_2O_3 (see Table 46). An interesting pattern
has been established in the distribution of additive elements
(MgO, TiO_2, and Fe_2O_3), the amount of which sequentially increases
with increase in the amount of Cr_2O_3 (Fig. 39), reaching 1.3% for
TiO_2, and 5.1% for Fe_2O_3. The presence of these additives does
not enable us to identify the color of the most chromiferous varieties.
We have noted that a special search for vanadium, by analogy with
eskolaite, has not been successful, and the order of the amount of
V_2O_3 in the most chromiferous varieties may only be assessed at
less than 0.1%. The compositions of this series have been crystallized not only in equilibrium with chrome-kyanite, but also have
been developed in the alteration products of such kyanites.

The compositions described were first defined for natural minerals.
Since extreme end members of this series (corundum and eskolaite)
are known for the natural system, and the current investigations
have established a break in miscibility, we may term the intermediate compositions, chrome-corundum (with a Cr_2O_3 content of
less than 40%) and alumo-eskolaite (with an Al_2O_3 content of less
than 35%).

§18. THE CHROME-SPINELS

The chrome-spinels are typical minerals of the kimberlite pipes
and have been recorded in the great majority of these pipes in

TABLE 47. Chemical Composition of Minerals of the Al_2O_3 - Cr_2O_3 Series from the 'Zagadochnaya' Pipe

Oxides	Z-54	Z-56	Z-51a	Z-51b	Z-51/1	Z-51/2	Z-51/3
SiO_2	0,07	0,07	—	—	—	—	0,05
TiO_2	0,09	0,10	0,32	0,28	0,24	1,19	1,31
Al_2O_3	92,3	93,5	55,1	58,4	60,6	26,3	15,5
Cr_2O_3	4,77	7,06	40,7	37,7	34,9	64,4	75,5
Fe_2O_3	0,50	0,48	1,86	1,62	1,41	4.58	5,11
MgO	0,14	0,04	0,10	0,10	0,08	0,31	0,34
Total	97,87	100,20	98,08	98,10	97,23	96,78	97,91
Ti	0,001	0,001	0,005	0,004	0,004	0,021	0,024
Al	1,922	1,895	1,312	1,374	1,422	0,715	0,438
Cr	0,068	0.097	0,651	0,595	0,550	1,170	1,431
Fe^{3+}	0,007	0,006	0,028	0,024	0,022	0,080	0,092
Mg	0,004	0,001	0,002	0,002	0,002	0,011	0,012
Total of cations	2,011	2,000	1,998	1,999	2,000	1,997	1,997
Cr-component	3,4	4,9	32,6	29,8	27,5	58,6	71,7
Fe-component	0,3	0,3	1,4	1,2	1,1	4,0	4,6
Ti-component	0,1	0,1	0,3	0,2	0,2	1,1	1,2
N_γ	1,787	1,807	1,930—1,968			—	—

Yakutia (Bobrievich et al., 1959, 1964; Rovsha and Ilupin, 1970; Ukhanov, 1970). They are known in the form of separate grains of varied morphology in kimberlite concentrate, in xenoliths of deep-seated ultramafic rocks, and also in the form of inclusions in diamonds. Up until very recently, data on the composition of the chrome-spinels from kimberlites were extremely limited, but they also indicated a significant variation (Rovsha and Ilupin, 1970; Ukhanov, 1970).

The use of the microprobe has revealed great possibilities in investigating the composition of individual small grains of chrome-spinels from xenoliths. The original materials for our work include about 40 new analyses of chrome-spinels (see Tables 10-12, 30, 34, and 48), which has more than doubled the amount of factual information. These data indicate still more significant variations in the composition of the chrome-spinels from kimberlites, and a number of features have been identified for the first time. Such features include the discovery of a large amount of TiO_2, reaching 6.5% in a chromite from Sample Uv-126, which is associated with a chrome-rich pyrope in a xenolith of peridotite (see Table 48), chrome-spinel xenocrysts containing a wide range of TiO_2 up to 12% (Sobolev et al., 1975), and also the discovery of a series of chromite compositions with variable amount of the magnetite component, reaching 25 mol %, which is almost equivalent to that in the chrome-spinel, the composition of which has earned the title of a new mineral, donathite (Seeliger and Mücke, 1969). Spinels of similar composition have also been found in the ultramafic rocks of Bulgaria (Zhelyazkova-Panaiotova and Ivchinova, 1970).

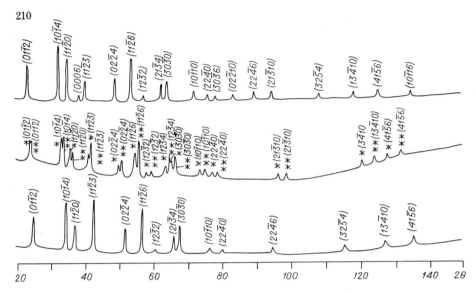

Fig. 38. Results of X-ray structural analysis of minerals of Al_2O_3 - Cr_2O_3 composition (URS-50-IM diffractometer, Cu-anitcathode): Upper and lower X-ray patterns are respectively Cr_2O_3 and Al_2O_3 standards; the middle pattern consists of two phases of an intermediate composition in the system Al_2O_3 - Cr_2O_3 (mineral, Sample Z-51); the phase, richer in Cr_2O_3, is denoted by one asterisk, and that richer in Al_2O_3, by two asterisks.

The presence of TiO_2 enables us to compare the chrome-rich spinels from the peridotite xenoliths with certain chrome-spinels found in lunar basalts (Haggerty et al., 1970; Agrell et al., 1970), and the variable content of magnetite may be regarded as an index of the activity of oxygen (Lindsley, 1963; Irvine, 1965, 1967).

Of all the chrome-spinels from the kimberlite pipes, the most unusual in composition are the chromites, enclosed in diamonds and known in intergrowths with polycrystalline diamond aggregate (Fig. 40). A comparison of the analytical results of about 50 such chromites (Sovolev et al., 1965), which comprise field III in Figure 40 with those of the chrome-spinels from ultramafic rocks (Pavlov et al., 1968), the compositions of which form field I, and with the chromites from meteorites (Bunch et al., 1967; Snetsinger et al., 1967) (field II), indicate clear differences between them. Whereas the most chromiferous spinels from the ultramafic rocks also partly overlap field III in Figure 40a, on the basis of the ratio of Al_2O_3 and Cr_2O_3 (see Fig. 40b), they are significantly different. Such differences demonstrate the absence of equivalents in composition in the chromites from diamonds amongst the known natural chrome-spinels.

A comparison of the compositions of all the chrome-spinels investigated from kimberlites (Fig. 41) emphasizes their exceptional variation. Two main trends in composition are clearly distinguishable: 1) a variable relationship in the amount of Al_2O_3 and Cr_2O_3 with a low constant amount of Fe_2O_3 and a low amount of TiO_2; and 2) a low amount of Al_2O_3, along with a variable amount of Cr_2O_3 and TiO_2. The first trend appears in the chrome-spinels, which

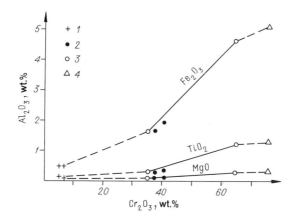

Fig. 39. Compositional features of minerals in the Al_2O_3 - Cr_2O_3 series: 1) from Samples Z-54 and Z-56; 2) from a zone of dissociation texture with lower Cr_2O_3 content (Sample Z-51); from coexisting zones of Sample Z-51. Compositional points for coexisting phases of Sample Z-51 have joined by continuous line.

belong to the series of peridotites with varying amount of Al_2O_3 (Betekhtin, 1937; Sokolov, 1946; Pavlov et al., 1968; Zimin, 1963). The chromites, associated with diamonds, occupy the extreme position in this group, being the most chromium-rich among the known natural chromites.

The second trend in change in composition is characteristic of a whole series of samples of chrome-spinels from zoned garnets, and from intergrowths in kimberlite (Sobolev, Khar'kiv et al., 1973), and the comparatively widespread distribution of spinels with variable increased content of the magnetite component has also been established for the concentrates of certain Yakutian kimberlite pipes (Sobolev et al., 1975).

The points for known analyses of mixtures of chrome-spinels from concentrates of kimberlite pipes are also located within the entire field of compositions, and the average compositions of the chrome-spinels from the concentrate from diamond-bearing pipes are richer in Cr_2O_3 than those from the non-diamond types (see Chapter VIII). An analysis of the numerous grains from concentrates has shown that in the 'Mir', 'Udachnaya', and 'Aikhal' pipes, practically all possible compositions of chrome-spinels are present, and 15-20% of all the grains studied correspond in composition to chromites, enclosed in diamond (Sobolev et al., 1975).

There is an interesting example of the discovery of two generations of chromite in an intergrowth with diamond (Sample MR-9), with the chromite of the first generation being equivalent to all the chromites, associated with diamonds, whereas the chromite of the second generation is distinguished by a larger amount of the magnetite component and titanium (see Fig. 41).

Thus, the chrome-spinels from the kimberlites are characterized by the exceptional variation in composition. Such variation may be associated, on the one hand, with the presence of spinels from xenoliths of ultramafic rocks of different composition, and on the other, with a change in the conditions of crystallization, most

TABLE 48. Chemical Composition of Chrome-Spinels from Kimberlites and Peridotite Xenoliths

Oxides	Peridotites								Kimberlites						
	Uv-126/ 'Udachnaya' pipe	A-70 'Aikhal' pipe	A-18 'Aikhal' pipe	Uv-161/ 'Udachnaya' pipe	O-802 'Obnazhennaya' pipe	Ud-3/ 'Udachnaya' pipe	A-101 'Aikhal' pipe	A-31 'Aikhal' pipe	2M 'Mir' pipe	5M 'Mir' pipe	6M 'Mir' pipe	4M 'Mir' pipe	3M 'Mir' pipe	Ut-1 'Udachnaya' pipe	Ut-2 'Udachnaya' pipe
SiO_2	0,49	0,13	0,07	0,14	0,36	0,36	0,16	0,34	0,45	0,41	0,42	0,41	0,45	0,10	0,02
TiO_2	6,44	2,82	0,96	4,59	1,16	0,15	1,30	0,16	0,11	0,11	1,30	0,81	0,41	0,13	0,17
Al_2O_3	8,27	6,39	21,1	8,97	10,7	15,6	4,70	9,79	38,1	37,2	23,0	23,4	13,9	9,88	5,44
Cr_2O_3	43,8	44,4	46,0	47,1	52,2	54,4	60,3	61,2	27,4	28,9	40,0	40,4	52,7	55,7	64,1
Fe_2O_3	9,66	10,4	3,6	7,6	4,5	0,0	5,3	0,0	2,5	2,2	3,0	3,2	2,6	7,9	4,0
FeO	17,8	22,1	12,1	20,2	18,7	15,0	17,7	15,4	10,4	11,0	14,7	14,0	16,4	15,4	15,2
MnO	0,14	0,30	0,15	0,36	0,32	0,22	0,29	0,28	0,15	0,19	0,24	0,22	0,29	0,32	0,30
MgO	13,9	9,80	15,6	11,1	10,8	12,0	10,5	10,9	17,7	17,2	14,2	14,4	12,4	11,4	11,7
Total	100,5	96,34	99,68	100,06	98,74	97,73	100,25	98,07	96,81	97,21	96,84	96,84	98,15	101,9	101,0
Si	0,016	0,005	0,005	0,004	0,012	0,012	0,006	0,011	0,013	0,012	0,013	0,013	0,015	0,003	0,001
Ti	0,156	0,077	0,022	0,115	0,029	0,004	0,032	0,004	0,002	0,002	0,031	0,019	0,010	0,003	0,004
Al	0,314	0,275	0,759	0,349	0,425	0,594	0,186	0,384	1,302	1,272	0,869	0,869	0,535	0,377	0,211
Cr	1,134	1,280	1,110	1,228	1,391	1,391	1,607	1,612	0,626	0,664	0,997	1,005	1,365	1,421	1,679
Fe^{3+}	0,236	0,286	0,081	0,190	0,114	0,000	0,134	0,000	0,055	0,048	0,072	0,075	0,065	0,192	0,099
Fe^{2+}	0,481	0,674	0,308	0,557	0,526	0,406	0,498	0,428	0,253	0,268	0,387	0,369	0,423	0,443	0,422
Mn	0,004	0,009	0,004	0,010	0,009	0,006	0,008	0,008	0,004	0,005	0,005	0,006	0,008	0,010	0,008
Mg	0,669	0,533	0,709	0,545	0,543	0,579	0,526	0,540	0,764	0,746	0,667	0,675	0,606	0,549	0,577
f	32,7	55,7	28,7	44,8	49,2	41,0	47,0	44,0	34,0	26,3	34,8	34,1	40,5	44,5	42,0
Cr-component	56,7	64,0	55,5	61,4	69,6	69,6	80,4	80,6	31,3	33,2	49,8	50,3	68,3	71,5	84,0
Al-component	15,7	13,8	38,0	17,5	21,4	29,7	9,3	19,2	65,1	63,6	42,8	43,5	26,8	18,9	10,6
Ulvöspinel	15,6	7,7	2,2	11,5	2,9	0,4	3,2	0,4	0,2	0,2	3,1	1,9	1,0	0,3	0,4
K_0	32,8	29,8	21,0	25,4	17,8	---	21,2	---	10,5	15,2	15,7	16,9	13,3	30,1	19,2

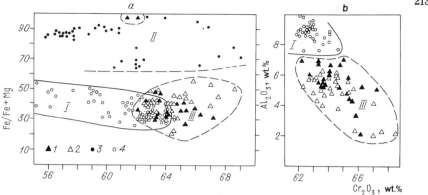

Fig. 40. Compositional features of chrome-rich spinels from diamonds
(I), from intergrowths with diamonds (2), from meteorites (3), and
from various ultramafic rocks and chromite deposits in the Kempirsai
pluton (4).

probably, in the temperatures and the activity of oxygen in the
kimberlite focus, which has led to significant variations in the
amount of the magnetite component present.

§19. ILMENITES

The ilmenites, or more precisely, the picroilmenites from the
Yakutian kimberlites occur both in the xenoliths of specific peridotites, and also in the form of monomineralic nodules in kimberlite.
The large number (about 120) of published analyses of ilmenites
(Milashev et al., 1963; Bobrievich et al., 1964; Frantsesson, 1968;
Vladimirov et al., 1971) demonstrate the substantial variation
in composition in the amount of MgO and in the addition of Fe_2O_3,
which significantly distorts the composition of the ilmenites from
the Stoichiometric pattern (Fig. 42). The representative material
on the ilmenites from the Yakutian kimberlites has enabled us to
identify certain patterns in their composition in individual pipes,
in particular in the well studied 'Udachnaya' and 'Mir' pipes
(Frantsesson, 1968).

Along with Fe_2O_3, an amount of Cr_2O_3 has been identified in the
ilmenite, reaching in a number of cases, 4% (Ilupin and Nagaeva,
1970), and in very rare cases even 10.7% (see Table 12). The results
of a study of individual monomineralic nodules of ilmenite have not
shown up any clear pattern of entry of the chromium additive (Ilupin
and Nagaeva, 1970; Blagul'kina, 1971).

A more definite answer may be provided by a study of the distribution of chromium in the ilmenites and associated minerals, such as
the garnets and chrome-spinels. The data in the present work have
identified a clear positive correlation between the amount of Cr_2O_3
in individual zones of a zoned garnet and the enclosed ilmenite
grains in the garnet. It is interesting that the maximum solubility
of chromium in ilmenite has been identified in a sample with a garnet
of variable (up to high) chrome-index, along with a very high amount
of Fe_2O_3 in the ilmenite and the coexisting chromite. This amount
in Sample M-41 is the maximum amongst the coexisting minerals of the

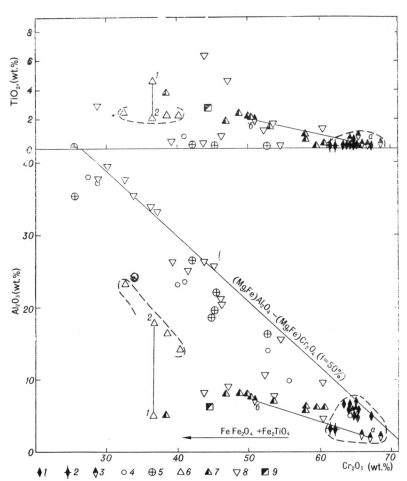

Fig. 41. Compositional features of chrome-spinels from kimberlites:
1) from diamonds; 2) from African diamond (Sample GL-3); 3) from
intergrowths with diamonds; 4) from kimberlite concentrate (individual grains); 5) from kimberlite concentrate (bulk analyses); 6)
from zoned garnet M-49 (points 1 and 2 respectively); 7) from
chromite-pyroxene-garnet intergrowths; 8) from peridotite xenoliths;
9) from peridotite xenolith A-70 ('Aikhal' pipe) in paragenesis with
ilmenite; *a, b*) compositional points for chromites of first and
second generation from intergrowth with diamond (Sample MR-9)
respectively.

kimberlites, and is probably a reflection of the highest oxidizing
conditions. At the same time, it is characteristic that in the
paragenesis of ilmenite with chromite in meteorites (Bunch et al.,
1970; Buseck and Kell, 1966) and in the lunar basaltoid rocks
(Haggerty et al., 1970), the solubility of chromium in the ilmenites
is extremely low.

Thus, the most probable reason for increase in the solubility of
chromium in ilmenites, associated with chromites (or chromium-rich

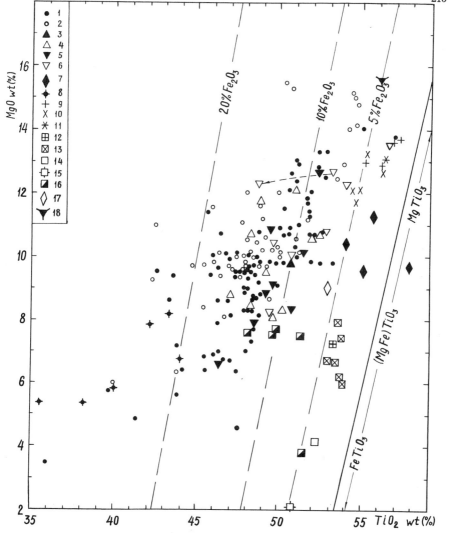

Fig. 42. Plot of MgO vs TiO_2 for ilmenites from various associations in kimberlite and experimentally synthesized ilmenites illustrating the divergence of ilmenites from the simple $FeTiO_3$ - $MgTiO_3$ solid solution. Rocks and parageneses for ilmenites: 1. Kimberlites of Yakutia; 2. Kimberlites from West Africa and South Africa; 3 - 4. Intergrowths with clinopyroxene, Yakutia and S. Africa; 5-6. From garnet peridotites or intergrown with garnets, Yakutia and S. Africa; 7. Intergrown with diamonds, Mir pipe, Yakutia; 8. Inclusions in zoned garnet, Mir pipe, Yakutia; 9 - 11. Ilmenites synthesized from pyrolite-40% olivine at T=1100, 1000, 950°C; 12 - 14. Ilmenites synthesized from olivine basanite at T=1100, 1000 and 900°C; 15. Synthesized ilmenite; 16. Megacrysts from basaltic lavas; 17. Inclusion from diamond, Mir pipe, Yakutia; 18. From diamondiferous pyrope peridotite, Udachnaya pipe, Yakutia: Data are: 1 - 16. From Green and Sobolev (1975); 17. From Sobolev et al., (1976); 18. From Pokhilenko et al., (1976).

pyropes) is an increase in P_{O_2}. In a reducing environment, with
such an amount of chromium in the melt, its solubility in ilmenite
is sharply diminished. It is also evident that the low content of
Cr_2O_3 in most of the ilmenites from the kimberlites of Yakutia
and Africa, which contain a significant amount of Fe_2O_3, indicates
their possible paragenesis with garnets, which contain almost no
chromium. It is necessary to arrange experimental investigations of
systems with titanium and chromium at high pressures and temperatures
(Lamprecht and Woermann, 1970) and varying P_{O_2}.

§20. RUTILE

Rutile is distributed in deep-seated xenoliths mainly in eclogites,
being a component part of the eclogites of different types. Only
individual rare occurrences of rutile are known in association with
chromium-bearing pyropes in the form of inclusions, and in this case,
the rutile contains up to 5% Cr_2O_3 (McGetchin and Silver, 1970).

In an unusual chromium-bearing eclogite paragenesis in association
with a chromium-bearing kyanite, we have found (in the form of an
inclusion in the kyanite) a rutile, containing about 10% Cr_2O_3. The
composition of this grain, about 50 microns in size, has been determined with the aid of the microprobe: TiO_2, 86.0%; Al_2O_3, 0.64%;
Cr_2O_3, 9.81%; Fe_2O_3, 0.62%; MgO, 0.003%; CaO, 0.03%; total, 97.14%.
Rutile of such composition has been identified for the first time
in natural materials.

The identification of Cr_2O_3 in rutiles from chromium-bearing
associations is a convincing reason for assigning the non-chromium
rutiles, found in the form of isolated grains and inclusions in
diamonds, to parageneses which do not contain Cr_2O_3, that is,
eclogitic types. In Table 35, four analyses of rutiles of this
type from intergrowths with diamonds are presented. In these
rutiles, the stable increased content of Fe_2O_3, according to experimental data (Witike, 1967), may be regarded as an indicator of
the temperature. Arising from these data, we may regard the rutiles
from intergrowths with diamonds as higher-temperature in nature
than those from the chromium-bearing kyanite with 0.62% Fe_2O_3.

It is likely that the investigation of the amount of iron and
aluminum in the rutiles from eclogites may provide additional
information for identifying the temperature of equilibrium of
various eclogites (Sobolev, Lavrent'ev and Usova, 1972). It must
be emphasized that, as a result of clarifying the definite role
of titanium-bearing minerals in association with diamond and the
special conditions of their crystallization (low activity of oxygen),
it is impossible to exclude the possibility of discovering under
these conditions the mineral $(Fe_{0.5}^{2+}Mg_{0.5})Ti_2O_5$ (armolcolite), found
in lunar basalts (Anderson et al., 1970[+].

[+]When the present work was sent to the press, a report appeared
(Haggerty, 1973) on the first discovery of armolcolite under terrestrial conditions in the Dutoitspan kimberlite pipe in South
Africa. It has been identified here in the form of individual
crystals, 25-50 microns in size, and also in a paragenesis with
rutile and picroilmenite.

CHAPTER VI

DISTRIBUTION PATTERNS OF COMPONENTS IN MINERALS OF VARIABLE COMPOSITION

§21. DISTRIBUTION OF MAGNESIUM AND IRON

An important feature of the metamorphic and magmatic rocks is the presence of phases of variable composition, with different elements simultaneously entering certain minerals. The compositions of the coexisting minerals are altered regularly and conjugately, which indicates the relative equilibrium state of the processes of mineral formation.

The most important isomorphous elements in the rock-forming minerals are iron and magnesium. The problem of the relationships between these components in the universally existing equilibrium solid solutions was tackled simultaneously by Korzhinsky (1936) for the metamorphic rocks and by Bowen and Schairer (1935) for equilibria with melts. For the relationships of the garnet-cordierite and biotite-orthopyroxene pairs in the metamorphic rocks and for the equilibrium of olivine with pyroxenes (experimentally), these authors demonstrated, on the basis of the phase rule, that the ratio of magnesium and iron in corresponding phases is mutually related in the presence of a degree of freedom and becomes definite in an invariant equilibrium.

Subsequently, actual examples and the theory of the problem were studied by a number of authors (Bartholome, 1962; Kretz, 1961; V. S. Sobolev, 1949; Marakushev, 1965, 1968; Sobolev, 1964a; Perchuk, 1967, 1970; etc.). In analyzing the relationships, use is most commonly made of distribution diagrams and the coefficient of distribution (K_D), calculated according to the formula

$$K_D = \frac{f_1(1 - f_2)}{(1 - f_1)f_2} \quad ,$$

where f_1 and f_2 are the iron-indices of the minerals compared.

During the study of the rock-forming minerals, a large amount of factual information has been collected on the relationships of iron and magnesium for a whole series of pairs, including those of minerals, which enter the peridotite and eclogite parageneses. However, its interpretation encounters a number of difficulties, mainly as a result of the variety of factors which influence the conditions of mineral formation, of which the principal ones may be considered the temperature, pressure, and chemical composition of the environment (Kretz, 1961; Marakushev, 1965, 1968; Banno, 1970; Sobolev, 1970; Perchuk, 1967, 1970; Dobretsov et al., 1970).

Garnet-clinopyroxene

In discussing the conditions of the crystallization of eclogites of different origin, there is significant interest in identifying the distribution patterns of magnesium and iron between garnets and clinopyroxenes. Wide variations in the values of K_D, have been established for the garnets and pyroxenes of different eclogites, and it has also been shown that the differences depend both on the effect of temperature, and on pressure and composition (Sobolev, 1964a; Marakushev, 1968; Banno, 1970; Perchuk, 1967, 1970.

However, the effect of temperature has been assessed as the more

217

significant (Marakushev, 1968; Banno, 1970; Perchuk, 1967, 1970).
It is true that attempts at a quantitative estimation of the effect
of temperature on the distribution of Mg and Fe (Perchuk, 1970)
cannot be regarded as successful, because the values obtained for
the deep-seated pyrope-bearing peridotites do not agree with those
based on the two-pyroxene geothermometer, which is well-founded
experimentally (Davis and Boyd, 1966). Further results were obtained
in experimental studies by Räheim and D. Green (1974).

We have compiled data on the distribution of Mg and Fe in garnets
and clinopyroxenes from ultramafic and basic associations of deep-
seated inclusions (Fig. 43). Since the major portion of the analyses,
used by us, has been obtained with the aid of the microprobe, the
overall, and not the specific iron-indices of the minerals, have
been taken into account. The diagram defines the extremely wide
range of values of K_D, the lower limit of which (0.2) is well maintained for the grospydites.

A comparison of the most magnesian pairs for the peridotites
demonstrates the extremely wide spread of the K_D values, both for
the known high-temperature peridotites from the 'Udachnaya' pipe,
and for other comparatively lower-temperature types. The inclusions
in diamonds and the diamond-bearing eclogites are characterized by
similar features of distribution of Mg and Fe. The data for the
pairs from the diamond-bearing eclogite and the inclusions of garnet
and pyroxene from the diamond in this same eclogite are of great
interest. Along with some difference in the iron index of the
associated minerals, and its increase in the minerals enclosed in
the diamond, the value of K_D is pratically unchanged, which together
with the complete similarity in composition, based on the amount of
other components, is an emphasis of the closeness of the temperature
conditions of crystallization of these pairs in the sample of diamond-
bearing eclogite, M-46.

Interesting examples, enabling us to estimate the effect of pressure
on the value of K_D, are the kyanite eclogites and grospydites from
the 'Zagadochnaya' pipe, on the one hand, and the diamond-bearing
kyanite eclogite from the 'Roberts Victor' pipe, on the other, for
which, based on a number of features, a significant difference in
pressure has been demonstrated (see §6). In Figure 43, it may be
seen that the K_D values for these pairs do not differ significantly,
which may be regarded as an argument in favor of the insignificant
effect of pressure in this case.

For the series of eclogites, including the diamond-bearing varieties,
and the inclusions in diamonds, the range of variations in the K_D
values agrees with the data obtained during a comparison of the
Ca/(Ca + Mg) ratio of the associated garnets and pyroxenes (see §22).

It is likely that the search for quantitative temperature relationships for the garnet-pyroxene pairs from deep-seated associations
will be carried out in future by comparisons with the data from
the experimentally well-founded two-pyroxene geothermometer.

Garnet-olivine

During consideration of the features of the conjugate change in
the Fe/(Fe + Mg) ratio in coexisting olivines and garnets (Sobolev,
1964; Dobretsov et al., 1970), it has been shown that the coefficient
of distribution changes during the passage from rocks of ultramafic
composition to those richer in iron, from approximately 0.5 to 1.0

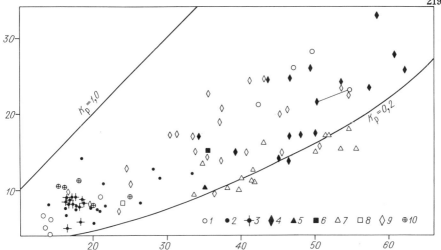

Fig. 43. Distribution diagram of iron index of garnets (along
horizontal axis) and clinopyroxenes in deep-seated inclusions: 1)
from diamonds; 2) from peridotites; 3) from porphyritic peridotites
in the 'Udachnaya' pipe; 4) from diamond-bearing eclogites; 5)
from a diamond-bearing kyanite eclogite; 6) from a diamond-bearing
corundum eclogite; 7) from kyanite eclogites and grospydites; 8)
from a corundum eclogite; 9) from eclogites; 10) from garnet-
pyroxene-chromite intergrowths in a kimberlite. The straight line
joins points of conjugate iron-index for the minerals from diamond-
bearing eclogite M-46 and from a diamond in this same eclogite.

in the most ferriferous compositions. In the present work we have
been interested in a detailed consideration of the distribution of
Fe and Mg in olivines and garnets of the magnesian associations.

Data from the study of the pyrope-peridotite associations both in
kimberlites and in solid outcrop make it possible to assess the nature
of the conjugate change in the compositions of the coexisting olivine
and garnet with quite a wide range of iron indices, essentially
embracing all possible fluctuations in the compositions of garnets
from peridotites, based on such indices from 12 up to 31%. The
points for the conjugate iron-index of 26 garnet-olivine pairs from
inclusions in diamonds, from peridotite-xenoliths in kimberlites,
and peridotites, known in solid outcrop (Meyer and Boyd, 1972; Sobolev
et al., 1970; O'Hara and Mercy, 1963; Fiala, 1965, 1966; Sobolev
and Lodochnikova, 1962; Mikhailov and Rovsha, 1966; Carswell, 1968a,
b) have been plotted on a distribution diagram (Fig. 44). In addition,
we have used some new data from peridotite xenoliths in the 'Udachnaya'
pipe.

In spite of the fact that the pairs investigated have been taken
from associations, formed under different conditions, the values of
the distribution coefficient (K_D) fluctuate over an extremely narrow
range: 0.33-0.52. For the field of magnesian compositions, this range
is in fact not wide, and has been controlled in part by possible
errors in determining the low concentrations of iron. For portion of
the intrusive peridotites of the Czech Massif, a certain increase
in K_D is associated, according to E. Fedyukova, with zonation of the

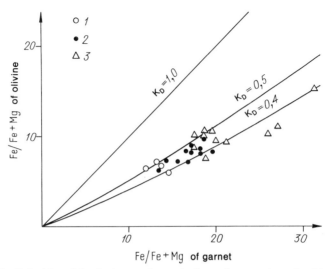

Fig. 44. Relationship between iron index of garnet and olivine in ultramafic parageneses: from diamonds (1), peridotite xenoliths (2), and peridotites of metamorphic complexes (3).

garnet, identified with the aid of the microprobe, and with increase in the iron index in the outer zone by 3-4%. Chemical analysis gives an average composition with a decreased iron index.

The available compositions of pyrope peridotites, associated with olivine, practically embrace the limits of variation in the composition of the magnesian garnets from kimberlites. Even allowing for variation in the values of K_D, it may be assumed that the iron index of the garnet, associated with olivine in diamond, cannot exceed 16-18%. Such an assemption agrees well with the factual data (see §9).

From the distribution diagram (see Fig. 44), another observation arises that the majority of the garnets known in the kimberlite concentrate, and rich in calcium and chromium (see §2), could not be formed along with diamond, because their iron index (about 25%) indicate possible association with olivine, containing 12-14% of fayalite.

Olivine-enstatite and Enstatite-clinopyroxene

The presence of a predominant temperature dependence in the distribution of Fe and Mg in the olivine-enstatite and enstatite-clinopyroxene pairs has been emphasized particularly for the high-temperature members (Marakushev, 1968). The value of K_D for these pairs approximates to 1.0, and sometimes the points even pass beyond the line of K_D = 1.0, creating an impression of extreme relationships. The results of experimental investigations indicate the absence of such a dependence (Medaris, 1969; Grover and Orville, 1969).

The data, presented in Figure 45, demonstrate that, with low values of the iron index, variations in the value of K_D may depend mainly on errors in analyses, and in the general case, the value of K_D may be regarded as somewhat less than 1.0.

Development of the two-pyroxene geothermometer (see §22) and the

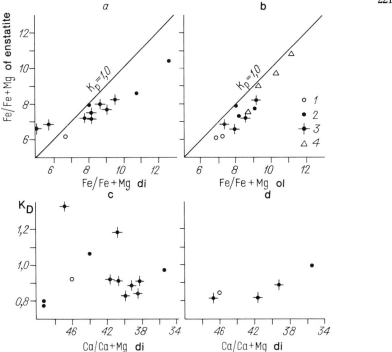

Fig. 45. Relationship between iron index of enstatites and diopsides (a); and enstatites and olivines (b) from pyrope peridotites, and association between the distribution coefficient of these values (K) and those of the calcium index of diopside (respectively c and d): 1) from diamonds; 2) from pyrope peridtites; 3) from pyrope peridotites of the 'Udachnaya' pipe; 4) from the pyrope peridotites of the Czech Massif.

available data on the associated minerals from peridotites, for which a wide range of equilibrium temperatures has been established, enable us to compare the values of K_D for the olivine-enstatite and enstatite-clinopyroxene pairs with the object of verifying the possible effect of temperature in different peridotites from kimberlite pipes. In this case, we shall not employ data on the pyrope peridotites of the metamorphic complexes, which, on the basis of a number of features, are regarded as lower-temperature on the whole as compared with the xenoliths (see §6). Such a comparison for the magnesian compositions (see Fig. 45) leads to no unequivocal conclusion, because the range of variations in the iron index of the coexisting minerals is too narrow, along with the very low absolute values.

§22. DISTRIBUTION OF Ca AND Mg IN GARNETS AND PYROXENES AND
THE EQUILIBRIUM TEMPERATURES OF ECLOGITES AND PERIDOTITES

In connection with the general objective of clarifying the distribution patterns of individual components in coexisting minerals and their possible dependence on temperature and pressure, there has been great interest in the study of the distribution of Ca and Mg

in the garnets and pyroxenes. This interest has been aroused to
a significant degree by the results of experimental investigations
in the system $CaMgSi_2O_6-Mg_2Si_2O_6$ (Boyd and Schairer, 1964; Davis
and Boyd, 1966), and also in the system $CaSiO_3-MgSiO_3-Al_2O_3$ (Boyd,
1970).

Data from an investigation into the nature of the miscibility
of the pyroxenes depending on temperature both at a pressure of 1 atm,
and at 30 kb have clearly demonstrated the possibility of using the
the composition of the clinopyroxene, associated with enstatite,
or more precisely, the Ca/(Ca + Mg) ratio as a geologic thermometer
(Boyd, 1969, 1970). An important argument in favor of such a conclusion arises from the similar nature of the solvus curves in the
system $CaMgSi_2O_6-Mg_2Si_2O_6$ for pressures of 1 atm and 30 kb (Fig. 46).

In spite of the unequivocal nature of the established relationship,
the appication of these results of the natural parageneses of deep-
seated ultramafic rocks requires a number of assumptions, based on
the assessment of the possible role of additional components, involved
in the composition of the natural minerals, namely: FeO, Cr_2O_3,
Na_2O and Al_2O_3. However, even without allowing for such an influence,
interesting results have been obtained on the estimate of the equilibrium temperature of crystallization of the clinopyroxenes from
xenoliths of pyrope peridotites and kimberlite concentrate (Boyd,
1969).

In fact, the presence of a low, and on average constant, 'background'
amount of these components in the deep-seated ultramafic rocks could
scarcely significantly shift the position of the solvus curve at $\underline{P}$ =
30 kb (see Fig. 46). At the same time, in order to verify such an
effect, results of actual experiments are necessary using compositions which contain these additives, and best of all, natural
compositions, with subsequent determination of the composition of
the end products with the aid of the microprobe.

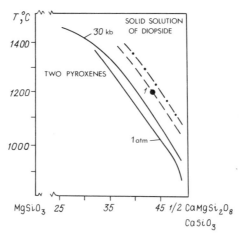

Fig. 46. The diopside solvus in the system $CaMgSi_2O_6$ - $Mg_2Si_2O_6$:
at 1 atm (after Boyd & Schairer, 1964); at 30 kb (after Davis &
Boyd, 1966). Point 1) composition of diopside, in equilibrium with
enstatite and pyrope, obtained experimentally at P = 30 kb and T =
1200°C (Boyd, 1970); a) assumed solvus for system with Al_2O_3 at P =
30 kb; b) assumed solvus for system with Al_2O_3 at P = 50 kb.

In this respect, the results obtained by Boyd (1970) during an investigation of the system $CaSiO_3-MgSiO_3-Al_2O_3$ at $\underline{T}$ = 1200°C and $\underline{P}$ = 30 kb, are noteworthy; this system, along with the two-pyroxene association, includes an analogous association with garnet and corresponds more closely to the composition of natural minerals as compared with the system $CaMgSi_2O_6-Mg_2Si_2O_6$. In Boyd's investigation, the microprobe was used for the first time in the work of the Geophysical Laboratory, and with its aid, the compositions of many phases, obtained as a result of the experiments, were determined.

In a number of the analyzed products of various experiments, there has been particular interest in those which produced an equilibrium association of two pyroxenes without garnet and with garnet, with different amounts of Al_2O_3 in the original material. The results of calculating the composition of the clinopyroxene on the basis of Boyd's analyses, and also the features of the composition of the associated minerals, obtained in certain experiments (Boyd, 1970), are presented below.

Experiment No.	Ca/(Ca + Mg) in clino-pyroxene	Al_2O_3 in original composition	in clino-pyroxene	in ortho-pyroxene	in garnet
2	0.419	--	--	--	--
3	0.409	--	--	--	--
6	0.428	1.9	0.7	0.6	--
7	0.426	1.1	0.7	0.8	--
8	0.414	2.3	1.1	1.1	--
17	0.431	13.4	3.1	4.5	15.4[+]
18	0.431	7.1	2.8	3.7	15.0[+]

[+]Ca-component in garnet: in remaining experiments, garnet is absent

From these data, it follows that the lowest value of the Ca/(Ca + Mg) ratio, on average 0.414, is typical of the pyroxenes, obtained in two experiments in the absence of Al_2O_3 in the original composition. This figure is readily comparable to that of 0.40 for the clinopyroxene composition in the system $CaMgSi_2O_6-Mg_2-Si_2O_6$, at $\underline{P}$ = 30 kb and $\underline{T}$ = 1200°C (Davis and Boyd, 1966). With the appearance of Al_2O_3, the value of the Ca/(Ca + Mg) ratio increases a little, on average 0.423 for three experiments without garnet, and reaches 0.431 in equilibrium with garnet.

These data, along with the support of the earlier established dependence, significantly facilitate the application of the pyroxene geothermometer to the natural parageneses of deep-seated xenoliths, where Al_2O_3 is constantly recorded in the pyroxenes and garnet is typical. Allowing for these data, and on the position of the compositional points for clinopyroxene in equilibrium with garnet, we have plotted the provisional dashed line $\underline{a}$ in Figure 46, which evidently corresponds more closely to the diopside solvus with addition of Al_2O_3 for the equilibrium of two pyroxenes with garnet.

It should be noted that even before the appearance of direct experimental data on the system with Al_2O_3, the effect of such material was assumed by Boyd and taken into account more completely in a quantitative fashion by O'Hara (1967), who constructed a diagram on which the values of the Ca/(Ca + Mg) ratio and Al_2O_3 in the

pyroxenes were simultaneously plotted according to the temperature
and pressure. This diagram may be used for estimating the conditions
of formation of various peridotites, allowing for all the available
information on the composition of pyroxene, although, in our opinion,
the role of the FeO in the pyroxenes has been equated with that of
MgO on insufficient grounds.

In regard to the effect of FeO in clinopyroxene on the temperature of its crystallization in paragenesis with clinopyroxene, we
may take as an example, the results of an experimental study of a
sample of olivine nephelinite (Bultitude and Green, 1968),
investigated at $\underline{T}$ = 1200°C and $\underline{P}$ = 18 kb. The composition of the
pyroxene in this sample, determined on the microprobe, is markedly
different from that of the pyroxenes in ultramafic rocks, both in
amount of Al_2O_3 = 15.5%, and in iron index ($\underline{f}$ = 27%). Almost all
the aluminum here consists of Al^{IV}, since the amount of Na_2O is
extremely low (1.3%), the content of FeO is 8.2%. For this pyroxene,
Ca/(Ca + Mg) = 41%, which agrees well with data on the $CaMgSi_2O_6$-
$Mg_2Si_2O_6$ system. If we allow for FeO together with MgO, then Ca/
(Ca + Mg + Fe) = 34%, which is associated with the position of the
solubility curve for the ferriferous pyroxenes significantly below
that for the magnesian types.

Thus, for equilibrium with garnet, the position of the solvus is
shifted a little towards the side of increase in temperature. It
is evident that such a tendency must also be observed during further
increase in pressure up to 50-60 kb, at which diamond is stable.
Experimental data are not so far available for this field of pressure,
although by analogy with the relative position of the solvus at
pressures of 1 atm and 30 kb, we may assume its apparently insignificant further displacement in the field of higher temperatures.
Consequently, the position of the solvus of the system with garnet
may be recorded to the right of the dashed line $\underline{a}$ plotted by us on
the diagram in Figure 46 (hypothetical curve $\underline{b}$).

Isolated confirmed examples of the equilibrium of two pyroxenes
with garnet at particularly high pressures are the associations, analyzed
from diamonds in the 'Mir' pipe (Samples Av-34 and Av-14). In both
cases, the garnets and clinopyroxenes are characterized by very
similar compositions, only insignificantly distinguished by the amount
of chromium present. The value of the Ca/(Ca + Mg) ratio = 46%
suggests, according to Figure 46, that the equilibrium temperature
of this association is of the order of 1150-1200°C. This is probably
the lowest equilibrium temperature of crystallization for the field
of stability of the diamond. It is evident that the lherzolitic
association may be regarded as eutectic.

It should be specified that enstatite inclusions in diamonds are
of great rarity and the specific dunite-harzburgite association
(olivine + pyrope + chromite + enstatite) is most widely distributed
in the stability field of the diamond. Such an association may
be regarded as the most high-temperature type.

Turning to the characteristics of the equilibrium temperatures
of the extremely widespread xenoliths of the shallower, non-diamond,
zone of the mantle, we note that a significant contribution to the
study of pyroxene compositions from peridotite xenoliths and kimberlites has appeared in the work of Boyd (Boyd, 1969; Boyd and Nixon,
1970). A study of more than 60 chromium-bearing pyroxenes from
kimberlites has enabled us to define their compositions more precisely,
and to recognize the marked predominance of calcic diopsides (with

Ca/(Ca + Mg) = 44-50%) over the subcalcic types (Ca/(Ca + Mg) = 32-36%). The latter, according to Boyd, comprise in all only 10% of all the pyroxenes examined. But even these isolated samples, the first of which (E-3) was described in the Lesotho kimberlites as long ago as 1963 (Nixon et al., 1963), has provided grounds for assuming significant variations in equilibrium temperatures in deep-seated xenoliths (from 900 to 1400°C). However, inadequately representative material and, in particular, the exceptional rarity of the subcalcic pyroxenes, have led Boyd to the erroneous hypothesis of a possible immiscibility gap between the subcalcic and calcic pyroxenes. A detailed investigation of the pyroxenes in deep-seated xenoliths from the Yakutian pipes has shown that all the compositions form a continuous series and increase in temperature, apparently partially, may (see Fig. 29) be regarded as a function of depth. Such a conclusion has been reached on the basis of a study of the pyroxenes from the perido-tites in the 'Udachnaya' pipe (see §1), and has been emphasized by a number of other features of the composition of the minerals in these xenoliths.

In spite of the different value of the Ca/(Ca + Mg) ratio in the pyroxenes from the peridotites (from 32 to 50), the amount of the calcium component, and consequently the analogous Ca/(Ca + Mg) ratio in the associated garnet, remains approximately constant.

If the pyroxene compositions on the basis of the Ca/(Ca + Mg) ratio are only changed conjugately in·accordance with temperature for the associated garnets and pyroxenes of the peridotite paragenesis, then a conjugate change on the basis of the calcium index in garnet and pyroxene is observed in the eclogite (Bimineralic) association. The lower limit of the equilibrium temperature for the eclogitic pairs containing garnet with low calcium-index, compared with such a peridotitic association, may be estimated directly from the calcium index of the pyroxene. However, the eclogites in most cases are characterized by garnets with increased calcium-index. Consequently, the Ca/(Ca + Mg) ratio in the pyroxene is also changed towards an increase.

During the consideration of a series of eclogites of different composition, their temperature of equilibrium may only be estimated in the case when the entire series under consideration has been formed under similar conditions of temperature and pressure, and the effect of such components as Na_2O may also be assessed as slight for indiv-idual associations. During observation of these conditions, the compositions of the associated garnets and pyroxenes, based on the amount of calcium (Ca/(Ca + Mg)) must be located along the univariant line, which may be regarded as a vector reflecting the direction of conjugate change in the composition of the minerals, formed at similar temperatures.

In order to check this hypothesis, we found it most convenient to use the compositions of minerals from 16 samples of diamond-bearing bimineralic eclogites, of which 15 were chosen from the 'Mir' pipe (see §8). The variations in the calcium-index of the garnets from these xenoliths embrace almost the entire range of possible garnet compositions from eclogites. The presence of diamond in all the samples enables us to assign them to the most deep-seated types, possibly also formed in a relatively narrow temperature range.

For this selection, we calculated the correlation coefficient (ρ) between the calcium-index of the garnet and pyroxene, which is apparently extremely high (ρ = 0.90). Such a value for ρ demonstrates

the validity of the hypothesis on the narrow temperature range of
formation of the diamond-bearing eclogites. Calculation of the
equations for a straight-line regression makes it possible to plot
an average univariant line (Fig. 47), most precisely reflecting
the change in conjugate composition of the garnets and pyroxenes
in the selected series of eclogites. The intersection of the
composition of the pyroxenes from the peridotites with the temperature scale makes it possible to estimate the average temperature
of equilibrium crystallization of the diamond-bearing eclogites
at about 1150-1200°C.

The validity of the approach selected has also been checked
against a set, chosen from garnet-pyroxene pairs, taken from eclogites, which do not contain diamonds, from the most different kimberlite pipes, both Yakutian and African. In this case, it is logical
to assume that the various eclogites were formed over a significantly
greater temperature range. In fact, a check of the association
between the calcium-index of the garnet and pyroxene for this set
(19 pairs) has given a correlation coefficient of = 0.68, which
is significantly different from that of 0.90 for the pairs from
the diamond-bearing eclogites.

Thus, on the basis of the above considerations, we shall accept
the position and slope of the univariant line, which reflects the
change in conjugate compositions of minerals of eclogites as a
basis for constructing a diagram of the dependence of temperature of
equilibrium crystallization of eclogites and peridotites on the
calcium-index of the coexisting garnets and pyroxenes (see Fig. 47).
Parallel to the calculated line (for the diamond-bearing eclogites),
we have plotted lines with a temperature interval of 100°C.

When plotting the points of conjugate calcium-index for garnet
and clinopyroxene, we have started with the proven constancy of the
garnet composition in the paragenesis with two pyroxenes, considering
the main cause of the spread of the points for these garnets, on
the basis of calcium content, to be the additives of Cr_2O_3 and Na_2O
in the system. The accumulated data enable us to make a quantitative
estimate of the degree of this effect (see §1). In order to plot
the position of the projection of the line of the pyroxene solvus
on the temperature scale, we shall accept the value of the average
amount of the calcium component in the garnets from peridotites
(Sobolev, 1964), and consequently the average value of the Ca/(Ca +
Mg) ratio on the basis of these data.

The plotted points for the conjugate compositions of the minerals
from various eclogites in the kimberlite pipes, and also the garnet-
pyroxene pairs, associated with diamond in inclusions and intergrowths
(see Fig. 47), enable us to compare their equilibrium temperatures
and to estimate the extremely wide temperature variations, and also
to assign the diamond-bearing associations to the highest-temperature
types. The temperatures of these associations may be accepted as
even higher, if we assume a shift in the solvus curve for the pyroxenes
of the $CaSiO_3-MgSiO_3-Al_2O_3$ towards increase in temperatures during
further increase in pressure (see Fig. 46). The displacement of
the temperature scale, using such an assumption, evidently does
not exceed a limit of 50-100°C.

For the diamond-bearing eclogites and associations with diamond,
the temperatures, estimated with allowance for possible verification
for pressure (see Fig. 47), embrace the range of 1150-1300°C. It
must be emphasized that the typical pyrope peridotites and eclogites

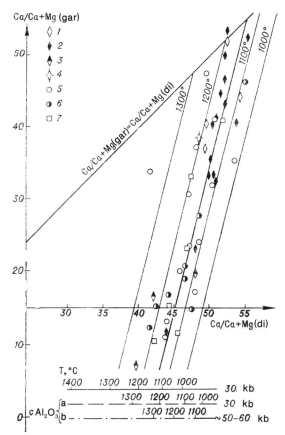

Fig. 47. Association between calcium index (Ca/(Ca + Mg)) of garnets (gar) and clinopyroxenes (di): 1) from diamonds; 2) from diamond-bearing eclogites; 3) from intergrowths with diamonds; 4) from diamond 57/a with inclusions of wehrlitic paragenesis; 5) from eclogite xenoliths; 6) from eclogites in 'Roberts Victor' pipe; 7) from garnet-pyroxene-chromite intergrowths from kimberlites, denoted by curves of diopside-enstatite solvus for P = 30 and 50-80 kb (see Fig. 46) Ca/(Ca + Mg) (di) = field of calcium-index of pyroxenes in the bipyroxene peridotites also.

of the kimberlite pipes, which do not contain diamonds, are also extremely high-temperature rocks: their equilibrium temperatures may be estimated as being within 1000-1350°C for the eclogites, and, as noted above, 900-1400°C for the peridotites. At the same time, in far from all cases, the increase in temperature may not be considered as a direct function of depth. A graphic support for this is certain samples of spinel peridotites from the 'Obnazhĕnnaya' pipe, for example, Sample 334/586 (Milashev et al., 1963). The relatively low pressure of formation of this rock is indicated by the presence of spinel and the increase amount of Al_2O_3 (5.69%) in the associated enstatite. The temperature of equilibrium of this sample, determined on the basis of Ca/(Ca + Mg) = 37%, reaches 1250°C. A similar temperature also characterizes eclogite 0-166

from this same pipe. There is no doubt that with further accumulation of data, based on parageneses of deep-seated xenoliths, the
range of defined equilibrium temperatures for rocks of different
depth may be extended even further, mainly towards the side of
increase.

The estimate of equilibrium temperature cited for the peridotites
and eclogites from the kimberlite pipes, on the basis of which the
pyrope geothermometer has been set up (Boyd and Schairer, 1964;
Davis and Boyd, 1966), significantly differs from those, based on a
study of the distribution of calcium between pyroxene and garnet,
by analogy with its distribution between garnet and plagioclase
(Perchuk, 1967, 1970). Although in the original treatment (Perchuk,
1967), the temperature values, obtained in this way, were also
distributed by the author over all the eclogite parageneses, in his
later work (1970) he pointed out that such a distribution coefficient
has a meaning only in those cases, when even a single sodium mineral
(amphibole or plagioclase) is present in the rock. This substantially
limits the possibility of the geothermometer, suggested by Perchuk,
and the figures presented by him for the eclogites without amphibole
and plagioclase, and even for the pyrope peridotites (Perchuk, 1967),
can scarcely be regarded as well-founded, arising from the limits
placed by the author himself (Perchuk, 1970). Consequently, the
conclusions about the possible low temperature values in the upper
mantle, of the order of 300°C (Perchuk, 1967) have little foundation
in fact, as also are those (about 650°C) in his later publications
(Perchuk, 1970).

§23. DISTRIBUTION OF Cr AND Al

Chromium is an extraordinarily characteristic element of the
ultramafic rocks. A. G. Betekhtin has been occupied in an
especially detailed study of its behavior in the minerals of ultramafic rocks, emphasizing that " . . . in the magmatic phase proper,
chromium as an element is not inclined to enter into silicates as
an isomorphous addition, but on the other hand, forms compounds of
the spinel type. At lower temperatures, the behavior of chromium
changes: it enters the silicates and its amount in them is frequently expressed in units of weight percent" (Betekhtin, 1937).
This problem also remains extremely apposite for the inclusions in
kimberlites, as has already been repeatedly recorded, but here it
stands on another plane. The spinel in these rocks passes to a
second arrangement or disappears completely, and chromium is concentrated in the silicates, that is, an increase in pressure in this
case is to a certain degree equivalent to a lowering in temperature
(Sobolev and Sobolev, 1967).

The results of a study of various pyrope-bearing ultramafic rocks
have enabled us to recognize two clear types of distribution of chromium in the rocks, with its overall similarity in the minerals (Fiala,
1965; Sobolev and Sobolev, 1967). The uneven nature of the distribution of chromium in the pyrope peridotites of the Czech Massif
and certain xenoliths from kimberlites, and also in the xenoliths
of complex composition, analogous to that described from the
'Obnazhënnaya' pipe (Sobolev and Kuznetsova, 1965), depends on various
causes. For the intrusive pyrope peridotites of Czechoslovakia,
where the primary-magmatic nature of the garnet is clear, the amount
of chromium, as shown by Fiala (1965), is inversely to that of aluminum

in the rock. Apparently, this cause is associated with the change in the amount of Cr and Al in the peridotites, in which garnet also has a magmatic origin, for example, in the 'Udachnaya' pipe. For a xenolith of complex composition, the amount of Al_2O_3 in the rock changes relatively little, but the amount of Cr_2O_3 is markedly altered, which is associated with the primary precipitation of spinels during crystallization differentiation.

Garnet-clinopyroxene

Even until quite recently, information on the pyropes, which contain a substantial amount of chromium, has been limited to isolated descriptions (Nixon et al., 1963; Fiala, 1965), and these garnets with Cr_2O_3, although also associated with clinopyroxenes, have been studied separately from them. Data on the distribution of Cr and Al in the garnets and clinopyroxenes have been limited to parageneses, the garnets of which contained not more than 10% of uvarovite.

A detailed study of deep-seated inclusions, especially in the porphyritic peridotites from the 'Udachnaya' pipe and unusual inter-growths in kimberlite, which, besides garnet, contain pyroxene and chromite, has provided data on the amount of Cr and Al in pairs of minerals with a chrome-index (Cr/(Cr + Al)) of over 30% in the garnets (Fig. 48). In addition, in order to determine the effect of pressure and temperature on the distribution of Cr and Al, there has been interest in the study of Sample T7/317 from the peridotites of the Czech Massif, with a garnet, containing 7.6% Cr_2O_3 (see Table 4), which was kindly presented to us by J. Fiala.

The first data from such investigations seemingly indicated an increase in the distribution coefficient (K_D) for Cr and Al in the high-chromium pairs (Sobolev, 1970). However, the new data have demonstrated that the values of K_D for pairs with variable chromium-index fall in the range of 0.3-0.5, and the distribution may be regarded as approaching the ideal state. It is likely that neither pressure nor temperature exerts any substantial influence on change in the value of K_D. This may be judged from a comparison of the relative nature of the distribution of Cr and Al for parageneses of known different pressures (and temperatures), namely, inclusions and intergrowths with diamonds, the porphyritic peridotites of the 'Udachnaya' pipe, and the peridotites of the metamorphic complexes.

At the same time, a clear tendency towards expansion of the field of chromium-bearing garnets with increase in pressure has been observed, which may be judged from the known limits of variation in the maximum amount of Cr_2O_3 for garnets from parageneses of different pressures. We may probably compare such kind of dis-tribution of Cr and Al with that of Fe and Mg in the associated garnets and cordierites, where the value of K_D is similar for parageneses with the most different iron index and conditions of formation, but the limiting amount of MgO depends on the pressure.

Garnet-chromite and Clinopyroxene-chromite

As noted above, the stability field of the spinels markedly con-tracts with increase in pressure, and there is simultaneous expansion of the field of chromium-bearing garnets. This pattern is clearly reflected in the nature of the distribution of Cr and Al between the garnets and spinels (see Fig. 48d). For the pyrope peridotites

230

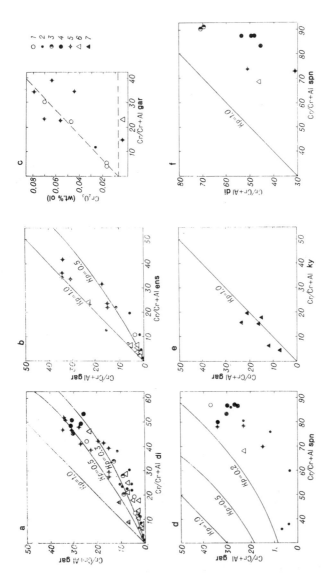

Fig. 48. Distribution features of Cr and Al in coexisting minerals: *a*) garnet-diopside; *b*) garnet-enstatite; *c*) olivine-garnet; *d*)garnet-spinel; *e*) garnet-kyanite; *f*) diopside-spinel. Points of conjugate chrome-index of coexisting minerals: 1) from diamonds; 2) from peridotites; 3) from intergrowths with diamonds; 4) from garnet-pyroxene-chromite intergrowths; 5) from porphyritic peridotites of the 'Udachnaya' pipe; 6) from pyrope peridotites of the metamorphic complexes; 7) from grospydites.

from the 'Udachnaya' pipe at increased pressures, and even more so
for the paragenesis with diamond, the garnet is associated with a
spinel having a very high $Cr/(Cr + Al)$ ratio of about 80%, and the
value of K_D for all the known pairs is less than 0.2. The distribution of Cr and Al in the clinopyroxene-spinel pairs (see Fig. 48f)
is almost of the same kind, although K_D for these pairs is significantly greater than for those with garnet. This is explained by
the greater chromium index of the pyroxene as compared with that
for the garnet.

Garnet-enstatite

The distribution of Cr and Al for the garnet-enstatite pair is
different (see Fig. 48b). In view of the very small amounts of
Cr_2O_3 and Al_2O_3 in the enstatites, significant variations in the
values of $Cr/(Cr + Al)$ are possible, resulting from analytical
errors. More reliable results have been obtained by using the
microprobe. Variations in the value of K_D are quite significant
(from 0.5 to 1.0). However, in spite of the presence of these
fluctuations, this distribution enables us quite precisely to estimate
the chromium-index of the garnet from the value of $Cr/(Cr + Al)$ of
the associated enstatite, which is especially important during the
reconstruction of the parageneses of indidIual inclusions in diamonds.

Garnet-kyanite

The paragenesis of chromium-bearing garnets with kyanite is a
particularly rare case. We have already noted the uneven nature
of the distribution of chromium in the kyanites in some samples,
and even in individual grains (Sobolev, Zyuzin and Kuznetsova, 1966;
Sobolev and Sobolev, 1967; Sobolev et al., 1968), and the closeness
of the value of K_D to 1.0. The new information emphasizes these
conclusions (see Fig. 48e). Consequently, not only the chromium-
index of the kyanite may be regarded as a function of pressure,
which is supported by experiments (Seifert and Langer, 1970) carried
out after the discovery of these kyanites and the proposed hypothesis
on limited miscibility of Cr and Al in them (Sobolev et al., 1968).
Garnet, known to be associated with kyanite, may probably serve as an
analogous indicator.

Garnet-olivine

The most interesting features here include the appearance of
chromium (Cr^{2+}) in the olivines of the chromium-bearing rocks (see
ᴙ15). This is associated, as has been clearly shown for the lunar
basalts, with a significant decrease in oxygen fugacity.

A positive correlation has been demonstrated in the lunar rocks
between the amount of chromium in the olivines and in the rocks,
indicating similar conditions of crystallization. In this respect,
a very interesting pattern has also been observed in the minerals
enclosed in diamond. The clear association between the amount of
chromium in the olivines and the chromium-index of the associated
pyropes from inclusions in diamonds of different deposits (see
Fig. 48c) enables us not only to stress the common pattern of
conditions (activity of oxygen), but also to reconstruct the nature
of the paragenesis of the small number of inclusions of olivine in

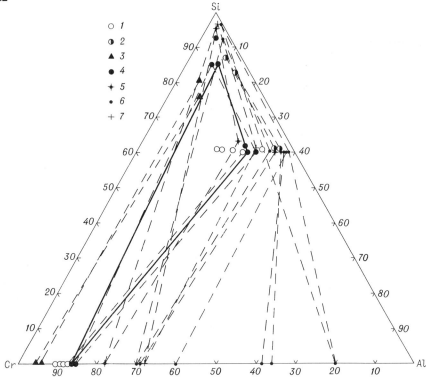

Fig. 49. Association between compositions of spinels, garnets, and clinopyroxenes in deep-seated inclusions: 1) from diamonds; 2) from intergrowths with diamonds; 3) from intergrowths with diamonds (clinopyroxene-chromite paragenesis); 4) from garnet-pyroxene-chromite intergrowths; 5) from peridotite xenoliths in the 'Udachnaya' pipe; 6) from peridotite xenoliths; 7) from peridotites of the Czech Massif.

diamonds with a low content of chromium, with low-chrome pyropes. A substantially more diffuse picture is typical of the relationships of the amount of chromium in the pairs from the peridotites from the 'Udachnaya' pipe, which is possibly associated with some variations in oxygen fugacity.

In conclusion, let us dwell on some common features of the association of compositions of the chromium-bearing associated minerals from deep-seated parageneses: spinels, garnets, and clinopyroxenes (Fig. 49). In this figure, a displacement in the garnet-spinel and clinopyroxene-spinel connodes has been clearly defined towards the side of the chrome-spinels for the parageneses of higher pressures, especially the peridotite xenoliths and inclusions in diamonds. A significant increase in the chromium-index of garnet, along with a contraction in the spinel field has been recorded for the bimineralic associations, and here, garnets are perhaps possible, only a little less rich in chromium than in those identified, which contain about 50% of knorringite.

The compositions of the coexisting rocks will be projected into the upper portion of the triangle, and the Cr/(Cr + Al) ratio will change considerably mainly as a result of the low and variable amount of Al_2O_3. At lower pressures, the compositions, relatively rich in chromium, fall to the left of the displaced phase triangle, that is, into the field of the spinel-pyroxene rocks, and only the rocks with a low Cr/(Cr + Al) ratio are moved into the field of pyroxene-garnet rocks and into the phase triangle pyroxene-spinel-garnet. At higher pressures, as a result of displacement of the phase triangle, the field of trimineralic and garnet-pyroxene parageneses expands. This picture is somewhat complicated owing to the effect of Na_2O. Increase in the amount of sodium in the clino-pyroxene, defined in Figure 49 by an increase in the amount of Al + Cr, leads to breakdown of the chromium-bearing garnet (analogous to the breakdown of the Mg-Ca-garnet in the grospydites) with the formation of a paragenesis of pyroxene, rich in sodium and chromium, with chromite. The reaction has the form

$$3Mg_3Cr_2Si_3O_{12} + Na_2O = 2NaCrSi_2O_6 + 2MgCr_2O_4 + 2Mg_2SiO_4 + 3MgSiO_3$$

The position of the mineral compositions of the coexisting parageneses is shown on the basis of actual examples (see Fig. 49).

CHAPTER VII

THE PROBLEM OF THE COMPOSITION OF THE UPPER MANTLE

§24. THE MINERAL COMPOSITION AND FACIES OF
METAMORPHISM OF THE UPPER MANTLE

Of all the data on the structure and composition of the subcrustal
regions of the Earth, the greatest interest for petrology centers
round the data obtained by direct methods of investigation of subcrustal matter. The existence of fragments of this matter, brought
to the surface in the form of xenoliths in volcanic structures,
now gives rise to almost no doubts. Representatives from the
greatest depths are the xenoliths in kimberlite pipes, and it is
not by chance that the first discoveries of these exotic formations,
especially a xenolith of a diamond-bearing eclotite (Bonney, 1899),
have served as the focal point for the development of an hypothesis
on the eclogite layer of the Earth (Fermor, 1913).

Data on the composition of deep-seated xenoliths from both the African
and Yakutian pipes indicate the predominance of ultramafic material.
Amongst this material, the most widely distributed are the pyrope
lherzolites, the monotonous composition of which has been the cause
of their assignment in some cases to the primary material of the
mantle (O'Hara, 1970). Similar lherzolites, only spinel-bearing,
predominate in the higher horizons of the mantle. Along with the pyrope
lherzolites, there is widespread distribution of websterites and pyroxenites, often interlayered with lherzolites. The pyroxenites,
coexisting with the lherzolites, which is confirmed by the presence
of composite xenoliths are already rocks of the eclogite type and
form a transition to typical eclogites. The latter, in turn, are characterized by exceptional variation in composition, up to rocks
extraordinarily rich in aluminum.

Thus, investigations of the xenoliths demonstrate that the substrate from which they were extracted, has a very complex composition
being a mixture of peridotites and eclogites, with the former
predominating in quantity. These very compositional features of the
xenoliths have served as a basis for creating the pyrolite model
of the upper mantle (Ringwood, 1958), which assumes that the undifferentiated upper mantle must consist of a mixture of peridotite
and eclogite (basalt). Several variants of such a mixture have
been proposed, beginning with four parts of peridotite and one of
basalt (Ringwood, 1962) and ending with three parts of peridotite
and one of basalt (Ringwood, 1966). In later works (Ringwood,
1970), a wider tolerance for the compositional limits of hypothetical
pyrolite (within the basalt peridotite ratio of 1:1 to 1:4) has
been assumed, and wide variations in the composition of the pyrolite
in the upper mantle have been suggested.

However, the results from a study of xenoliths, formed both within
the stability field of graphite and that of diamond, strongly suggest
that, in all known cases, the material of the mantle has been
differentiated. This is convincingly indicated by data from the
diamond-bearing xenoliths, the composition of which varies from
dunite-harzburgites to corundum eclogites, and also the practically
analogous results from an investigation of the mineral associations
from inclusions in diamonds. If we take only the ultramafic compositions, then with depth (comparing diamond and non-diamond compositions)

234

there is a substantial increase in density in the very aluminum-poor
lherzolites, which contain chromium-rich pyropes, and the harzburgite-
dunites with pyrope. The variety of eclogite compositions, containing
diamonds, also indicate significant differentiation of the upper
mantle, lying below the boundary of diamond stability. These data
are well correlated with modern ideas of geophysicists on the complex
heterogeneous nature of the upper mantle (Subbotin et al., 1968).

Thus, it may be stressed that there are no features of primary
undifferentiated material of the mantle below the platforms, at
least to depths of 200 km. This does not agree with the variant of
the upper-mantle model (Ringwood, 1969), according to which undifferentiated material of the mantle occurs below the continents,
sometimes at a depth of 100-150 km. This hypothesis is based not
only on a study of the minerals, associated with diamond (in particular, from the eclogite association) from one region (Yakutia), but
also from other areas where diamond deposits are known: Africa,
Venezuela, and the Urals.

It is noteworthy that, in general, the relationship between the
minerals of the peridotite and the eclogite association, based on
inclusions in diamond, suggest for the xenoliths, the predominance
of ultramafic material. Thus, Yefimova's calculations (1961)
demonstrate that the orange garnets of the Yakutian deposits comprise
only 2.4% of the total number of garnet inclusions in the diamonds.
These figures for the inclusions are scarcely significantly altered
with more detailed counts, although it must be allowed that the
proportion of eclogitic material, belonging directly to the mantle,
must be significantly greater than the stated estimate, since a
substantial portion with ultramafic paragenesis of the inclusions,
undoubtedly crystallized in the depth of the kimberlite focus.

The composition of the diamond-bearing eclogites and minerals
of eclogite paragenesis for each actual region may serve as a special
indicator of the composition, and to a certain degree, the extent
of differentiation of the deep-seated parts of the upper mantle.
The possibility of such consideration is associated with the fact
that the diamond-bearing eclogites in a number of deposits, for
example, in Yakutia, are extremely rich in Na_2O, judging by the
pyroxene compositions. An analogous feature has also been identified
for the diamond-bearing kyanite eclogites in the 'Roberts Victor'
pipe (Switzer and Melson, 1969). This, together with the variation
in the composition of the eclogites themselves, may indicate a
significantly greater degree of differentiation of the portion of
the upper mantle, lying below the boundary of diamond stability,
support for which is found in the significant role of harzburgites
and/or dunites here.

Along with the predominance of the peridotite association over
the eclogite association in Yakutia and in the African deposits, it
may be considered as convincingly established that the Uralian diamonds
(Sobolev et al., 1971) show a marked predominance of the eclogitic
association over the peridotitic association, based on the inclusions in the diamonds. Such an anomaly may be explained either by
the great persistence of the diamonds carried in the eclogite xenoliths, or (as is more likely) by a different proportion of eclogitic
and peridotitic material in the mantle below the Urals (in the region
of unexposed solid-rock deposits). The latter hypothesis is real,
since according to the proportion of xenoliths, individual cases
are known of marked predominance of eclogites over peridotites both

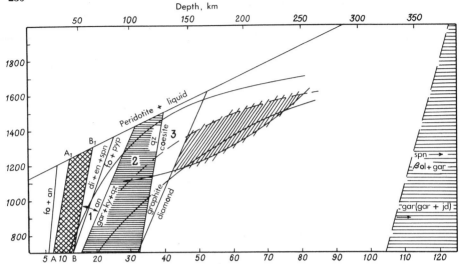

Fig. 50. Diagram of the facies of metamorphism of the upper mantle:
AA_1B_1B) range of eclogitization for basalts of different composition;
field of crystallization of natural diamonds indicated by cross-
hatching; the position of the field of phase transition of olivine
(Ringwood & Major, 1966) and formation of garnet with deficiency of
aluminum (Ringwood, 1967) by wide-spaced horizontal lines in the
right-hand portion of the diagram; 0) geothermal gradient for
oceanic regions; P_{cm}) geothermal gradient for the shields (Clark &
Ringwood, 1964); assumed geothermal gradient during period of crystallization of diamonds indicated by dashed line; assumed solidus
with allowance for influence of Na_2O and 2-3% H_2O indicated by lower
continuous line (1 - subfacies of garnetized peridotites, 2 - grospydite subfacies, 3 - subfacies of coesite eclogites).

in Africa (the 'Roberts Victor' pipe) and in Yakutia (the 'Zagadochnaya'
pipe) which may be a reflection of horizontal variations in the upper
mantle at the appropriate depths.

In establishing the compositional features of the deep-seated
xenoliths, brought up from the upper mantle, we shall in the present
work consider individual xenoliths and parageneses, in equilibrium
with reliably identified diamond (i.e., the deepest types), and formations which do not contain diamond, but perhaps containing graphite.
Accordingly, we may recognize two fundamental facies of garnet-
bearing rocks of the upper mantle: 1) graphite-pyrope (C); and 2)
diamond-pyrope (D).

Within the shallower facies (graphite-pyrope), it appears possible,
on the basis of experimental data and features of actual parageneses,
to recognize three subfacies: 1) garnetized peridotite; 2) grospydite;
and 3) coesite eclogite (see §6). The first subfacies consists
mainly of xenoliths with clearly defined reaction formation of the
garnet from spinel and pyroxenes. This may include most of the
rocks found in the 'Obnazhennaya' pipe and a number of others. The
boundary of the grospydite subfacies is defined by the breakdown
reaction of anorthite into grossular + kyanite + quartz (Boettcher,
1970) (Fig. 50). Recognition of the last subfacies is to a signifi-

cant degree provisional, since coesite eclogites have not been found in the mantle. This subfacies consists of hyperbasites only (the specific pseudoporphyritic peridotites of the 'Udachnaya' pipe, for which, based on a number of features (a constant amount of K_2O in the pyroxenes, and a very large amount of Cr_2O_3 in the garnets) we may assume a pressure of more then 30 kb (see §1). The absence of diamond in these rocks restricts the stability field of these peridotites on the basis of pressure.

When assigning the xenoliths to a particular facies it is necessary to take account of the probabiliby of the random absence of diamond in the xenoliths assigned to the deeper, diamond-bearing facies. In this case, valuable additional information about the depth of the xenoliths may be obtained arising from the amounts of certain trace-elements in the minerals--sodium in the garnets (Sobolev and Lavrent'ev, 1971) and potassium in the pyroxenes, the increased amounts of which are associated with increase in pressure (see §§12 and 13).

In recognizing the facies of diamond-bearing pyrope peridotites and eclogites (diamond-pyrope), we may in the same way accurately recognize the association of minerals, which crystallized below the isobaric surface of diamond stability (Leipunsky, 1939), that is, at depths of not less than 120 km. It is significantly more difficult to assess the complete volume of material, required to define this facies, that is, concerning the maximum depth of the xenoliths, known in the kimberlite pipes. Experimental data, available at present, cannot help to establish the complete volume of the facies, beacuse in the 50-100 kb range, the features of phase transitions in the minerals, corresponding in composition to those of the peridotites and eclogites, are lacking. An exception is the discovery of a phase transition of TiO_2 in the pressure range of 40-120 kb (Bendeliani et al., 1966). However, the comparatively low temperatures of transition do not permit us to use it as a geologic barometer. It is also impossible to use the transition of coesite into a rutile-like modification of silica (Stihov and Popova, 1961) both as a result of the absence of free silica in the rocks of the upper mantle, and also as a result of the very high pressures, at which the transition is achieved. In spite of this, several features of the mineral composition may be used for this purpose. The most promising in this respect is the dependence of the entry of chromium into the silicate structure (in polymineralic parageneses), and also of the disseminated amounts of sodium in garnets and of potassium in pyroxenes, on pressure. There are already several experimental data for the latter. The demonstrated absence of potassium in the pyroxenes, which have crystallized in the potassium-bearing system at pressures of up to 30 kb (Erlank and Kushiro, 1970), enables us to assign the pyroxenes with an accurately established amount of K_2O to higher-pressure formations. This is emphasized by the identification of K_2O (up to 0.3%) in the pyroxenes, which have been synthesized in the pressure range of 40-100 kb (Shimizu, 1971).

There is interest in establishing the temperatures for a series of actual parageneses with diamond, in particular the lherzolite paragenesis (1150-1200°C), probably close to the lower limit of the temperatures of formation of the diamond. At the same time, the predominance of high-temperature associations (harzburgite-dunite) of the order of 1300-1400°C has been established, and it may be greater. These data enable us to discern the PT field of crystallization. Some assumptions, which it is necessary to make, conform with the current ideas of the role of H_2O in the mantle (see §25). We have recognized the field of crystallization of diamonds and diamond-bearing parageneses (see Fig. 50), limited by the assumed line of the solidus with P_{H_2O} P_{total}, allowing for the effect of Na_2O, and also the assumed line for the geothermal gradient during the period of diamond formation. The entire field indicated, which has the form of a lens on the diagram, also extends beyond the limits of diamond stability, and may probably characterize a zone of partial melting in the mantle, corresponding to the low-velocity layer (Gutenberg, 1953; Magnitsky, 1968; Subbotin et al., 1968; Magnitsky, 1971; Lambert and Wyllie, 1970; Wyllie, 1970; Anderson and Sammis, 1970).

The significant differentiation of the upper-mantle material, available for investigation through xenoliths, is emphasized not only by the data from examination of the xenoliths in the kimberlite pipes, but also by information from xenoliths in alkaline basaltoids, where both lherzolites, and also websterites and pyroxenites have been recorded (Denisov, 1970; Yamaguchi, 1964; Jackson, 1968; Kuno, 1969; Kutolin, 1970). Moreover, samples of complex composition with coexisting garnetized websterite and spinel lherzolite (Reid and Frey, 1971) have been identified amongst the xenoliths of the Hawaiian volcanoes that are analogous to those of the kimberlite xenoliths. We cannot concur with the interpretation of these authors, that the spinel lherzolite has been formed as a residual after melting-out of the basaltic liquid from a pyroxenite. Such forma-tions are similar to the xenoliths of the 'Obnazhennaya' pipe, and define, in all probability, that same narrow range of conditions, when eclogitization of rocks, richer in silica (websterite), outstrips the initiation of garnetization of the spinel lherzolites (see Fig. 50).

We emphasize once more that the degree of differentiation of the mantle, probably increases with depth, as may be judged from a comparison of the series of xenoliths in the alkaline basaltoids, and also the garnet-bearing rocks with graphite or with diamond. Judging by the nature of the composite xenoliths which have banded gneiss-like structures, the differentiated upper mantle evidently has a stratified nature with a predominance of peridotites, with individual seams and lenses of eclogites (Fig. 51).

At present, there are direct data, permitting an approximate estimate of the age of the differentiation of the upper mantle. These are results of the absolute-age determination of lherzolite xenoliths from basalts in Victoria (Australia) based on lead-isotope ratios (2000-2500 m.y.) (Kleeman and Cooper, 1970), and also dating by approximate methods of a zircon from a pyrope peridotite from the basaltoid pipes of the Minusinsk Depression (1700 m.y.) (Kryukov, 1968). Similar figures also have been obtained for five xenoliths from the 'Roberts Victor' pipe (Manton and Tatsumoto, 1969) with variations between 1700 and 2400 m.y. Results of dating by other methods, especially the K-Ar method, also indicate a marked age difference between a xenogenic phlogopite from kimberlites (about 1000 m.y.) and one from the groundmass of kimberlites, the age of which approximately corresponds to the age of injection of kimber-lites of an actual pipe (Sarsadskikh et al., 1966). A Precambrian age has been established in a similar way for a xenolith of pyrope websterite (about 700 m.y.) from the 'Obnazhennaya' pipe (Firsov

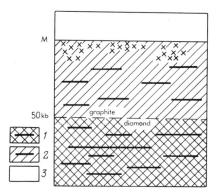

Fig. 51. Diagram of construction of upper mantle below the Siberian Platform. 1) rocks of facies of diamond-bearing peridotites and eclogites with minor eclogites; 2) rocks of facies of pyrope peridotites and eclogites with graphite; 3) zone enriched in phlogopite (V.S. Sobolev, 1972). *II*) Mohorovicic surface.

and Sobolev, 1964), for which the discovery of a belemnite (Milashev and Shul'gina, 1959) has demonstrated a Cretaceous age. These so far limited data indicate that differentiation of the mantle occurred during the early Precambrian and evidently, during quite a narrow time interval.

In conclusion, let us dwell on the features of the material composition of the Mohorovicic boundary, about which there are two principal hypotheses; it is the boundary between the basic and ultramafic rocks (Vinogradov, 1959), or between basic rocks and eclogites with an eclogite layer of significant thickness (Lovering, 1958; Belousov, 1960). Hypotheses have also been put forward about the coexistence of both variants (V. S. Sobolev, 1962, 1964a; Sobolev and Sobolev, 1964; Wyllie, 1963). Objections have also been raised to this interpretation of the Moho boundary (Afanas'ev, 1968).

Information from deep seated inclusions in kimberlites and basalts indicate that the Moho boundary may lie in the field of different facies for the continents and oceans. For the former, it lies in the field of the spinel peridotites on their boundary with eclogitic rocks, that is, it may belong to the granulite facies (Sobolev and Sobolev, 1964). For the latter, it corresponds to the amphibolite facies, and where the upper part of the mantle consists of eclogites, there must be a layer of transitional rocks with increased wave velocities, which have been revealed in a number of places, above the boundary (Belousov, 1968). This latter requires further checking with the aid of geophysical methods.

§25. THE ROLE OF H_2O IN THE UPPER MANTLE

At the present time, there is considerable discussion about the behavior and role of water in the upper mantle, in which respect each new identification of water-bearing minerals in definite deep-seated parageneses arouses great interest. Of particular importance here are the parageneses with diamond, being the deepest known. The undoubted involvement of water in magma is indicated by the presence of phlogopite in kimberlites, and there is interest in

identifying its stability field in the mantle. Experimental investigations have shown that phlogopite may be stable up to pressures
of 66 kb (Markov et al., 1966; Kushiro et al., 1967). A special
search for phlogopite inclusions in diamonds has so far not met
with success, and it will probably be difficult to expect such
discoveries in future, because if they were discovered in the
stability field of the diamond, the phlogopite would belong to the
most widely distributed inclusions in the diamond. Phlogopite
has been accurately identified in intergrowths with diamond in
only two cases. In the first (Boyd and Nixon, 1970), it has been
recorded in an intergrowth with diamond along with chrome-diopside,
without any details of the characteristics. In the second case,
it occurs in an intergrowth with a polycrystalline aggregate diamond
(MR-9) from the 'Mir' pipe (see Table 34), and a clear association
between the phlogopite and later chromite, markedly distinguished
in composition from the chromite syngenetic with diamond (see §11)
in this same sample, has definitely been demonstrated.

Some authors regard the phlogopite in the peridotite xenoliths,
associated with unaltered olivine, as primary (Dawson and Powell,
1969), which is not, however, supported by data from numerous
xenoliths. Without excluding the possibility of such an equilibrium
paragenesis in the uppermost portions of the mantle, we regard the
rare cases of phlogopite, present in peridotite xenoliths, as a
result of a later phlogopitization. Cases are also known of phlogopite in eclogite xenoliths, but usually in the peripheral parts. Most
probably, phlogopite may be regarded as a mineral, typical of the
uppermost portion of the upper mantle (Kushiro, 1969), where it
coexists with such water-bearing minerals as the amphiboles (Oxburgh,
1964).

Besides phlogopite and the amphiboles, there are other minerals
of interest, which are possible concentrators of water in the upper
mantle. Of special interest here is titanoclinohumite, reliably
identified in the form of inclusions in pyrope in association with
olivine and ilmenite in a kimberlite dike at Moses Rock (Arizona,
USA) (McGetchin and Silver, 1970). Although this mineral has not been
found in xenoliths, its paragenesis with ilmenite, makes its discovery
in diamonds unlikely. It has been found in the Yakutian kimberlites
in the form of individual xenocrysts (Voskresenskaya et al., 1965;
Shchelchkova and Brovkin, 1969). Its paragenesis here is unknown,
but it may most likely be found in association with ilmenite and
pyrope. A single described occurrence of this mineral in unusual
peridotites in crustal conditions in Kazakhstan (Yefimov, 1961),
diagnosed in error as titanolivine, is of interest. It is associated
in these peridotites with pyrope ($\underline{f}$ = 30%).

In every case, the role of titanoclinohumite as a concentrator of
water in the mantle can scarcely be significant at great depths,
since the adequately representative sampling of the composition
of the mantle through xenoliths and inclusions in diamond, indicate
its limited distribution.

Hypotheses have also been proposed on the content of water in
normal silicates of the garnet and pyroxene type, which form the
rocks of the mantle. In these phases, the substitution $(OH)^4$ $(SiO)^{4-}$
similar to the hydrogarnets (Fyfe, 1970; Sclar, 1970), has been
proposed. A reflection of such substitution may be the deficiency
of Si in minerals, analyzed with the aid of the microprobe. Relying
on the adequately representative data on the composition of the

pyroxenes, associated with diamonds, we may affirm that the features of such a substitution are lacking. Consequently, we may confidently speak of the absence of hydrosilicates in that field where typical water-bearing minerals are absent (in the stability field of diamond). Of course, in this case it is impossible to deny the possibility of the presence of water in the pyroxenes in amounts of the order of 0.1-0.5%, which is sometimes recorded in chemical analyses, but the role of the substitution indicated is probably very insignificant, which is well-defined in the compositions of the particularly deep-seated clinopyroxenes, containing almost no Al^{IV}.

Thus, water is probably present in the mantle in the dissolved from in the interstitial melt and in the fluid in equilibrium with the melt, significantly lowering the solidus temperature of the mantle rocks (Kadik and Khitarov, 1970; Khitarov et al., 1968; Lambert and Wyllie, 1970; Wyllie, 1970, 1971; Kushiro and Aoki, 1968).

§26. THE ROLE OF Na_2O, K_2O, AND Cr_2O_3 IN THE UPPER MANTLE

The results of investigations of xenoliths indicate the significant role of Na_2O in the deep-seated clinopyroxenes, which are the principal concentrators of sodium. The presence of pyroxenes in the diamond-bearing eclogites and in diamonds, containing 7-9 and even up to 11% Na_2O (which exceeds 70% jadeite component), demonstrates that some rocks of the upper mantle may contain more than 5% Na_2O. Data on the composition of the pyroxenes from eclogites (see Table 42) indicate that the average eclogite of the mantle contains about 2.5% Na_2O. Such information enables us to reconsider some individual conclusions, reached during experiments on the melting of mantle eclogites in a dry system (O'Hara and Yoder, 1967), in particular, concerning the higher temperature of the commencement of melting of the mantle eclogites (1550°C) as compared with that of the pyrope lherzolites (1450°C), since the pyroxene of the eclogite, used in these experiments, contains half the amount of Na_2O as compared with the average pyroxene (and consequently, the eclogite). As may be judged from a comparison with the P and T values, obtained during the melting of an alkaline olivine basalt (Ringwood and Green, 1966), the temperature of melting of the 'average' mantle eclogite must be approximately 150-250°C lower, giving, as a result of the ratio of melting temperatures of eclogites and peridotites, the inverse of those presented by O'Hara and Yoder (1967).

During the description of the mineralogy of the diamond-bearing eclogites (Sobolev and Kuznetsova, 1966), we suggested that some types of basalts may be obtained not by partial but by complete melting of rocks of the diamond-bearing eclogite type in the mantle. This applies not only to the diamond-bearing types, but also to the eclogites of the mantle in general.

Under especially high-pressure conditions in the upper mantle, Na_2O is probably redistributed from the pyroxenes into the garnets, which may be assessed from the reliable data on the amount of Na_2O up to 0.22% in the garnets (Sobolev and Lavrent'ev, 1971). During the formation of garnet with a deficiency of aluminum, a small amount of pure jadeite will be found in equilibrium with it (Ringwood, 1970).

In regard to the possible mineral-concentrators of K_2O in the mantle, the most common ones (phlogopite and amphibole) are concentrated only in the uppermost parts of the mantle (see §25). The very rare amphibole, richterite, with a significant amount of K_2O (about 5%) is clearly associated with phlogopite and diopside in the unusual xenoliths from kimberlite pipes (Erlank and Finger, 1970).

The results of our investigations, illustrated in §13, definitely indicate the possible entry of K_2O into the clinopyroxenes under especially deep-seated conditions, up to an amount of 0.27-0.3% in completely fresh pyroxene, enclosed in diamond. However, it is noteworthy that in certain pyroxenes from diamonds, K_2O is absent completely, which cannot be associated with the composition features of the pyroxene itself, but indicates the heterogeneity of the upper mantle with respect to the content of K_2O.

There is particular interest in the extraction of K_2O from the pyroxene in diamond-bearing eclogite, which has been demonstrated by the content of K_2O in the pyroxene from diamond in this eclogite, which is four times greater than that in the pyroxene of the rock itself (see §13), along with a similar amount of all the remaining elements. Such extraction may occur only after capture of the inclusions by the diamond, which grew in a melt of eclogitic composition and protected the inclusion from further alterations.

Data on the amount of K_2O in the pyroxenes depend in turn on experiments based on the verification of the solubility of potassium in the pyroxenes (depending on pressure) for pressures above 30 kb, because in the range up to 30 kb there is already a negative result (Erlank and Kushiro, 1970). For pressures of 40-100 kb, confirmation of the existance of such a relationship has been obtained (Shimizu, 1971).

It is possible that the data on the solubility of potassium in the pyroxenes, as for sodium in the garnets, in dependence on pressure, may be used for defining more precisely the position of the field of crystallization of natural diamonds on the basis of pressure, and consequently, for defining the boundaries of depth of the material of the mantle, carried to the surface by the kimberlites. The stability field of the sodium-potassium jadeites probably belongs to the deep-seated portions of the upper mantle, and the pyroxenes, which contain large amounts of potassium, may be synthesized apparently at pressures of more than 200 kb.

Let us dwell briefly on the possible role of the chromium-bearing minerals in the upper mantle. Since increase in pressure favors redistribution of chromium from the spinels into the silicates (Sobolev and Sobolev, 1967), we may expect magnesian garnets, especially rich in chromium, in rocks with a deficiency in aluminum, that is, with an increased $Cr/(Cr + Al)$ ratio, at great depths in the mantle. At the maximum depths for the nucleation of the kimberlite foci, which we have estimated at approximately 200 km, the limiting amount of the magnesium-chromium component (knorringite) in the garnets is about 50 mol % (Sobolev, Lavrent'ev et al., 1973a). In compositions, similar to those of dunites, which contain about 0.5 wt % Cr_2O_3 (as has been suggested in the composition of the average peridotite of the upper mantle in most of the existing models), at depths of 300 km and more (P more than 100 kb) up to 3% garnet may be present, approximating in composition to a pure knorringite. Enstatite and chromite will be unstable under these conditions.

In the unusual chromium-bearing eclogites, such as those identified in the 'Zagadochnaya' pipe, a chromium-rich kyanite may be present. In the stability field of the diamond, kyanite will dissolve up to

50 mol % of chromium component. Under deeper-seated conditions, at pressures over 100 kb, the chromium analog of kyanite may be stable.

§27. THE REGIME OF OXYGEN IN THE UPPER MANTLE

In connection with the overall conditions of mineral formation in the upper mantle, there is interest in clarifying the features of the regime of oxygen under particularly deep-seated conditions. An estimate of the value of the partial pressure of oxygen is fundamentally possible, during the study of minerals containing iron, and the compositions most sensitive to change in oxygen potential are those of the complex oxides, the spinels and ilmenites.

Experiments have established the relationship between the degree of oxidation of the equilibrium Fe-Ti oxides and the interrelated values of the oxygen potential and the temperature (Lindsley, 1963). It must be noted, however, that the possibility of such an estimate is associated with the need for the number of assumptions, especially the level of influence of amounts of Mg, Cr, and Al. Because Cr and Al are present in the form of the spinel component, it is suggested that they be allowed for as 'ulvöspinel' (Lindsley, 1963).

The estimate of the value of the partial pressure of oxygen under high-pressure conditions in the mantle may be sufficiently accurate only when exemplified by mineral associations, containing free carbon in the form of diamond. Theoretical calculations (Korzhinsky, 1940; French, 1966; Dobretsov et al., 1970) have shown that an increase in pressure must favor the oxidation of the silicates in paragenesis with free carbon and a gas phase mainly of H_2O-CO_2 composition, that is, it will contribute to an increase in P_{O_2}. A similar point of view on the regime of oxygen during the crystallization of kimberlite has been expressed by O'Hara (O'Hara and Mercy, 1963). An increase in temperature must operate in the opposite direction, and the dependence on temperature is expressed much more intensely. However, this applies only to the carbon-bearing rocks, for which so far there are few data.

New data on the composition of the chromites, associated with diamonds (Meyer and Boyd, 1972; Sobolev et al., 1971), both in the form of inclusions (see Table 30), and in intergrowths with diamonds (see Table 34), indicate clearly defined features, especially the very low content of Fe^{3+}, that is, a lower P_{O_2} value. This feature may be regarded as reliably established, since amongst about 50 analyzed chromites, associated with diamonds from various regions (see Fig. 40), there is not one exception to this rule. At the same time, for the chromites from the kimberlites, a trend has been clearly established towards an increase in the amount of the magnetite component from 3-5% in association with diamonds, up to 26% in parageneses, without carbon with all intermediate compositions, and in these chromites from intergrowths with chrome-rich garnets and clinopyroxenes (see §2), the overall Cr/(Cr + Al) ratio remains constant in all the samples. Judging by the paragenesis of Cr-garnet + clinopyroxene, identical to that in diamonds, these intergrowths are formed, on the basis of pressure, under similar conditions with diamonds, but with lowering of temperature. Such a feature has also been identified in Sample MR-9, where chromite of two generations is present in intergrowth with a polycrystalline aggregate of diamond (see Table 34), and the late chromite, more oxidized and enriched in titanium, is developed in the form of very small segregations (5–10 microns) in clear association with the phlogopite of this intergrowth, which is also later. This indicates that diamond is preserved as the temperature falls and the partial pressure of oxygen increases.

This evolution may have been associated with two factors: lowering of the crystallization temperature, as is seen for the series of chromites, extending from the stability boundary of carbon, that is, the lowering limit. The absence of combustion of diamond is no proof of its stability in this PT field, because the precipitated carbon is extremely inert and is readily preserved without change. It is impossible, however, to exclude the probability that the formation of diamond has already taken place on the boundary of the stability field of CO_2 in the upper mantle, that is, CO_2 will not be in equilibrium with the fluid, but rather hydrocarbons, as must be the case in the deeper portions of the upper mantle (Dobretsov et al., 1970).

The hypothesis concerning the lowering of the partial pressure of oxygen in the stability field of diamond is also favored by the composition of the ilmenites from intergrowths with diamonds, which are almost devoid of Fe^{3+} (see §10). And finally, an additional confirmation is the stable amount of Cr_2O_3 in the olivines from diamonds and diamond-bearing kimberlites (in the form of Cr^{2+}), which reflects reducing conditions (Haggerty et al., 1970).

These features approximate the compositions of minerals, associated with diamonds, and with the compositions of equivalent minerals in the lunar basalts, formed under significantly more reducing conditions and completely devoid of Fe^{3+}. Such features, as the amount of Cr in the olivines, being an exceptional rarity in terrestrial ultramafic rocks, have been expressed very clearly in the chromium-bearing basalts on the moon. In two samples, a regular association has been observed between the amount of Cr_2O_3 in the olivines (0.2 and 0.4%) and the amount of Cr_2O_3 in the rocks (respectively 0.32 and 0.53%) (Bell, 1971; Haggerty and Meyer, 1971). In the olivines from the diamonds, with a significantly smaller amount of Cr, there is also a very clear positive correlation with the chromium-index of the associated pyrope. In addition to the olivines, associated with diamonds, certain peridotites from the 'Udachnaya' pipe and also the meimechites, bear features of especially reducing conditions of formation. Thus, for the formations of the upper mantle, the most reliable representatives of which are the parageneses with diamonds, clear features of the lowering of the partial pressure of oxygen have been established, as compared with the values, identified for the basalts (Anderson, 1968; Sato and Wright, 1966), but they are lower than for the lunar rocks (Sato and Helt, 1971).

For the magmatic series of the platform, a peculiar means of development was established quite a long time ago (V. S. Sobolev, 1937), with typical rapid increase in the iron index of the femic minerals. Experimental investigations (Osborn, 1959) have shown that this feature indicates the lower oxygen potential. It is possible that such lowering of the oxygen potential is also typical of the deeper portions of the mantle below the platforms, favoring the release of free carbon, and at an appropriate depth, crystallization of diamond.

§28. FEATURES OF THE PARTIAL MELTING OF THE MATERIAL
OF THE UPPER MANTLE

At the present time, there is extensive discussion of the question
of the nature of the low-velocity layer in the upper mantle (Gutenberg,
1953), and many investigators suggest the existence of a layer of
partial melting (Magnitsky, 1968, 1971; Wyllie, 1970).
If such rocks with a non-discrete interstitial melt really exist,
they must possess specific structures in those cases where the xenoliths of the rocks will be brought to the surface. Such features
must be especially well expressed in the xenoliths, where eruption
is very rapid. The search for such features must be the object of
special investigation. However, we may at present speak of certain
features, recorded during a study of the mineralogy of the xenoliths.
We must first of all dwell on the features of the grospydites
and the kyanite eclogites. We have already recorded that between
the grains of the primary minerals there is an aggregate of secondary
products, crystallized under lower-pressure conditions, the composition of which involves a basic plagioclase and secondary corundum
(Sobolev et al., 1968), and there are also secondary hydrous minerals.
A special investigation of the secondary products has been carried
out on the material of the kyanite eclogites of the 'Roberts Victor'
pipe, including a diamond-bearing sample (Switzer and Melson, 1969).
The identification of a glass in the interstices between the mineral
grains, a glass enriched in Na_2O, interpreted as a result incongruent
melting of omphacite, is of special interest. The unaltered nature
of the garnet and kyanite emphasizes that melting mainly affected
the omphacite, in which, besides the glass, new formation of plagioclase has also been recorded. In kyanite, it is true, a newly-formed
fibrous mineral has been recorded, determined after analysis with
the aid of the microprobe as mullite. This mineral undoubtedly
crystallized under conditions of lowering of pressure.
We must specially emphasize the features of clear cataclasis of
the rocks at depths of more than 100 km. Independently of the
discussion of the causes of crushing of the rocks, even involving
plate tectonics (Boyd, 1973), this feature of the very deep-seated
rocks is of considerable interest to the tectonists and such rocks
deserve a further wide-ranging study. The features of partial melting in the garnet peridotites are the kelyphitic rims. Already
during the description of the garnet peridotite from the Czech
Massif (Sobolev and Lodochnikov, 1962), it has been recorded that
the structure of these rims probably indicates the formation of an
intergranular melt. In the xenoliths of peridotitic composition
from the kimberlites, the kelyphitic rims have been significantly
thickened towards the contact with the kimberlite in association
with melting after the capture of the xenoliths. At the same time,
in the inner portions of the xenoliths, often in completely fresh
samples, the kelyphitic rims retain a constant width, depending on
the type of xenolith. In the case of the Czech peridotites, it is
most likely that the cause of melting has been associated with
injection into a zone of lower pressures.
Although kelyphitic rims are also typical in xenoliths in the
pseudoporphyritic peridotites, they are absent in the granular rocks
of the 'Obnazhennaya' pipe, which have clearly been caught up from
the higher horizons of the mantle. In the diamond peridotites,
rims around garnet are lacking, that is, the formation of the kimber-

litic foci have already been associated with local causes, either
with increase in pressure of water or CO_2, or with lowering of the
melting point owing to the influence of volatiles.

Thus, certain features of melting have been identified for the
rocks caught up in depth, approximately corresponding to the depth
of the low-velocity layer during the type of formation of the kimberlites. It must be affirmed, however, that the information presented
only records the means of searching for direct features of traces
of melting in eclogite xenoliths and other deep-seated rocks.

+

+   +

In emphasizing once more that the upper mantle has been significantly differentiated, we shall assume that the deepest known material
corresponds to a pressure of 50-60 kb, that is, to depths of the
order of 200 km. So far, it is impossible to regard as confirmed
that the deeper formations could reach the surface in the kimberlite
pipes. This applies to the interpretation of the well-known pyroxene-
ilmenite intergrowths in kimberlites (Williams, 1932; Bobrievich
et al., 1964) as breakdown textures of an assumed high-pressure
phase (Dawson and Reid, 1970), namely, an unusual garnet, in the form
of which it has been possible to crystallize a mixture at pressures
of 100 kb (Ringwood and Lovering, 1970). Since the regular intergrowths of pyroxene and ilmenite are interpreted in another way
(crystallization from a melt), and garnets of such unusual composition (or breakdown textures corresponding to them) have not been
found in diamonds, the above hypotheses may be regarded as poorly
founded. In contrast to the assumption that such intergrowths are
dissociation textures, there is also the presence of similar textures
in the xenoliths of ilmenite-bearing peridotites in the kimberlite
pipes of Yakutia.

In regarding the upper mantle as having been differentiated
and possessing a layered nature, we shall not dwell on the possible
interpretation of the primary material (Vinogradov, 1959, 1961;
Harris et al., 1967; O'Hara, 1970; etc.), which in every case must
be distinguished in composition from the well-known deep-seated
xenoliths, but rather on the features of the mechanism, which led
to differentiation. It is possible that such a mechanism may be
melting, similar in type to zone melting (Vinogradov and Yaroshevsky,
1965; Vinogradov et al., 1970; Magnitsky, 1968; Yaroshevsky, 1968).
In order to solve this very complex problem in finality, serious
theoretical and experimental investigations are necessary. Further
detailed study of the deep-seated material may provide considerable
assistance in such investigations, as a result of which it may be
possible to compile an even more complete idea about the composition
and construction of the upper part of the mantle, at least, down to
depths of the order of 200 km.

CHAPTER VIII

THE MINERALOGICAL CRITERIA OF THE DIAMOND-BEARING KIMBERLITES

In a series of numerous attempts to establish the characteristics
of the diamond-bearing nature of the kimberlites, the principal
place is occupied by the search for an association between the features
of the chemical composition of the kimberlites and the content of
diamonds in them. The possibility of the existence of such a relationship arises from the hypothesis about the crystallization of
all diamonds in a kimberlite melt, and the tendency towards such
an association has been suggested by a number of authors, who have
studied kimberlites (Wagner, 1909; Williams, 1932; Bobrievich et al.,
1959a; Milashev, 1965; Dobretsov et al., 1966, 1972). Although a
great deal of work has been carried out in this direction, which has
provided certain definite recommendations (Milashev, 1965; Dobretsov
et al., 1966, 1972), this problem has encountered numerous difficulties and is still far from resolution. The most significant
difficulty is the intense depletion of kimberlites in xenogenic
material, portion of which is of a deep-seated nature, and part
belongs to the surrounding rocks, and also a significant change in
the composition of the kimberlite as a result of later processes.

As a possible feature of diamond-occurrence, we have also employed
the principal minerals of the kimberlites, constantly present in the
kimberlite pipes, in particular, pyrope, the importance of which
as a satellite of diamond in the kimberlites had already been recorded
in the materials for the study of the South African kimberlites
(Williams, 1932; Sobolev, 1951).

A study of the inclusions in diamonds by X-ray methods (see §9)
has stressed the importance of pyrope as a satellite of diamond,
and has revealed an assemblage of minerals, which form simultaneously
with diamond. This has enhanced the interest in searching for an
association between the amounts of satellites of diamond (pyrope,
chromite, and olivine), identified in inclusions, and such a widely
distributed mineral in the kimberlites as picroilmenite (although
it has not so far been found as inclusions), and the content of
diamonds. The results of these researches appeared to be negative
(Bobrievich et al., 1964), with the exception of the discovery of
a tendency towards a positive correlation between the amount of
chromite and that of diamond in the diamond-rich pipes (Dobretsov
et al., 1966).

In regard to pyrope, although it has also been regarded as the
principal satellite of diamond and the measurement of the pyrope
content is one of the principal methods of exploring for the kimberlite pipes of Yakutia (Sarsadskikh, 1958), there has been no success
in finding any associations between its amount and the diamond
content. Moreover, in a number of pipes, which contain a significant
quantity of pyrope, there have been very few diamonds or they are
completely absent.

Such extreme cases of differences in the kimberlites, with or without
diamonds, but rich in pyrope, are reflected in the recognition of the
diamond and pyrope subfacies of the kimberlites (Milashev, 1965).
However, it must be stressed that, even in the pipes that belong to
the diamond subfacies, the minerals, actually syngenetic with diamond,
in particular pyrope, have been depleted of a much larger amount of
similar minerals, which have no relation to the diamond. In this

respect, using Milashev's terminology (1965), we must reject the
concept of a pure diamond subfacies of the kimberlites, because in
all the diamond pipes, even the richest, minerals of both the diamond
and the pyrope subfacies are manifested.

The first determination of the composition of a pyrope, enclosed
in diamond (Meyer, 1968b), allowed the plotting of a sharp compositional boundary between this chrome-pyrope, poor in calcium, and the
large amount of pyrope garnet, present in the concentrate of the
heavy fraction of the kimberlite. We shall not dwell on the compositional features of the garnets and other minerals, enclosed in
diamonds, which have been described in §9, but we shall once again
emphasize only the principal features of their differences from
similar minerals of the kimberlites:

the pyropes, with marked predominance of a calcium-poor, chromium
variety, are characterized on average by a lower iron-index;

the pyrope-almandine garnets are distinguished from those of similar
composition by the presence of a stable amount of Na_2O (0.10-0.20%);

the olivines, with decreased iron index, are distinguished from
the majority of those from ultramafic rocks by the presence of a
constant amount of Cr_2O_3;

the chromites are distinguished by a large amount of the Cr-
component (more than 80%) or more than 62% Cr_2O_3 and low Al_2O_3 (less
than 7%) from all the chromites of the ultramafic rocks of terrestrial
origin, and they are comparable in composition only with chromites
from meteorites;

the enstatites are characterized in most cases by an increased
value of the ratio $Cr_2O_3/(Cr_2O_3 + Al_2O_3)$;

the chrome-diopsides possess less clearly distinguishing features,
though in a large number of samples, there is a tendency towards
enrichment in Cr_2O_3 and Na_2O as compared with analogous pyroxenes
from peridotites.

Besides the minerals enclosed in diamond, there is valuable information to be obtained from the compositions of mineral-intergrowths
with diamonds. Garnets have been identified in the intergrowths,
that are analogous to the inclusions in the diamonds, and the chromites
from the intergrowths, as exemplified by the 'Mir' pipe, are without
exception equivalent in composition to those of the inclusions in
diamonds. Unique compositions have been discovered among the clinopyroxenes, which are markedly enriched in the ureyite component
(up to 46% $NaCrSi_2O_6$).

Thus, the similarity in the compositions of many inclusions and the
appropriate minerals from intergrowths with diamonds, may also, with
great probability, be regarded as being in equilibrium with diamond
aggregates or seemingly with their outer zones. This conclusion
is important for assessing the relationship between diamond and
such a mineral as ilmenite, which up till now has not been found
as inclusions, but is known only in the form of intergrowths with
diamonds (Harris, 1968). The first ilmenites from intergrowths
with a monocrystal and aggregate of diamond were investigated in
the 'Mir' pipe (Sobolev et al., 1971). A comparison between their
composition and that of the ilmenites from the kimberlites of Yakutia
and Africa (more than 120 analyzed samples) has shown clear differences, expressed in the almost complete absence of Fe^{3+} (Fig. 42),
which agrees with the general compositional features of the minerals,
associated with diamonds. It is evident that an insignificant
amount of picroilmenite with very small amount of Fe^{3+}, and con-

sequently, with a large amount of TiO$_2$, is crystallized in the last phases of formation of diamond in equilibrium with chromite, olivine, and pyrope in a medium of ultramafic composition. The marked difference in the composition of the ilmenite, associated with diamond, serves as yet a further conformation of the clear typomorphism of the compositions of the minerals, syngenetic with diamond, and the necessity for their demarcation from the xenogenic minerals, and also the later products of crystallization of kimberlite.

The principal features listed, which distinguish the inclusions in diamond from analogous minerals of the kimberlites and xenoliths, can be traced for the syngenetic inclusions in diamonds of the most varied deposits (South Africa, Ghana, Sierra Leone, Venezuela, Brazil, Yakutia, and the Urals). In some of the diamond-bearing regions listed, such as Brazil, Venezuela and the Urals, no solid-rock sources for the diamonds have so far been found, although the nature and features of the inclusions indicate a similarity in the conditions of formation of diamonds and the composition of the diamond-forming medium for all the deposits. This important conclusion, reached on the basis of a comparison between diamonds from the African deposits and Venezuela (Meyer and Boyd, 1972), has been convincingly confirmed by the results of a study of inclusions in the Yakutian and Uralian diamonds (Sobolev, Lavrent'ev et al., 1969; Sobolev, Botkunov, Lavrent'ev and Pospelova, 1971).

In this respect, the question arises as to the possibility of using the features of the differences in the minerals enclosed in diamonds from those of the peridotite xenoliths and kimberlite in an attempt to find minerals which crystallized simultaneously with diamond, in the concentrates from the kimberlite pipes. Such an attempt has aroused considerable interest, because the accurate identification of minerals, analogous in composition to the syngenetic inclusions in diamonds, may be the correct answer to the question on the possibility of the presence of diamonds in an actual kimberlite pipe, even without doing the appropriate work based on sampling. In addition, the discovery of minerals that crystallized simultaneously with diamond, may be used in the search for quantitative association, with the diamond-capacity of individual pipes (Sobolev, 1971).

Of the most widely distributed minerals in diamonds, olivine is seemingly the least satisfactory for use as a mineralogic indicator, owing to its widespread distribution and similarity in composition in different types of ultramafic rocks. However, the search for olivines, which contain chromium, may be of significant interest in determining the facies assignment of a particular pipe and in clarifying the quantitative amount of relict olivine (which remained during the formation of the kimberlite melt) and the olivine, which crystallized in the stability field of diamond. In the case of the investigation of a large quantity of olivine grains from individual pipes, it is possible to identify differences not only in the amount of chromium, but also in iron index, even amongst the most magnesian varieties.

It is apparently impossible to use enstatite and clinopyroxene as mineralogical indicators of diamond-capacity mainly because of their significant rarity in diamonds as compared with other minerals and less clearly defined features, which would distinguish them from the corresponding minerals of the ultramafic xenoliths.

Arising from the features of the predominant paragenesis of diamond (dunite) it may be suggested that, in the primary diamond-forming medium, the main minerals, which crystallized in equilibrium with diamonds, were olivines, pyrope, and chromite. Olivine, as already stated, does not possess clearly defined typomorphic compositional features, and moreover, it is in most cases completely altered. Judging by the quantitative relationships between diamond and garnet and chromite in individual known xenoliths from dunites and by analogy with the amount of diamond in already known, quite numerous xenoliths of diamond-bearing eclogites, it may be considered that the original amount of diamonds in the rock during the period of their crystallization was significantly less than 0.1 wt %. The amount of minerals, associated with them, that is, typical mineral-satellites, such as garnet and chromite, must exceed the amount of diamonds by at least two orders, and perhaps sometimes more.

Thus, it is necessary in more detail to consider the question of the possibility of finding in the kimberlite concentrate such minerals as chromite and garnets, corresponding in composition to the inclusions in diamonds. In this respect, let us dwell on the features of the quantitative-mineral composition of the concentrate of the heavy fraction of the kimberlite and the quantitative relationships of these minerals in the heavy fraction.

As is known, in almost all pipes fragments of pyrope-bearing ultramafic rocks, and also eclogites and various crystalline schists, are present in variable amounts. Crushing of xenoliths of these rocks during enrichment contributes to the entry of a significant amount of garnet into the concentrate. A substantial percentage of the grains and fragments of garnets, present in the binding mass of the kimberlite, is also genetically associated with the garnet-bearing inclusions.

The most reliable example for assessing the amount of minerals in the heavy fraction, which are potentially associated with diamond, is the 'Mir' pipe, for which results have been published on the mineralogical examination of seven bulky samples (Vasil'ev et al., 1968). In the concentrate of the heavy fraction, obtained from a serpentinized kimberlite of the upper horizons, olivine is practically absent. The exceptional rarity of fragments of garnet-bearing crystalline schists (Bobrievich et al., 1959a, 1964) indicates the predominant amount of garnet alone in the heavy fraction, genetically associated with deep-seated xenoliths (peridotites and eclogites), and with kimberlite itself. From the available information, without involving data on ilmenite and other later minerals, we may reach the conclusion as to the quite constant average ratio of chrome-spinels and pyrope in all seven samples. The former comprise approximately 0.2-0.3 wt % of the amount of pyrope. If we turn to data on other diamond-bearing regions and the pipes of Yakutia (Bobrievich et al., 1964), we may reach the conclusion that there is an overall significant predominance of pyrope over chromite in regions of diamond-bearing pipes (see below).

Regions	Pyrope	Chromite	Chromite/(chromite + pyrope), %
Aldan	0.72	0.18	20
Malo-Botuoba	0.19	0.001	0.5
Daldyn-Alakit	0.11	0.002	2
Verkhne-Muna	0.09	0.01	10
Sredne-Olenëk	0.06	Rare signs	1
Nizhne-Olenëk	0.22	0.05	19

The average amount of pyrope and chrome-spinels in kimberlites from certain diamond-bearing regions of Yakutia (in wt %) (Bobrievich et al., 1964) gives some idea of the quantitative relationships of the chrome-spinels and pyrope in the kimberlites. In half of the regions under consideration, the amount of chrome-spinels is only 0.5-2.0% of their total amount together with pyrope, along with very low amounts of the minerals themselves. However, in some regions, for example, in the Aldan and the Nizhne-Olenek areas, the relative and absolute amounts of chrome-spinels markedly increase. For the Nizhne-Olenek area, in particular, such an increase may be explained by the widespread distribution of xenoliths of ultramafic rocks, containing chrome-spinels, including unusual pyrope-spinel ultramafic rocks, for instance, in the 'Obnazhennaya' (Sobolev and Sobolev, 1964) and 'Slyudyanka' (Lutts, 1965) pipes.

The figures given, although obtained on the basis of small samples, apparently correctly reflect the order of the ratio of pyrope and chrome-spinels, because the results of bulk sampling of the 'Mir' pipe (Vasil'ev et al., 1968) give similar values.

Judging from the data on inclusions in diamonds, chromite occurs in diamonds significant more often than pyrope. From the figures obtained on the ratio of chromite to pyrope in the concentrate of the heavy fraction of kimberlites from the various diamond-bearing regions of Yakutia, it may be suggested that the chromite, syngenetic with diamond, will be discovered in the kimberlite concentrate much less commonly than the corresponding pyrope. At the same time, it must be noted that although there are indications of a direct correlation between the amount of chromite and diamond content (Dobretsov et al., 1966), a number of data on the anomalously large amount of chromite in certain regions of development of kimberlites are in opposition to these conclusions (Rovsha and Ilupin, 1970). In fact, even in the rich pipes, the correlations between the overall content of any mineral of the heavy fraction, especially chromite, and the diamond content, may be complicated by the presence of xenogenic chromite. We have provisionally checked this for the 'Mir' pipe, the chromite of which, in spite of an overall small amount, consists of a mixture of chrome-spinels of different composition. Individual grains, selected at random from the enrichment concentrate of a particular bulk sample from the 'Mir' pipe, kindly presented to us by G. V. Zol'nikov, and analyzed with the aid of the microprobe, are characterized by very wide fluctuations in composition with respect to Al_2O_3 (from 15 to 40%) and Cr_2O_3 (from 57 to 30%).

An analysis for aluminum and chromium in about 160 grains of chrome-spinels from this pipe has confirmed their wide fluctuations in composition within the 'Mir' pipe (with 22 to 67% Cr_2O_3) and has revealed that about 12% of all the grains studied are identical in composition to the inclusions in diamonds (Sobolev, Pokhilenko et al., 1975). This indicates the significantly greater variation in composition of the chrome-spinels within the 'Mir' pipe as compared with that suggested earlier on the basis of X-ray data (Bobrievich et al., 1964). We have also identified large amounts of 'diamond' chromite in the 'Aikhal' and 'Udachnaya' pipes (about 20%). In the weakly diamond-bearing pipes, only isolated grains of almost pure chromite (less than 1%) have been found. Thus, the first results of detailed investigations of the chromite concentrates confirm the validity of the stated hypotheses on the special importance of chromite composition as an exploration indicator for diamond-bearing pipes (Sobolev, 1971a).

In considering the data on the composition of chromites, associated
with diamonds, which are already quite representative (more than 50
analyses along with some of our new data), we may conclude that in
no single case does the amount of Cr_2O_3 fall below 61-62% (see
Fig. 40). In the overwhelming majority of samples there is more
than 63% Cr_2O_3 (see §18). In some published bulk analyses of chrome-
spinels from concentrates of different pipes, extremely different
amounts of Cr_2O_3 have been found: 28% in the 'Aldan' dike, which
contains no diamonds, and 42-52% in the diamond-bearing pipes.
Even on the basis of this conclusion, it is possible with sufficient
assurance, to consider that the concentrate from the 'Aldan' dike
lacks especially chromium-rich spinels, which are potential satellites
of diamond. The new data on the composition of chromites from
diamond require very careful checking of the representative material
based on the chrome-spinels from the pipe, for the possibility of
an unequivocal solution about the presence of chromites in the
concentrate, which are syngenetic with diamonds. The necessity for
such verification is seen from the above example from an investigation of the individual grains of chrome-spinels from the 'Mir'
pipe. The wide variations in composition are not associated here
with morphologic features, as suggested by Rovsha and Ilupin (1970).
If we compare the data on the average amounts of Cr_2O_3 and Al_2O_3,
calculated on the basis of six determinations of the composition
of spinel grains from the 'Mir' pipe concentrate, with the already
averaged data from analyses of chromite concentrates, then as a
result, we may reach two completely definite conclusions: the six
grains examined do not embrace the entire variation in compositions
of the chrome-spinels from the 'Mir' pipe; the selection did not
include the most chromium-rich spinels, that is, those syngenetic with
diamond. The presence of a significant amount of almost pure chromites
in the concentrates of the heavy fraction from kimberlites has already
been identified by us in the 'Udachnaya', 'Aikhal', and 'Mir' pipes.
    An additional study of the composition of chrome-spinels of dif-
ferent morphology (a separate analysis of a selection of chromites
of octahedral habit and of complex morphology) for the 'Udachnaya'
pipe (Sobolev, Pokhilenko et al., 1975) emphasizes that the composition of the chrome-spinels does not depend on their morphology
(Frantsesson, 1968), but also establishes that an investigation of
about 100 individual grains of chrome-spinels from concentrates
is completely adequate and representative for their reliable average
characterization within an entire pipe.
    During an investigation of a chromite concentrate with the object
of diagnosing the chromite, syngenetic with diamond, the completion
of a partial analysis of the chrome-spinels for Cr and Al or for Cr
only, was sufficient. The chromite, syngenetic with diamond, is
readily diagnosed from the following features: low content of
Al_2O_3 in association with a high chromium-index (Cr/(Cr + Al) 80%)
or a large amount of Cr_2O_3 62%.
    In this respect, it is necessary to dwell in greater detail on
the compositional features of the chromium-rich garnets from the kimberlites. Until recently, actual information on such garnets was
limited to isolated records (see §2). However, during the study
of the mineralogy of deep-seated xenoliths from the 'Udachnaya'
pipe (Sobolev, 1970), a series of garnet compositions was found
with refractive indices reaching 1.790. A detailed study of the
physical properties (n and a_o) of this series of garnets with focus
on the most chromium-rich compositions has revealed several com-

positions (see Fig. 4), similar to those of garnets, enclosed in
diamonds (low-calcic types). It is possible that the xenoliths
containing these garnets and differing in textural features from
the porphyritic peridotites of the 'Udachnaya' pipe, also belong
to potentially diamond-bearing varieties analogous to the xenoliths
from the 'Aikhal' pipe (V. S. Sobolev, Nai et al., 1969). The
compositional points for the garnets studied from the peridotite
xenoliths have been plotted on a diagram (Fig. 52) based on the
amounts of Cr_2O_3 and CaO; here, we have plotted the compositional
points of inclusions of chrome-pyropes from diamonds in different
parts of the globe based on the data from Tables 27, 28, and 32.
For these garnets (31 analyses), the average amount of Cr_2O_3 =
9.5 ± 2.5%, and CaO = 1.96 ± 0.7%. In addition to these data, in
Figure 52, we have used data on chrome-garnets from peridotites
of samples E ≃ 10 (Nixon et al., 1963) and T7-317 (Fiala, 1965),
which respectively contain 7.5 and 6.85% Cr_2O_3, and also our new
data based on Sample T7-272 from the peridotites of the Czech Massif

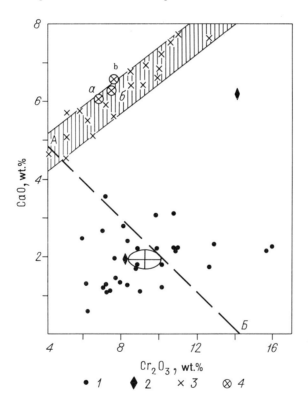

Fig. 52. Amounts of Cr_2O_3 and CaO in magnesian garnets: 1) from
diamonds; 2) from xenoliths of diamond-bearing serpentinites; 3)
from peridotite xenoliths of the 'Udachnaya' pipe; 4a, c) from
peridotites of the Czech Massif; 4b) from peridotite E-10 from
Lesotho (South Africa). Compositional field of garnets from
lherzolites cross-hatched; lower portion of diagram shows confidence
ellipse of average amount of CaO and Cr_2O_3 in garnets from diamonds
(p = 0.05).

with 7.6% Cr_2O_3. The compositional points for garnets from xenoliths
of diamond-bearing peridotites from the 'Aikhal' pipe (V. S. Sobolev,
Nai et al., 1969), one of which is completely analogous to those
enclosed in diamonds, have been plotted individually. The compositional point of another garnet lies at a significant distance from
the overall field of compositions, but its position is clearly defined
below the limits of the linear field of compositions of garnets
from peridotites (lherzolites).

As is well-known from a study of garnets from peridotites, they
are distinguished by an exceptionally constant amount of calcic
component (see §12). The experimental data also emphasize that the
composition of the garnet from the lherzolite association is practically independent of pressure. Nevertheless, it has been successfully demonstrated that increase in the amount of calcium in these
garnets takes place with increase in the amount of chromium (Sobolev
et al., 1969), which is also well supported by the data in Figure 52.

It is well-known that the pyropes, enriched in chromium, change
their color from reddish-violet to green with change in illumination
(Bobrievich et al., 1959a, 1964; Gnevushev et al., 1959). The nature
of this phenomenon (the alexandrite effect) has been interpreted
from a study of isomorphous mixtures of Al_2O_3 and Cr_2O_3 (Grum-
Grzhimailo, 1946), in which reddish color has been identified for a
small amount of Cr_2O_3 and green for a large amount. Such a change
in color through reddish-violet with an intermediate amount of Cr_2O_3
(30-50%) occurs as a result of a shift in the absorption bands into
the red field of the spectrum during increase in the amount of Cr_2O_3
(Grum-Grzhimailo, 1946; Grum-Grzhimailo and Rovsha, 1960). As a
result of a change in illumination, the color changes in some samples
from a pinkish-violet in evening light to a greenish-pink during
daylight. The alteration in the nature of the color is associated
not only with the chromophore itself, but also with accompanying
ions; in particular, the amount of CaO favors the appearance of a
greenish color (Sobolev, 1949).

These conditions are well illustrated by garnets, which contain
variable amounts of Ca and Cr. Thus, in certain chromium-bearing
calcic garnets from grospydites (Sobolev, Kuznetsova and Zyuzin,
1966), an alteration in color during change in illumination occurs
with an amount of 1-2% Cr_2O_3, and garnets containing 4-5% Cr_2O_3
are already tinted green in evening light. In low-calcium garnets
(inclusions in diamonds), the color is unaltered even with an amount
of about 8% Cr_2O_3 and changes gradually, passing into a bluish-green
during daylight in the range of 10-16% Cr_2O_3. However, even at
the maximum known amount of Cr_2O_3 (16%), these garnets have a reddish-
violet color in evening light (n = 1.790). In the garnets of intermediate type (garnets from peridotites) with CaO = 4.5-7.5% (depending
on the amount of chromium), an intense change in color with change
in illumination has been observed with an amount of Cr_2O_3 or 4%.
This tendency is approximately reflected in the line AB on Figure 52:
on the right of the line are compositions with intense manifestation
of the alexandrite effect, and to the left, those with weak manifestation or lacking it. Thus, to the left of the line lie the
majority of compositional points of garnets from diamonds, which
may serve as a diagnostic feature in searching for such garnets
in a concentrate. The principal recommendation may be summarized
as follows: the normal chrome-pyropes from xenoliths with n 1.750
will be distinguished by an intense manifestation of this effect.

Grains of pyropes with n͟ 1.750 and weak manifestation of the
alexandrite effect may be assigned to the low-calcium type.

A study of garnet concentrates from the 'Mir' and 'Aikhal' pipes,
using the above-noted method of diagnosis of low-calcium garnets
rich in chromium, with subsequent checking of the composition of
selected grains using the microprobe, has revealed garnets in these
pipes, analogous to inclusions in diamonds (several percent of the
total content of garnets). Thus, the amount of these unusual garnets
in the diamond-bearing pipes is constantly high. In the poorly
diamond-bearing pipes, they have also been identified, but their
content is extremely low and comprises 0.1 to 0.5%. However, in all
cases, the amount of such garnets is many times greater than that
of the green chromium- and calcium-rich garnets in the concentrates,
which once more emphasizes that the rarity of green garnets is associated in the main, not with particularly high pressures (Sarsadskikh,
1970), but with the rarity of the appropriate compositions (Sobolev,
1971a͟). The garnets, rich in the knorringite component and actually
formed at pressures, considerably exceeding 30 kb (Malinovsky et al͟.,
1973), were earlier simply passed over in the concentrates (Sobolev,
Botkunov and Kuznetsova, 1969). The chrome-pyropes, poor in calcium,
have now been found in the concentrates of both the Yakutian (Sobolev,
Lavrent'ev et al͟., 1973a͟) and the South African kimberlites (Gurney
and Switzer, 1973). A specially assembled standard collection of
garnets with variable amount of calcium and chromium is of great
assistance in the visual diagnosis of such garnets.

For the green pyrope-uvarovite garnets, additional verification
of the amount of iron is necessary, since the majority of such grains
have an iron index greater than that of garnets from diamond (20-40%)
and in a number of cases they are present in the concentrate of the
non-diamondiferous pipes (Sobolev, Kuznetsova and Zyuzin, 1966).

It must be emphasized that the sole use of color differences in
pyropes for assigning them to types syngenetic with diamonds and
shallower (Sarsadskikh, 1970) is clearly inadequate. A noteworthy
example of this is the clear difference in composition of the lilac
pyropes, enclosed in diamonds, from those of similar color, studied
in many ultramafic xenoliths and in kimberlite concentrate, and
even in peridotites, known in solid outcrop (Fig. 52). An overall
increase in the amount of Cr_2O_3 in the pyrope, identified by change
in color, may only be regarded as an additional feature of the
diamond-bearing pipes in combination with other features. This
may apply only to rare, very chromium-rich garnets, containing
more than 8-10% Cr_2O_3, and by no means to all the chromium-bearing
pyropes with 1-5 and even 7% Cr_2O_3, occurrences of which are known
both in many non-diamondiferous pipes and in the pyrope peridotites
of the Czech Massif (Fiala, 1965). The inadequacy of the use of
color as a feature of depth is emphasized by the coexistence of
orange and lilac pyropes, markedly differing in amount of Cr_2O_3
in various zones of a single sample, which is controlled by features
of chemistry (Fiala, 1965; Sobolev and Sobolev, 1967) and not
pressure.

In addition to the search for chrome-pyropes in the concentrate,
interest may also be centered on the search for pyrope-almandine
garnets, which have entered the kimberlite as a result of crushing
of xenoliths of diamond-bearing eclogites. Although such garnets
are also similar in composition to those of the normal eclogites,
a recognized stable feature (the amount of Na_2O) may be used for

their diagnosis (Sobolev and Lavrent'ev, 1971), and individually selected grains from the concentrate may be checked for sodium on the microprobe.

Thus, on the basis of data from a study of inclusions in diamonds, stable features of compositional differences have been recognized in most minerals, syngenetic with diamond, from those of analogous minerals of deep-seated xenoliths and from the concentrate of the heavy fraction of the kimberlites. We have checked these features for two kimberlite pipes ('Udachnaya' and 'Obnazhennaya'), which are respectively diamond-bearing and diamond-free. On the basis of an investigation of a large number of garnets from concentrate and from xenoliths of both pipes (more than 300 in each), we have established the maximum amount of Cr_2O_3 in the pyropes, and also in the chrome-spinels. Olivine grains from kimberlite have been investigated with the object of finding the amount of chromium. The results of the investigations are presented below.

Diagnostic features	'Udachnaya'	'Obnazhennaya'
Maximum amount of Cr_2O_3 in pyrope, wt %	18.7	4.0 (6.4)*
Presence of calcium-poor chrome-pyropes	Present	Not found
Maximum amount of Cr_2O_3 in chrome-spinels, wt %	66.0	52.2
Average amount of Cr in olivines (in the form of Cr_2O_3), wt %	0.038 ± 0.020 ($\underline{N}$ = 25) fo = 92.5	0.007 ($\underline{N}$ = 10) fo = 92.3

*According to Ukhanov (1970).

As mineralogical criteria for the occurrence of diamonds, the use is recommended of compositions of olivine, chromite, and garnet, which have been genetically associated with diamond, the discovery of which in the concentrate of the heavy fraction of kimberlite and placers, unequivocally enables us to solve the problem of the presence of diamonds in actual pipes, and may also serve as a basis for identifying quantitative patterns.

These recommendations may be completely applied to the diamond-bearing regions of Yakutia, but in the case of the Uralian placers, the use of mineral-satellites is significantly complicated owing to the rarity of pyrope and the impoverishment in chromite of the ultramafic type. Comparative investigations of the inclusions in the Uralian and Yakutian diamonds suggest the significant predominance of the eclogite association over the ultramafic type in the Uralian diamonds (Sobolev et al., 1971). Consequently, as a basis for work in the search for mineral-satellites in Uralian diamonds, along with the search for chrome-pyrope and chromite, syngenetic with diamond, in the concentrates, we may place the search for eclogitic garnets, containing sodium. Undoubtedly, the possibility of using such a method may be supported by special investigations.

CONCLUSIONS

The principal results of the investigations, partially summarized in Chapters VI and VII, are as follows.

The deep-seated inclusions in kimberlites have been separated on the basis of conditions of formation into two main groups relative to the boundary of the graphite-diamond phase transformation: a) xenoliths, formed in the pressure range of 20-50 kb, and b) xenoliths and minerals, formed at higher pressure. This separation may be defined not only by the presence or absence of diamond, but also by a number of compositional features of the minerals in xenoliths, which enable us to make further subdivisions within the first group.

For the facies of the pyrope peridotites and eclogites with graphite (graphite-pyrope), various parageneses of ultramafic type (peridotites) and basic (eclogites, including kyanite and corundum types, and grospydites) have been described, and a scheme for their further subdivision into subfacies, based on pressure, has been recorded. The unusual peridotites from the 'Udachnaya' pipe, formed according to a number of features at the highest pressures (more than 30 kb), have been separated into a subfacies, provisionally termed the coesite subfacies.

For the diamond-bearing facies (diamond-pyrope), in addition to a number of eclogite xenoliths, including kyanite and corundum types, identification and study have for the first time been made of xenoliths of diamond-bearing peridotites of the harzburgite-dunite type, which in paragenesis and composition of minerals are markedly distinguished from the predominant lherzolites in the kimberlite pipes.

The assemblages of inclusions in diamonds and the intergrowths of minerals in a single diamond, revealed for the first time, comprise equilibrium parageneses. The following facts are proof of this important proposition: 1) the grains of one mineral, enclosed in a single diamond crystal, have identical composition during wide variations in composition in different diamond crystals; 2) the composition of minerals in intergrowths from diamond is identical to that of individual inclusions from the same diamond; and 3) the distribution features of the components in coexisting minerals are analogous to the established patterns for equilibrium parageneses of rocks.

A study of the minerals, associated with diamond in xenoliths, inclusions, and intergrowths, has enabled us to recognize a number of stable parageneses, belonging to the ultramafic type (with olivine) and the basic type (without olivine). The following parageneses have been identified for the first type: harzburgite-dunite (markedly predominant), lherzolite, and wehrlite (with and without pyrope). For the second type, we have the eclogite paragenesis (garnet + omphacite), with kyanite eclogites and corundum eclogites.

In contrast to the inclusions, two generations of minerals are often observed in the intergrowths with diamonds: 1) those growing in equilibrium with diamond; and 2) those later than diamond, crystallizing in the kimberlite during a significant change in conditions (fall in temperature, etc.), which is especially well illustrated by the recognized generations of chromite.

It has been emphasized that increase in pressure extends the stability field of the sodic pyroxenes and especially the magnesian garnets. For the garnets, stability of the continuous series of diamond-

pyrope-grossular compositions, beginning with $\underline{f}$ = 30%, has been
demonstrated, and increase in the iron index and pressure increases
the stability of the intermediate members.

It has been shown that with increase in pressure, there are
marked increases in the limits of substitution of aluminum by
chromium and consequently, redistribution of chromium from spinels
into silicates. This leads to the formation of new varieties of
garnets of the pyrope-knorringite and pyrope-uvarovite series (with
up to 50% chromium component), chrome-kyanite (up to 20% chromium
component), and other new varieties of minerals. A study of systems
with Cr_2O_3 must give basic information for estimating absolute
pressures of the deep-seated associations.

The behavior and relationship of titanium to chromium are extremely
complex and deserve further study. In the harzburgite-dunite paragenesis (most chromium-rich), the garnets and chromites are very poor
in titanium. However, in the lherzolite paragenesis, the chrome-
garnets, and often the chromites also, simultaneously contain a
considerable amount of TiO_2, and chromium-rich ilmenites also appear,
with the ore minerals on the basis of compositional features approximating to those of the lunar basalts. In the deep-seated eclogites
and peridotites, the amount of TiO_2 in the garnets (replacing aluminum
and chromium) is clearly associated with high temperature, and dissociation is frequently observed with the formation of oriented
needles of sphene or ilmenite. A significant portion of the TiO_2
apparently goes into the kimberlites at the expense of xenogenic,
less deep-seated minerals.

The significant role of Na_2O in many deep-seated associations
has been demonstrated. This component leads to the formation of
omphacitic pyroxenes at the expense of garnet--the breakdown of
intermediate garnets with the formation of an omphacite-kyanite
paragenesis--the breakdown of chrome-pyropes with the formation
of a paragenesis of clinopyroxene (containing up to 45% ureyite)
and chromite, and also to a shift in composition of other characteristic points on the paragenetic diagrams, especially to a lowering
of the calcium index of garnet, in equilibrium with two pyroxenes.
In the most deep-seated associations, the marked solubility of
sodium in garnets has been established for the first time.

The significant isomorphism of K_2O (up to 0.3%) has also been
convincingly established in the clinopyroxenes, probably formed at
pressures of more than 30 kb. It has been shown that the pyroxenes
are concentrators of K_2O at such pressures, when phlogopite and
amphibole are already unstable.

With the aid of the bipyroxene geothermometer, the equilibrium
temperatures for rocks of the non-diamond facies (from 900 to $1400^{\circ}C$)
have been established, and also the lower temperature limit for the
facies of diamond-bearing rocks ($1150^{\circ}C$). The upper limit for this
facies has been estimated at close to $1400^{\circ}C$ and somewhat above.

It has been emphasized that water-bearing minerals are absent
in the association with diamond and they evidently appear only in
the uppermost parts of the mantle. However, water, although at very
low partial pressure, is present in the fluid (P_{H_2O} 0.1), substantially lowering (by $200-300^{\circ}C$) the melting temperature. Some features,
indicating the possibility of the existence of an intergranular melt
in the low-velocity layer have been demonstrated, which supports
the conclusions of geophysicists. Kimberlite foci evidently develop
under conditions of local increase in temperature or increase in partial
pressure of water.

From the low degree of oxidation of iron in the spinels and ilmen-

ites, it has been shown that the oxidation potential decreases with
depth (in the diamond-bearing parageneses), such potential here lying
in the range between those of the terrestrial and the lunar basalts.
As a result of this, certain minerals approximate in compositional
features to those of the Moon. In particular, the olivines are
characterized by the appearance of chromium (Cr^{2+}).

The mineral compositions of the deep-seated xenoliths and inclusions in diamonds reflect the features of high-pressure conditions,
typical of the upper mantle below the platforms, at depths between
70 and 200 km. A study of these formations has shown significant
differentiation of the deep-seated material, the degree of which
even increases with depth.

The construction of paragenetic diagrams and the study of their
changes with alteration in pressure enable us, on the basis of
individual minerals in concentrates and inclusions in diamonds
(especially pyroxenes, garnets, kyanite, and spinel), to determine
the type of paragenesis, which makes it possible to make practical
use of the conclusions reached for separating the active satellites
of diamond from similar minerals and for separating the diamond-
bearing nature of the pipes on the basis of mineralogical criteria.
The most important minerals that may be used as criteria for diamond
occurrences, and primarily the depth of nucleation of the kimberlite focus, are the magnesian-chrome-garnets, poor in calcium,
the compositions of which may be determined on the basis of physical
properties from the diagram compiled by us, the most chrome-rich
spinels, and in part, the amount of chromium in the olivines.

Quite early on, cataclased pyrope peridotites were described among
the Yakutian kimberlites. Their wide distribution and the existence
of gradual transitions into porphyritic peridotites with a crystallized groundmass, have been demonstrated. It has been established
that these xenoliths belong to an extremely deep-seated type (the coesite subfacies of the graphite-pyrope facies) and to the highest-
temperature formations. Their various quantitative relationships
with the shallower peridotites in different pipes indicate the
selective nature of the transport of the xenoliths. Analogous
rocks have been identified by Boyd (1973) in the Lesotho kimberlites,
and they have been assigned to a deeper-seated type, that is, to
the diamond-pyrope facies. One of the elements of the petrographic
lattice, which determines the position of these xenoliths on the PT
diagram, is, according to F. R. Boyd, the amount of Al_2O_3 in the
orthopyroxenes. However, with a small amount of Al_2O_3 (1%), this
method does not give sufficiently precise results. A more satisfactory
barometer is evidently the limiting amount of magnesian-chrome
(knorringite) component in the garnet. Further experimental investigations are necessary here for calibration purposes.

On the basis of a study of the minerals and parageneses of these
rocks F. R. Boyd has further extended the conclusions about their
cataclasis during plate movements, and he associates the formation
of the most deep-seated kimberlite foci with the release of heat
as a result of such movements. A number of facts, the principal
of which is the marked difference in the composition of the minerals,
accurately assigned to the diamond-pyrope facies (syngenetic inclusions in diamonds), from that of the minerals of the cataclased
peridotites, contradict this hypothesis. This hypothesis, however,
is undoubtedly of interest. The very fact of the formation of unusual
mylonites at great depths in the upper mantle must attract the
serious attention of tectonists, and such rocks deserve further
profound investigation.

BIBLIOGRAPHY

AFANAS'EV, G. D., 1970: The boundary of the crust and upper mantle, in Kora i verkhnyaya mantiya Zemli (The Crust and Upper Mantle), pp. 14-24. Nauka Press, Moscow.

AGRELL, S. O. et al., 1970: Observations on the chemistry, mineralogy and petrology of some Apollo 11 lunar samples. Proc. Apollo-11 lunar sci. Conf., 1, pp. 93-128. Pergamon Press, Oxford.

ALBEE, A. L. and CHODOS, A. A., 1969: Minor element content of coexistent Al_2SiO_5 polymorphs. Am. J. Sci., 267, pp. 310-316.

ANDERSON, A. T., 1968: The oxygen-fugacity of alkaline basalt and related magmas, Tristan da Cunha. Am. J. Sci., 266, pp. 704-727.

ANDERSON, A. T., BUNCH, T. E., CAMERON, E. N. et al., 1970: Armalcolite: a new mineral from the Apollo-11 samples. Proc. Apollo-11 lunar sci. Conf., 1, pp. 55-63. Pergamon Press, Oxford.

ANDERSON, D. L. and SAMMIS, C., 1970: Partial melting in the upper mantle. Phys. earth planet. Inter., 3, pp. 41-50.

BANNO, S., 1967: Effect of jadeite components on the paragenesis of eclogitic rocks. Earth planet. Sci. Lett., 2, pp. 248-254.

BANNO, S., 1970: Classification of eclogites in terms of physical conditions of their origin. Phys. earth planet. Inter., 3, pp. 405-421.

BANNO, S., KUSHIRO, I. and MATSUI, Y., 1963: Enstatite from a garnet peridotite inclusion in kimberlite. J. geol. Soc. Japan, 69, pp. 157-159.

BARTOLOME, P., 1962: Iron-magnesium ratio in associated pyroxenes and olivines. Bull. geol. Soc. Am., Buddington Vol., pp. 1-20.

BARTOSHINSKY, Z. V., 1960: Diamonds from an eclogite in the 'Mir' kimberlite pipe. Geologiya Geofiz. Novosibirsk, No. 6, pp. 129-131.

BARTOSHINSKY, Z. V., KHAR'KIV, A. D., BOTKUNOV, A. I. and SOBOLEV, N. V., 1973: New data on diamonds from eclogites in the 'Mir' pipe. Geologiya Geofiz. Novosibirsk, No. 5, pp. 108-112.

BAUER, J., 1966: Inclusions in garnets of ultrabasic and granulitic rocks in the northern tract of the Bohemian Massif. Kristallinikum, 4, pp. 11-18.

BAUER, M. and SPENCER, L. J., 1904: Precious Stones. Griffin, London.

BELL, P. M., 1971: Analysis of olivine crystals in Apollo-12 rocks. Yb. Carnegie Instn. Wash., 69, pp. 228-229.

BELOUSOV, V. V., 1960: The development of the globe and tectogenesis. Sov. Geol., No. 7, pp. 3-27.

BELOUSOV, V. V., 1968: Some general problems of the development of the tectonosphere (crust and upper mantle), in Kora i verkhnyaya mantiya Zemli (The Crust and Upper Mantle), pp. 5-13. Nauka Press, Moscow.

BELYANKIN, D. S., LAPIN, V. V. and SHUMILO, I. M., 1941: New data on the optics and chemistry of corundum. Dokl. Akad. Nauk SSSR, 30, pp. 738-741.

BENDELIANI, N. A., POPOVA, S. V. and VERESHCHAGIN, L. F., 1966: The synthesis of a new modification of titanium dioxide, stable at high pressures. Geokhimiya, No. 5, pp. 499-501.

BERG, G. W., 1968: Secondary alterations in eclogites from kimberlite pipes. Am. Miner., 53, pp. 1336-1346.

BETEKHTIN, A. G., 1937: The Shordzha chromite-bearing peridotite massif (in Transbaikalia) and the origin of deposits of chrome

ironstone in general. in Khromity SSSR (Chromites of the USSR),
Vol. 1. Moscow.

BLAGUL'KINA, V. A., 1971: The composition of ilmenite from kimberlites. Zap. vses. miner. Obshch., 100, pp. 194-198.

BOBRIEVICH, A. P. and SOBOLEV, V. S., 1957: Eclogitization of
pyroxene crystalline schists from the Archean complex. Zap.
vses. miner. Obshch., 86, pp. 3-17.

BOBRIEVICH, A. P. and SOBOLEV, V. S., 1962: The kimberlitic association of the northern part of the Siberian Platform, in Petrografiya Vostochnoi Sibiri (The Petrography of Eastern Siberia),
Vol. 1, pp. 341-416. Akad. Nauk SSSR Press, Moscow.

BOBRIEVICH, A. P., KALYUZHNYI, V. A. and SMIRNOV, G. I., 1957:
Moissanite in the kimberlites of the East Siberian Platform.
Dokl. Akad. Nauk SSSR, 115, pp. 1189-1192.

BOBRIEVICH, A. P., SMIRNOV, G. I. and SOBOLEV, V. S., 1959b: An
eclogite xenolith with diamonds. Dokl. Akad. Nauk SSSR, 126,
pp. 637-640.

BOBRIEVICH, A. P., SMIRNOV, G. I. and SOBOLEV, V. S., 1960: The
mineralogy of xenoliths of a grossular-pyroxene-kyanite rock
(grospydite) from the Yakutian kimberlites. Geologiya Geofiz.
Novosibirsk, No. 3, pp. 18-24.

BOBRIEVICH, A. P., BONDARENKO, M. N., GNEVUSHEV, M. A. et al.,
1959a: Almaznye Mestorozhdeniya Yakutii (The Diamond Deposits
of Yakutia). Gosgeoltekhizdat, Moscow.

BOBRIEVICH, A. P., PANKRATOV, A. A., KOZLOV, I. T. et al., 1960:
The petrography of the kimberlites in the basins of the Rivers
Oleněk and Muna on the Siberian Platform. Mater. po Geol. i
polez. Iskop. YaASSR, 3, pp. 54-124.

BOBRIEVICH, A. P., ILUPIN, I. P., KOZLOV, I. T. et al., 1964:
Petrografiya i mineralogiya kimberlitovykh porod Yakutii (The
Petrography and Mineralogy of the Kimberlitic Rocks of Yakutia).
Nedra Press, Moscow.

BOETTCHER, A. L., 1970: The system $CaO-Al_2O_3-SiO_2-H_2O$ at high
pressures and temperatures. J. Petrology, 11, pp. 337-379.

BONNEY, T. G., 1899: The parent-rock of the diamond in South Africa.
Geol. Mag., dec. 4, 6, pp. 309-321.

BONNEY, T. G., 1901: Additional notes on boulders and other rock-
specimens from Newlands diamond mines, Griqualand-West. Proc.
R. Soc., 67, pp. 475-484.

BOTKUNOV, A. I., 1964: Some distribution patterns of diamonds in
the 'Mir' pipe. Zap. vses. miner. Obshch., 93, pp. 424-435.

BOWEN, N. L. and SCHAIRER, J. F., 1935: The system $MgO-FeO-SiO_2$.
Am. J. Sci., ser. 5, 29, pp. 151-217.

BOYD,JR., F. R., 1960: The quartz-coesite transition. J. geophys.
Res., 65, pp. 749-756.

BOYD, JR., F. R., 1960: Aluminous enstatite. Yb. Carnegie Instn
Wash., 59, pp. 49-52.

BOYD, JR., F. R., 1969: Electron-probe study of diopside inclusions
from kimberlite. Am. J. Sci., 267A, pp. 50-69.

BOYD, JR., F. R., 1970: Garnet peridotites and the system $CaSiO_3-MgSiO_3-Al_2O_3$. Spec. Pap. Miner. Soc. Am., 3, pp. 63-75.

BOYD, JR., F. R., 1973: A pyroxene geotherm. Geochim. cosmochim.
Acta, 37, pp. 2533-2546.

BOYD, JR., F. R. and ENGLAND, J. L., 1964: The system enstatite-
pyrope. Yb. Carnegie Instn Wash., 63, pp. 157-161.

262

BOYD, JR., F. R. and NIXON, P. H., 1970: Kimberlite diopsides. Yb. Carnegie Instn Wash., 68, pp. 324-329.

BOYD, JR., F. R. and SCHAIRER, J. F., 1964: The system $MgSiO_3$-$CaMgSi_2O_6$. J. Petrology, 5, pp. 275-309.

BOYD, JR., F. R., BELL, P. M. and FINGER, L., 1971: Pyroxenes and olivines from Oceanus Procellarum. Abstr. Lunar Sci. Conf., 1, p. 148.

BROWN, G. M., EMELEUS, C. H., HOLLAND, J. G. and PHILLIPS, R., 1970: Mineralogical, chemical and petrological features of Apollo-11 rocks and their relationship to igneous processes. Geochim. cosmochim. Acta., 34, Suppl. 1, Vol. 1, pp. 195-219.

BRYHNI, I., BOLINGBERG, H. J. and GRAFF, P. R., 1969: Eclogites in quartzo-feldspathic gneisses of Nordfjord, West Norway. Norsk geol. Tidsskr., 49, pp. 194-225.

BULTITUDE, R. J. and GREEN, D. H., 1968: Experimental study at high pressures on the origin of olivine nephelinite and olivine melilite nephelinite magmas. Earth planet. Sci. Lett., 3, pp. 325-337.

BUNCH, T. E., KEIL, K. and SNETSINGER, K. G., 1967: Chromite composition in relation to chemistry and texture of ordinary chondrites. Geochim. cosmochim. Acta, 31, pp. 1569-1582.

BUNCH, T. E., KEIL, K. and OLSEN, E., 1970: Mineralogy and petrology of silicate inclusions in iron meteorites. Contr. Mineral. Petrol., 25, pp. 297-340.

BUSECK, P. R. and KEIL, K., 1966: Meteoritic rutile. Am. Miner., 51, pp. 1506-1515.

BYKOVA, Yu. M. and GENSHAFT, Yu. S., 1972: A synthesis of chromium garnets of the pyrope-knorringite series. Geokhimiya, No. 10, pp. 1291-1294.

CARSWELL, D. A., 1968a: Picritic magma-residual dunite relationships in garnet peridotite at Kalskaret near Tafiord, South Norway. Contr. Mineral. Petrol., 19, pp. 27-124.

CARSWELL, D. A., 1968b: Possible primary upper mantle peridotite in Norwegian basal gneisses. Lithos, 1, pp. 322-355.

CARSWELL, D. A. and DAWSON, J. B., 1970: Garnet-peridotite xenoliths in South African kimberlite pipes and their petrogenesis. Contr. Mineral. Petrol., 25, pp. 163-184.

CHERKASOV, Yu. A., 1957: The use of 'focal screening' in measuring refractive indices by the immersion method, in Sovremennye metody mineralogicheskogo issledovaniya gornykh porod, rud i mineralov (Modern Methods of Mineralogic Investigations of Rocks, Ores, and Minerals), pp. 184-207. Moscow.

CHINNER, G. A., BOYD, JR., F. R. and ENGLAND, J. L., 1960: Physical properties of garnet solid solutions. Yb. Carnegie Instn Wash., 59, pp. 76-78.

CHINNER, G. A., SMITH, J. V. and KNOWLES, C. R., 1969: Transition-metal contents of Al_2SiO_5 polymorphs. Am. J. Sci., 267A, pp. 96-113.

CHUKHROV, F. V. and BONSHTEDT-KUPLETSKAYA, E. M. (Eds), 1967: Mineraly (Minerals). Vol. 2, pt. 3. Nauka Press, Moscow.

CLARK, S. P. and RINGWOOD, A. E., 1964: Density distribution and constitution of the mantle. Rev. Geophys., 2, pp. 35-88.

COES, L., 1955: High pressure minerals. J. Am. ceram. Soc., 38, p. 298.

COHEN, L. H., ITO, K. and KENNEDY, G. C., 1967: Melting and phase relations in anhydrous basalt to 40 kbars. Am. J. Sci., 265, pp. 475-518.

COLEMAN, R. G., LEE, D. E., BEATTY, L. B. and BRANNOCK, W. W., 1965:
Eclogites and eclogites: their differences and similarities.
Bull. geol. Soc. Am., 76, pp. 483-508.

CORSTORPHINE, G. S., 1908: The occurrence in kimberlite of garnet-
pyroxene nodules carrying diamonds. Trans. geol. Soc. S. Afr.,
10, pp. lxi-lxiv.

DAVIDSON, C. F., 1966: Diamantiferous diatremes. Econ. Geol., 61,
pp. 786-790.

DAVIDSON, C. F., 1967: The so-called "cognate xenoliths" of kimberlite, in Wyllie, P. J. (Ed.), Ultramafic and Related Rocks,
pp. 269-278. Wiley, New York.

DAVIS, B. T. C. and BOYD, JR., F. R., 1966: The join $Mg_2Si_2O_6$ -
$CaMgSi_2O_6$ at 30 kilobars pressure and its application to pyroxenes
from kimberlites. J. geophys. Res., 71, pp. 3567-3576.

DAWSON, J. B., 1968: Recent researches on kimberlite and diamond
geology. Econ. Geol., 63, pp. 504-511.

DAWSON, J. B. and POWELL, D. G., 1969: Mica in the upper mantle.
Contr. Mineral. Petrol., 22, pp. 233-237.

DAWSON, J. B. and REID, A. M., 1970: A pyroxene-ilmenite intergrowth
from the Monastery Mine, South Africa. Contr. Mineral. Petrol.,
26, pp. 296-301.

DEER, W. A., HOWIE, R. A. and ZUSSMAN, J., 1964: Rock Forming
Minerals, Vol. 1. Longmans, London.

DENISOV, E. P., 1970: Inclusions of ultramafic rocks in alkaline
basaltoids--a possible indicator of the composition of the upper
mantle, in Problemy stroeniya zemnoi kory i verkhnei mantii
(Problems of the Structure of the Crust and Upper Mantle),
pp. 160-171. Nauka Press, Moscow.

DOBRETSOV, N. L., KHAR'KIV, A. D. and SHEMYAKIN, M. L., 1966: The
use of multi-dimensional statistical analysis for solving prognosis
problems based on the example of diamond occurrences in kimberlites. Geologiya Geofiz. Novosibirsk, No. 8, pp. 15-22.

DOBRETSOV, N. L., ZUENKO, V. V. and KHAR'KIB, A. D., 1972: Factors
and types of diamond occurrences in the kimberlitic pipes of
Yakutia (on the basis of a statistical treatment of the data).
Geologiya Geofiz. Novosibirsk, No. 7, pp. 31-39.

DOBRETSOV, N. L., REVERDATTO, V. V., SOBOLEV, V. S., SOVOLEV, N. V.
and KHLESTOV, V. V., 1970: Fatsii metamorfizma (The Facies of
Metamorphism). Nedra Press, Moscow.

DOBRETSOV, N. L., KOCHKIN, Yu. N., KRIVENKO, A. P. and KUTOLIN, V. A.,
1971a: Porodoobrazuyushchie pirokseny (The Rock-Forming Pyroxenes).
Nauka Press, Moscow.

DOBRETSOV, N. L., LAVRENT'EV, Yu. G., POSPELOVA, L. N., SOVOLEV,
N. V. and SOBOLEV, V. S., 1971b: Features of the mineralogy and
origin of the eclogite-glaucophane complexes (as exemplified by
the South Urals). Geologiya Geofiz. Novosibirsk, No. 7, pp. 3-15.

DUNCUMB, P. and REED, S. J. B., 1968: The calculation of stopping
power and backscatter effects in electron probe microanalysis,
in Quantitative electron probe microanalysis. Spec. Publs Natn.
Bur. Stand., No. 298, pp. 133-154.

ERLANK, A. J., 1969: Microprobe investigation of potassium distribution in mafic and ultramafic nodules. EOS Trans. Am. geophys.
Un., 50, p. 343.

ERLANK, A. J. and FINGER, L. W., 1970: The occurrence of potassic
richterite in a mica nodule from the Wesselton kimberlite, South
Africa. Yb. Carnegie Instn Wash., 68, pp. 320-324.

ERLANK, A. J. and KUSHIRO, I., 1970: Potassium contents of synthetic pyroxenes at high temperatures and pressures. Yb. Carnegie Instn Wash., 68, pp. 233-236.

ESKOLA, P., 1921: On the eclogites of Norway. Skr. VidensksSelsk Christiania, I. Mat.-naturv. Kl., 8.

FERMOR, L. L., 1913: Preliminary note on garnet as a geological barometer and on an intra-plutonic zone in the Earth's crust. Rec. geol. Surv. India, 2, pp. 41-47.

FIALA, J., 1965: Pyrope of some garnet peridotites of the Czech Massif. Kristallinikum, 3, No. 4, pp. 55-74.

FIALA, J., 1966: The distribution of elements in mineral phases of some garnet peridotites from the Bohemian Massif. Kristallinikum, 4, No. 4, pp. 31-53.

FIRSOV, L. V. and SOBOLEV, N. V., 1964: The absolute age of a xenolith of eclogite from the 'Obnazhennaya' kimberlite pipe. Geologiya Geofiz. Novosibirsk, No. 10, pp. 74-77.

FORBES, R. B., 1965: The comparative chemical composition of eclogite and basalt. J. geophys. Rev., 70, pp. 1515-1521.

FRANK-KAMENETSKY, V. A., 1964: Priroda strukturnykh primesei v mineralakh (The Nature of Structural Additives in Minerals). Leningr. Gos. Univ. Press.

FRANTSESSON, E. V., 1968: Petrologiya kimberlitov (The Petrology of the Kimberlites). Nedra Press, Moscow.

FRANTSESSON, E. V. and PROKOPCHUK, B. I., 1968: Kimberlites--the tectonomagmatic facies of the alkaline-ultramafic association of the platforms, in Vulkanizm i tektogenez (Volcanism and Tectogenesis), pp. 159-164. Nauka Press, Moscow.

FRENCH, B. M., 1966: Some geological implications of equilibrium between graphite and a C-H-O gas phase at high temperatures and pressures. Rev. Geophys., 4, pp. 223-253.

FUTERGENDLER, S. I., 1956: An examination of inclusions in diamonds by the method of X-ray structural analysis. Zap. vses. miner. Obshch., 85, pp. 568-569.

FUTERGENDLER, S. I., 1958: X-ray investigation of solid inclusions in diamonds. Kristallografiya, 3, pp. 494-496.

FUTERGENDLER, S. I., 1960: An X-ray study of the solid inclusions in Uralian and Yakutian diamonds. Trudy VSEGEI, n.s., 40, pp. 73-87.

FUTERGENDLER, S. I., 1965: A case of overgrowth of rutile by diamond crystals. Geologiya Geofiz. Novosibirsk, No. 1, pp. 172-173.

FUTERGENDLER, S. I., 1969: An X-ray investigation of natural diamond with an inclusion of garnet in the form of crystallized silicate melt. Kristallografiya, 14, pp. 217-230.

FUTERGENDLER, S. I., and FRANK-KAMENETSKY, V. A., 1961: Oriented ingrowths of olivine, garnet, and chrome-spinel in diamonds. Zap. vses. miner. Obshch., 90, pp. 230-236.

FYFE, W. S., 1970: Lattice energies, phase transformations and volatiles in the mantle. Phys. Earth planet. Inter., 3, pp. 196-200.

GELLER, S., 1967: Crystal chemistry of the garnets. Z. Kristallogr., 125, pp. 1-47.

GILLER, Ya. L., 1962: The X-ray diagnosis of garnets in Rentgenografiya mineral'nogo syr'ya. Vyp. I (The X-ray Characters of Mineral Raw Materials. Part I), pp. 79096. Gosgeoltekhizdat, Moscow.

GINZBURG, I. V., 1970: Review of pyroxene systematics, in Mineraly

bazitov v svyazi s voprosami petrogenezisa (The Minerals of
Basic Rocks in Relation to Problems of Petrogenesis), pp. 5-39.
Nauka Press, Moscow.

GNEVUSHEV, M. A. and BARTOSHINSKY, Z. V., 1959: The morphology
of Yakutian diamonds. Trudy yakutsk. Fil. sib. Otd., Akad.
Nauk SSSR, sb. 4, pp. 74-92.

GNEVUSHEV, M. A. and FUTERGENDLER, S. I., 1965: Traces of magmatic
melt in diamonds. Geologiya Geofiz. Novosibirsk, No. 2, pp.
155-157.

GNEVUSHEV, M. A. and NIKOLAEVA, E. S., 1958: Inclusions of olivine
and pyrope in Yakutian diamonds. Miner. Sb. l'vovsk. geol.
Obshch., 12, pp. 440-442.

GNEVUSHEV, M. A. and NIKOLAEVA, E. S., 1961: Solid inclusions in
the diamonds from Yakutian deposits. Trudy yakutsk. Fil. sib.
Otd. Akad. Nauk SSSR, 6, pp. 97-105.

GNEVUSHEV, M. A., GOMON, G. O. and CHERNENKO, A. I., 1958: The
effect of the amount of chromium in pyrope on the height of maxima
on the spectral absorption curves. Zap. vses. miner. Obshch., 87,
pp. 85-89.

GNEVUSHEV, M. A., KALININ, A. I., MIKHEEV, V. I. and SMIRNOV, G. I.,
1956: The change in cell dimensions of garnets depending on
composition. Zap. vses. miner. Obshch., 85, pp. 472-490.

GODOVIKOV, A. A., 1968: The CaO:MgO:FeO ratio and the mineral
composition of eclogites. Dokl. Akad. Nauk SSSR, 179, pp. 668-
671.

GOLDSCHMIDT, V., 1897: Kristallographische Winkeltabellen. Berlin.

GREEN, D. H. and RINGWOOD, A. E., 1967: The genesis of balsaltic
magmas. Contr. Mineral. Petrol., 15, pp. 103-190.

GREEN, T. H., 1967: An experimental investigation of sub-solidus
assemblages formed at high pressure in high-alumina basalt, kyanite
eclogite and grosspydite compositions. Contr. Mineral. Petrol.,
16, pp. 84-114.

GRIGOR'EV, D. P., 1961: Ontogeniya mineralov (The Ontogeny of
Minerals). L'vov Univ. Press.

GROVER, J. E. and ORVILLE, P. M., 1969: The partitioning of cations
between coexisting single and multi-site phases with application
to assemblages: orthopyroxene-clinopyroxene and orthopyroxene-
olivine. Geochim. cosmochim. Acta, 33, pp. 205-226.

GRUM-GRZHIMAILO, S. V., 1946: The relationship between color and
the change in lattice parameters in the system Al_2O_3-Cr_2O_3.
Report dedicated to D. S. Belyankin. Akad. Nauk SSSR, Moscow.

GRUM-GRZHIMAILO, S. V., 1953: The properties of isomorphous mixtures
of Al_2O_3-Cr_2O_3. Trudy Inst. Kristallogr., 8, pp. 27-34.

GRUM-GRZHIMAILO, S. V. and ROVSHA, V. S., 1960: The color of the
satellite minerals of the diamond. Trudy VSEGEI, n.s., 40,
pp. 57-64.

GRUM-GRZHIMAILO, S. V. and UTKINA, E. I., 1963: The possibility
of determining the amount of chromium in ruby by the optical
method. Trudy Inst. Kristallogr., 8, pp. 99-110.

GURNEY, J. J. and SWITZER, G. S., 1973: The discovery of garnets
closely related to diamonds in the Finsch pipe, South Africa.
Contr. Mineral. Petrol., 39, pp. 103-116.

GURNEY, J. J., SIEBERT, J. C. and WHITE-COOPER, G. G., 1969: A
diamondiferous eclogite from the Roberts Victor Mine. Spec.
Publs geol. Soc. S. Afr., 2, pp. 351-357.

GURNEY, J. J., MATHIAS, Morna, SIEBERT, J. C. and MOSELEY, G., 1971:

Kyanite eclogites from the Rietfontein kimberlite pipe, Mier Coloured Reserve, Gordonia, Cape Province. Contr. Mineral. Petrol., 30, pp. 46-52.

GURULEV, S. A., KOSTYUK, V. P., MANUILOVA, M. M. and RAFIENKO, N. I., 1965: The discovery of a blue diopside in Siberia. Dokl. Akad. Nauk SSSR, 163, pp. 443-446.

GUTENBERG, B., 1951: Internal Constitution of the Earth. (2nd Ed.). Dover, New York.

HAGGERTY, S. E., 1973: The chemistry and genesis of opaque minerals in kimberlites. Abstr. Vol., Int. Confer. on Kimberlites, 147-150. Capetown.

HAGGERTY, S. E. and MEYER, H. O. A., 1971: Apollo 12: opaque oxides. Abstr. Vol., 2nd Lunar Sci. Confer., p. 98.

HAGGERTY, S. E., BOYD, JR., F. R., BELL, P. M., FINGER, L. W., and BRYAN, W. B., 1970: Opaque minerals and olivine in lavas and breccias from Mare Tranquillitatis. Geochim. cosmochim. Acta, 34, Suppl. 1, Vol. 1, pp. 513-538.

HARIYA, Y. and KENNEDY, G. C., 1968: Equilibrium study of anorthite under pressure and high temperature. Am. J. Sci., 266, pp. 193-203.

HARRIS, J. W., 1968a: The recognition of diamond inclusions. Pt. 1. Syngenetic mineral inclusions. Indust. Diam. Rev., 28, No. 334, pp. 402-410.

. . . . 1968b: The recognition of diamond inclusions. Pt. 2. Indust. Diam. Rev., 28, No. 335, pp. 458-461.

HARRIS, P. G., REAY, A. and WHITE, I. G., 1967: Chemical composition of the upper mantle. J. geophys. Res., 72, pp. 6359-6369.

HARTMAN, P., 1954: On oriented olivine inclusions in diamond. Am. Miner., 39, pp. 624-625.

HARTMAN, P. 1969: Can Ti^{4+} replace Si^{4+} in silicates? Mineralog. Mag., 37, pp. 366-369.

HENTSCHEL, H., 1937: Der Eklogite von Gilsberg im Sachsischen Granulitgebirge und seine metamorphen Umwandlungsstoffen. Tschermaks miner. petrogr. Mitt., 49, 1, pp. 42-88.

HOLM, J. L., KLEPPA, O. J. and WESTRUM, E. F., 1967: Thermodynamics of polymorphic transformations in silica. Thermal properties from 5 to 1070°K and pressure-temperature stability field for coesite and stishovite. Geochim. cosmochim. Acta, 31, pp. 2289-2307.

HOLMES, A. and PANETH, F. A., 1936: Helium ratios of rocks and minerals from the diamond pipes of South Africa. Proc. R. Soc., 154A, pp. 385-413.

HOWIE, R. A. and WOOLEY, A. R., 1968: The role of titanium and the effect of TiO_2 on the cell-size, refractive index, and specific gravity in the andradite-melanite-schorlomite series. Mineralog. Mag., 36, pp. 775-790.

ILUPIN, I. P. and KOZLOV, I. T., 1970: Zircon in kimberlites, in Geologiya, petrografiya i mineralogiya magmaticheskikh obrazovaniy severo-vostochnoy chasti Sibirskoy platformy (The Geology, Petrography, and Mineralogy of Magmatic Formations in the North-eastern Part of the Siberian Platform), pp. 254-266. Nauka Press, Moscow.

ILUPIN, I. P. and NAGAEVA, N. P., 1970: Chromium and nickel in an ilmenite from the Yakutian kimberlites, in Geologiya, petrografiya i mineralogiya magmaticheskikh obrazovaniy severo-vostochnoy chasti Sibirskoy platformy (The Geology, Petrography, and Miner-

alogy of the Northeastern Part of the Siberian Platform), pp.
288-300. Nauka Press, Moscow.

ILUPIN, I. P., LEVSHOV, P. P. and ROVSHA, V. S., 1969: The composition
and origin of the kelyphitic rims on garnet-pyrope in the Yakutian
kimberlites. Uchen. Zap. nauchno-issled. Inst. Geol. Arkt.
Regional'naya Geol., 16, pp. 45-52.

IL'VITSKY, M. I. and KOLBANTSEV, R. V., 1968: Paragenetic types
and a statistical analysis of the chemistry of olivines. Dokl.
Akad. Nauk SSSR, 179, pp. 1428-1431.

IRVINE, T. N., 1965: Chromian spinel as a petrogenetic indicator.
Pt. 1. Theory. Can. Jl earth Sci., 2, pp. 648-672.

IRVINE, T. N. and KENNEDY, G. C., 1967: Chromian spinel as a
petrogenetic indicator. Pt. 2.--Petrologic applications.
Can. Jl earth Sci., 4, pp. 71-103.

ITO, K. and KENNEDY, G. C., 1967: Melting and phase relations in
a natural peridotite to 40 kilobars. Am. J. Sci., 265, pp.
519-538.

ITO, K. and KENNEDY, G. C., 1968: Melting and phase relations in
the plane tholeiite-lherzolite-nepheline basanite to 40 kbars
with geological implications. Contr. Mineral. Petrol., 19,
pp. 177-211.

JACKSON, E. D., 1968: The character of lower crust and upper mantle
beneath the Hawaiian Islands. Int. geol. Congr., 23, Proc. 1,
pp. 135-150.

KADIK, A. A. and KHITARKOV, N. I., 1970: Influence of water on
melting of silicates at high pressure. Phys. Earth planet. Sci.,
3, pp. 343-347.

KHAR'KIV, A. D. and MAKOVSKAYA, N. S., 1970: Chromium and titanium
in garnets from Yakutian kimberlites. Dokl. Akad. Nauk SSSR, 193,
pp. 173-176.

KHAR'KIV, A. D., SOBOLEV, N. V. and CHUMIRIN, K. G., 1972: Inclusions
of chrome-diopside in zircon from kimberlites in the rocks of the
Malo-Botuoba region. Zap. vses. miner. Obshch., 101, pp. 431-433.

KHITAROV, N. I., KADIK, A. A. and LEBEDEV, E. V., 1968: The solubility
of water in a basalt melt. Geokhimiya, No. 7, pp. 763-772.

KLEEMAN, J. D. and COOPER, J. A., 1970: Geochemical evidence for
the origin of some ultramafic inclusions from Victorian basanites.
Phys. Earth planet. Inter., 3, pp. 302-308.

KNOWLES, C. R., SMITH, J. V., BENCE, A. E. and ALBEE, A. L., 1969:
X-ray emission microanalysis of rock-forming minerals. VII.
Garnets. J. Geol., 77, pp. 439-451.

KOLESNIK, Yu. N., and GULETSKAYA, E. S., 1968: Paragenetic types
of corundum. Geologiya Geofiz. Novosibirsk, No. 2, pp. 128-132.

KORZHINSKY, D. S., 1936: A paragenetic analysis of quartz-bearing,
calcium-poor crystalline schists of the Archean complex of the
South Baikal region. Zap. vses. miner. Obshch., 65, pp. 247-280.

KORZHINSKY, D. S., 1940: Factors in the mineral equilibria and the
mineralogic depth facies. Trudy Inst. geol. Nauk Akad. Nauk
SSSR, ser. petrogr., 12, No. 5, p. 99.

KUOVO, O. and VUORELAINEN, Y., 1958: Eskolaite, a new chromium
mineral. Am. Miner., 43, p. 1098.

KOVAL'SKY, V. V. and YEGOROV, O. S., 1964: The content of xenoliths
in explosive kimberlite breccias and a method of calculating it.
Geologiya Geofiz. Novosibirsk, No. 11, pp. 140-143.

KRASNOV, I. I., LUR'E, M. L. and MASAITIS, V. L. (Eds), 1966:
Geologiya sibirskoi platformy (The Geology of the Siberian Platform).
Nedra Press, Moscow.

268

KRESTEN, P., 1973: Kimberlitic zircons. Abstr. Vol., Int. Confer. on Kimberlites, pp. 191-194. Capetown.

KRETZ, R., 1961: Some applications of thermodynamics to coexisting minerals of variable compositions. J. Geol., 69, pp. 361-387.

KROPOTOVA, O. I. and FEDORENKO, B. V., 1970: The isotope composition of carbon in diamond and graphite from eclogite. Geokhimiya, No. 10, p. 1279.

KRYUKOV, A. V., 1968: Inclusions of pyrope peridotite as an indicator of the relation between kimberlites and alkaline basaltoids, in Kora i verkhnyaya mantiya Zemli (The Crust and Upper Mantle), pp. 141-145. Nauka Press, Moscow.

KUKHARENKO, A. A., 1955: Almazy Urala (The Diamonds of the Urals). Gosgeoltekhizdat, Moscow.

KUNO, H., 1969: Mafic and ultramafic inclusions in basaltic rocks and the nature of the upper mantle, in The Earth's Crust and Upper Mantle. Monogr. Am. geophys. Un., 13, pp. 507-513.

KUSHIRO, I., 1970: Stability of amphibole and phlogopite in the upper mantle. Yb. Carnegie Instn Wash., 68, pp. 245-247.

KUSHIRO, I. and AOKI, K., 1968: Origin of some eclogite inclusions in kimberlite. Am. Miner., 53, pp. 1347-1367.

KUSHIRO, I. and YODER, JR., H. S. 1965: The reactions between forsterite and anorthite at high pressures. Yb. Carnegie Instn Wash., 64, pp. 89-94.

KUSHIRO, I., SYONO, Y. and AKIMOTO, S., 1967a: Effect of pressure on garnet-pyrope equilibrium in the system $MgSiO_3$-$CaSiO_3$-Al_2O_3. Earth planet. Sci. Lett., 2, pp. 460-464.

KUSHIRO, I., SYONO, Y. and AKIMOTO, S., 1967b: Stability at high pressures and possible presence of phlogopite in the Earth's upper mantle. Earth planet. Sci. Lett., 3, pp. 197-203.

KUTOLIN, V. A., 1968: Petrochemical features of basalt of different associational types and the composition of the upper mantle of the Earth, in Kora i verkhnyaya mantiya Zemli (The Crust and Upper Mantle), pp. 90-97. Nauka Press, Moscow.

KUTOLIN, V. A., 1970: Ultrabasic nodules in basalts and the upper mantle composition. Earth planet. Sci. Lett., 7, pp. 320-322.

LAMBERT, I. B. and WYLLIE, P. J., 1970: Low-velocity zone of the Earth's mantle. Incipient melting caused by water. Science, N. Y., 169, pp. 764-766.

LAMPRECHT, A. and WOERMANN, E., 1970: Das System FeO-SiO_2-TiO_2 bei 1130°C und Drucken bis 30 kb. Naturwissenschaften, 57, No. 4, p. 191

LAVRENT'EV, Yu. G. and VAINSHTEIN, É. E., 1965: The effect of instrumental error on the precision and sensitivity of X-ray spectral analysis. Report 1. Zh. anal. Khim., 20, pp. 918-926.

LEBEDEV, V. I., 1959: Some results of a study of garnets from metamorphosed basic rocks and gneisses from the White Sea region. Vestn. leningr. Univ., 18, No. 3, pp. 5-20.

LEGG, C. A., 1969: Some chromite-ilmenite associations in the Merensky Reef, Transvaal. Am. Miner., 54, pp. 1347-1354.

LEIPUNSKY, O. I., 1939: Artificial diamonds. Uspekhi Khimii, 8, pp. 1519-1534.

LEONT'EV, L. N. and KADENSKY, A. A., 1957: The nature of the Yakutian kimberlite pipes. Dokl. Akad. Nauk SSSR, 115, pp. 368-371.

LINDSLEY, D. H., 1963: Equilibrium relations of coexisting pairs of Fe-Ti oxides. Yb. Carnegie Instn Wash., 62, pp. 60-66.

LOVERING, J. F., 1958: The nature of the Mohorovičić discontinuity. Trans. Am. geophys. Un., 39, pp. 947-955.

LUTTS, B. G., 1965: The eclogitization reaction in deep-seated rocks.²⁶⁹
  Geologiya rudn. Mestorozh., 7, pp. 18-30.
LUTTS, B. G. and MARSHINTSEV, V. K., 1963: A xenolith of garnet
  pyroxenite from the 'Mir' kimberlite pipe. Trudy yakutsk. Fil.
  sib. Otd. Akad. Nauk SSSR, vyp. 10.
McCONNELL, D., 1966: Propriétés physiques des grenats. -- Calcul
  de dimension de la maille unité à partir de la composition chimique.
  Bull. Soc. fr. Minér. Cristallogr., 89, No. 1, pp. 14-17.
McGETCHIN, T. R. and SILVER, L. T., 1970: Compositional relations in
  minerals from kimberlite and related rocks from the Moses Rocks
  dike, San Juan County, Utah. Am. Miner., 55, pp. 1738-1771.
MacGREGOR, I. D., 1964: The reaction 4 enstatite + spinel   forsterite
  + pyrope. Yb. Carnegie Instn Wash., 63, p. 157.
MacGREGOR, I. D., 1965: Stability fields of spinel and garnet
  peridotites in the synthetic system MgO-CaO-Al₂O₃-SiO₂. Yb.
  Carnegie Instn Wash., 64, pp. 126-134.
MacGREGOR, I. D., 1967: Model mantle compositions, in Wyllie, P. J.
  (Ed.), Ultramafic and Related Rocks, pp. 382-392.
MacGREGOR, I. D., 1970: The effect of CaO, Cr₂O₃, Fe₂O₃, and Al₂O₃
  on the stability of spinel and garnet peridotites. Phys. Earth
  planet. Inter., 3, pp. 372-377.
MacGREGOR, I. D. and CARTER, J. L., 1970: The chemistry of clino-
  pyroxenes and garnets of eclogite and peridotite xenoliths from
  the Roberts Victor Mine, South Africa. Phys. Earth planet. Inter.,
  3, pp. 391-397.
MAGNITSKY, V. A., 1968: Sloi nizkikh skorostei verkhnei mantii
  Zemli (The Low-Velocity Layer of the Earth's Mantle). Nauka
  Press, Moscow.
MAGNITSKY, V. A., 1971: Geothermal gradients and temperatures
  in the mantle and the problem of fusion. J. geophys. Res., 76,
  pp. 1391-1396.
MALINOVSKY, I. Yu., DOROSHEV, A. M. and GODOVIKOV, A. A., 1973:
  The stability of garnets of the pyrope-grossular-uvarovite series
  at T = 1200°C and P + 30 kb. Proc. 9th All-Union Confer. on
  Exper. Mineralogy, Irkutsk.
MANNING, P. G. and HARRIS, D., 1970: Optical absorption and electron
  microprobe studies of some high temperature andradites. Can.
  Mineralogist, 10, pp. 260-271.
MANTON, W. I. and MATSUMOTO, M., 1969: Isotopic composition of lead
  and strontium in nodules from the Roberts Victor Mines, South
  Africa. Los Trans. Am. geophys. Un., 50, p. 343.
MAO, H. K., 1971: The system jadeite (NaAlSi₂O₆)--anorthite (CaAl₂Si₂O₆)
  at high pressures. Yb. Carneigie Instn Wash., 69, pp. 163-168.
MARAKUSHEV, A. A., 1965: Problemy mineral'nykh fatsii metamorficheskikh
  i metasomaticheskikh gornykh porod (The Problems of the Mineral
  Facies of the Metamorphic and Metasomatic Rocks). Nauka Press,
  Moscow.
MARAKUSHEV, A. A., 1968: The effect of temperature on the orthopyroxene-
  clinopyroxene and orthopyroxene-olivine equilibria in Metasomatizm
  i drugie voprosy fiziko-khimicheskoi petrologii (Metasomatism
  and Other Problems of Physicochemical Petrology), pp. 31-52. Nauka
  Press, Moscow.
MARKOV, V. K., PETROV, V. P., DELITSIN, I. S. and RYABININ, Yu. N.,
  1966: The cornersion of phlogopite at high pressures and temperatures. Izv. Akad. Nauk SSSR, ser. geol., No. 6, pp. 10-20.
MARSHINTSEV, V. K., SHCHELCHKOVA, S. G., ZOL'NIKOV, G. V. and

VOSKRESENSKAYA, V. B., 1967: New data on moissanite from the Yakutian kimberlites. Geologiya Geofiz. Novosibirsk, No. 12, pp. 22-31.

MATHIAS, Morno and RICKWOOD, P. C., 1969: Ultramafic xenoliths in the Matsoku kimberlite pipe, Lesotho. Spec. Publs geol. Soc. S. Afr., 2, pp. 358-369.

MATHIAS, Morno and RICKWOOD, P. C., 1970: Some aspects of the mineralogy and petrology of ultramafic xenoliths in kimberlite. Contr. Mineral. Petrol., 26, pp. 75-123.

MATTHES, S., RICHTER, P. and SCHMIDT, K., 1970: Die Eklogitvorkommen des kristallinen Grundgebirges in NE-Bayern. III. Der Disthen (kyanit) der Eklogite und Eklogitamphibolite des Münchberger Gneisgebietes. Neues Jb. Miner. Abh., 113, pp. 111-137.

MEDARIE, JR., L. G., 1969: Partitioning of Fe^{2+} and Mg^{2+} between coexisting synthetic olivine and orthopyroxene. Am. J. Sci., 267, pp. 945-968.

MENYAILOV, A. A., 1962: Some patterns of distribution, structure, and formation of diamond basement deposits in Yakutia. Trudy yakutsk. Fil. sib Otd. Akad. Nauk SSSR, 8, pp. 5-18.

MEYER, H. O. A., 1968a: Mineral inclusions in diamonds. Yb. Carnegie Instn Wash., 66, pp. 446-450.

MEYER, H. O. A., 1968b: Chrome pyrope: an inclusion in natural diamond. Science, N. Y., 160, pp. 1446-1447.

MEYER, H. O. A. and BOYD, JR., F. R., 1972: Composition and origin of crystalline inclusions in natural diamonds. Geochim. cosmochim. Acta, 36, pp. 1255-1274.

MEYER, H. O. A. and BOYD, JR., F. R., 1970: Inclusions in diamonds. Yb. Carnegie Instn Wash., 68, pp. 315-320.

MEYER, H. O. A. and SVIZERO, D. P., 1973: Mineral inclusions in Brazilian diamonds. Abstr. Vol., Int. Confer. on Kimberlites, pp. 225-228.

MIKHAILOV, N. P. and ROVSHA, V. S., 1965: The effect of pressure on the paragenesis of ultramafic rocks. Dokl. Akad. Nauk SSSR, 160, pp. 1175-1178.

MIKHAILOV, N. P. and ROVSHA, V. S., 1966: Pyrope-bearing peridotites of the Bohemian Massif and their genesis. Kristallinikum, 4, pp. 87-107.

MIKHEENKO, V. N., VLADIMIROV, B. M., NENASHEV, N. I., and SEL'DISHEVA, E. B., 1970: A cobble of diamond-bearing eclogite from a kimberlite in the 'Mir' pipe. Dokl. Akad. Nauk SSSR, 190, pp. 1440-1443.

MIKHEEV, V. I. and KALININ, A. I., 1961: A comparison of the magnetic properties, density, and unit-cell parameters of ilmenites. Zap. leningr. gorn. Inst., 38, vyp. 2, pp. 73-98.

MILASHEV, V. A., 1960: Cognate inclusions in the 'Obnazhennaya' kimberlite pipe (basin of River Olenëk). Zap. vses. miner. Obshch., 89, pp. 284-299.

MILASHEV, V. A., 1965: The petrochemistry of the Yakutian kimberlites and factors in their diamond occurrences. Trudy nauchno-issled. Inst. Geol. Arkt., 139.

MILASHEV, V. A. and SHUL'GINA, N. I., 1959: New data on the age of the kimberlites of the Siberian Platform. Dokl. Akad. Nauk SSSR, 126, pp. 1320-1322.

MILASHEV, V. A., KRUTOYARSKY, M. A., RABKIN, M. I. and ÉRLIKH, É. N., 1963: The kimberlitic rocks and picritic porphyries of the northeastern part of the Siberian Platform. Trudy nauchno-issled. Inst. Geol. Arkt., 126.

MIRKIN, L. I., 1961: Spravochnik po rentgenostrukturnomu analizu polikristallov (Handbook for the X-ray Analysis of Polycrystals). Fizmatgiz, Moscow.

MITCHELL, R. S. and GIARDINI, A. A., 1953: Oriented olivine inclusions in diamond. Am. Miner., 38, pp. 136-138.

MOOR, G. G. and SOBOLEV, V. S., 1957: The problem of the Siberian kimberlites. Mineralog. Sb. l'vovsk. geol. Obshch., 11, pp. 369-371.

MRHA, J., 1900: Beitrage zur Kenntnis des Kelyphits. Tschermaks mineral. petrogr. Mitt., 19, pp. 111-143.

NALIMOV, V. V., 1960: Primenenie matematicheskoi statistiki pri analize veshchestva (The Use of Mathematical Statistics in Analyzing Information). Fizmatgiz, Moscow.

NEUHAUS, A., JUMPERTZ, E. and BREHNER, P., 1962: Über die Mischbarkeit von Korund und Cr_2O_3 in Hydrothermalversuch bei T bis 800°C. Proc. 3rd Congr. Europ. Federation Chem. Engrg. London (Kongressbericht).

NIXON, P. H. and HORNUNG, G. A., 1969: A new chromium garnet end member, knorringite, from kimberlite. Am. Miner. 53, pp. 1833-1840.

NIXON, P. H., KNORRING, O. von and ROOKE, J. M., 1963: Kimberlites and associated inclusions of Basutoland; a mineralogical and geochemical study. Am. Miner., 48, pp. 1090-1132.

NOCKOLDS, S. R., 1954: Average chemical compositions of some igneous rocks. Bull. geol. Soc. Am., 65, pp. 1007-1032.

ODINTSOV, M. M. and STRAKHOV, L. G., 1968: Trap and kimberlite pipes as an index of features of structural development of the continental crust of ancient platforms, in Vulkanizm i tektogenez (Volcanism and Tectogenesis) pp. 165-173. Nauka Press, Moscow.

O'HARA, M. J., 1967: Mineral parageneses in ultrabasic rocks, in Wyllie, P. J. (Ed.), Ultramafic and Related Rocks, pp. 393-402. Wiley, New York.

O'HARA, M. J., 1970: Upper mantle composition inferred from laboratory experiments and observations on volcanic products. Phys. Earth planet. Inter., 3, pp. 236-245.

O'HARA, M. J. and MERCY, E. L. P., 1963: Petrology and petrogenesis of some garnetiferous peridotites. Trans. R. Soc. Edinb., 65, pp. 251-314.

O'HARA, M. J. and MERCY, E. L. P., 1966a: Eclogites, peridotites and pyrope from the Navajo Country, Arizona and New Mexico. Am. Miner., 51, pp. 336-352.

O'HARA, M. J. and MERCY, E. L. P., 1966b: Exceptionally calcic pyralspite from South African kyanite eclogite. Nature, Lond., 212, pp. 68-69.

O'HARA, M. J. and YODER, JR., H. S., 1967: Formation and fractionation of basic magmas at high pressure. Scott. J. Geol., 3, pp. 67-117.

ORLOV, Yu. L., 1959: Syngenetic and epigenetic inclusions in diamond crystals. Trudy miner. Muz., 10, pp. 103-120.

OSBORN, E. F., 1959: Role of oxygen pressure in the crystallization and differentiation of basaltic magmas. Am. J. Sci., 230, pp. 57-71.

OXBURGH, E. R., 1964: Petrological evidence for the presence of amphibole in the upper mantle and its petrogenetic and geophysical implications. Geol. Mag., 101, pp. 1-19.

OZEROV, K. L. and BYKHOVER, N. A., 1936: Deposits of corundum and

272

kyanite in the Upper Timpton region of the Yakutian SSSR. Trudy
tsent. nauchno-issled. geol.-razv. Inst., 82.
PANKRATOV, A. A., 1960: Garnets from the Yakutian kimberlite pipes.
Izv. vost.-sib. Fil. Akad. Nauk SSSR, 2, pp. 52-60.
PAVLOV, N. V., KRAVCHENKO, G. G. and CHUPRYNINA, I. I., 1968:
Khromity Kempirsaiskogo plutona (Chromites of the Kempirsai
Pluton). Nauka Press, Moscow.
PERCHUK, L. L., 1967: The pyroxene-garnet equilibrium and the
depth facies of eclogites. Izv. Akad. Nauk SSSR, ser. geol.,
No. 11, pp. 41-86.
PERCHUK, L. L., 1970: Ravnovesiya porodoobrazuyushchikh mineralov
(The Equilibria of the Rock-Forming Minerals). Nauka Press,
Moscow.
PETROV, V. S., 1959: The genetic association between diamonds and
the carbonatites of kimberlites. Vestn. mosk. Univ., No. 2,
pp. 13-20.
PHILIBERT, J., 1963: A method for calculating the absorption correc-
tion in electron-probe microanalysis, in Pattee, H. H. et al. (Eds),
X-ray Optics and X-ray Microanalysis, pp. 379-392. Academic Press,
New York.
PONOMARENKO, A. I., PONOMARENKO, G. A., KHAR'KIV, A. D., and SOBOLEV,
N. V., 1972: New data on the mineralogy of inclusions in ilmenite
ultramafics from the kimberlite pipes of Western Yakutia. Dokl.
Akad. Nauk SSSR, 207, pp. 946-949.
PRINZ, M., MANSON, D. V., HLAVA, P. F. and KEIL, K., 1973: Inclusions
in diamonds: garnet lherzolite and eclogite assemblages. Abstr.
Vol., Int. Confer. on Kimberlites, Capetown.
RÅHEIM, A. and GREEN. D. H., 1974: Experimental determination of
the temperature and pressure dependence of the Fe-Mg partition
coefficient for coexisting garnet and clinopyroxene. Contr. Miner.
Petrol., 48, pp. 179-203.
REED, S. J. B., 1965: Characteristic fluorescence correction in
electron probe microanalysis. Brit. J. Appl. Phys., 16, p. 913.
REID, A. M., BROWN, R. W. and DAWSON, J. B., 1973: Diamond-bearing
eclogites. Abstr. Vol., Int. Confer. on Kimberlites, Capetown,
p. 271.
REID, JR., J. B. and FREY, F. A., 1971: Rare earth distribution in
lherzolite and garnet pyroxenite xenoliths and the constitution
of the upper mantle. J. geophys. Res., 76, pp. 1184-1196.
RICKWOOD, P. C., 1968: On recasting analyses of garnet into end-
member molecules. Contr. Mineral. Petrol., 18, pp. 175-198.
RICKWOOD, P. C. and MATHIAS, Morna, 1970: Diamondiferous eclogite
xenoliths in kimberlite. Lithos, 3, pp. 223-225.
RICKWOOD, P. C., MATHIAS, Morna, and SIEBERT, J. C., 1968: A study
of garnets from eclogite and peridotite xenoliths found in
kimberlite. Contr. Mineral. Petrol. 19, pp. 271-301.
RICKWOOD, P. C., GURNEY, J. J. and WHITE-COOPER, D. R., 1969: The
nature and occurrences of eclogite xenoliths in the kimberlites
of Southern Africa. Spec. Publs geol. Soc. S. Afr., 2, pp.
371-393.
RINGWOOD, A. E., 1958: Constitution of the mantle. 3. Geochim.
cosmochim. Acta, 15, pp. 195-212.
RINGWOOD, A. E., 1962: A model for the upper mantle. Pt. 1. J.
geophys. Res., 67, pp. 857-867; Pt. 2. idem, pp. 4473-4477.
RINGWOOD, A. E., 1966: The mineralogy of the mantle. in Advances
in Earth Sciences, pp. 257-417. MIT Press, Boston.

RINGWOOD, A. E., 1967: The pyroxene-garnet transformation in the Earth's mantle. Earth planet. Sci. Lett., 2, pp. 255-263.

RINGWOOD, A. E., 1969: Composition and evolution of the upper mantle. Geophys. Monogr. Am. geophys. Un., 13, pp. 1-17.

RINGWOOD, A. E., 1970: Phase transformations and the composition of the mantle. Phys. Earth planet. Inter., 3, pp. 109-155.

RINGWOOD, A. E. and GREEN, D. H., 1966: An experimental investigation of the gabbro-eclogite transformation and some geophysical implications. Tectonophysics, 3, pp. 383-427.

RINGWOOD, A. E. and LOVERING, J. F., 1970: Significance of pyroxene-ilmenite intergrowths among kimberlite xenoliths. Earth planet. Sci. Lett., 7, pp. 371-375.

RINGWOOD, A. E. and MAJOR, A., 1966: Synthesis of Mg_2SiO_4-$FeSiO_4$ spinel solid solutions. Earth planet. Sci. Lett., 1, pp. 241-245.

RINGWOOD, A. E. and MAJOR, A., 1971: Synthesis of majorite and other high pressure garnets and perovskites. Earth planet. Sci. Lett., 12, pp. 411-418.

ROVSHA, V. S. and ILUPIN, I. P., 1970: Chrome-spinels in Yakutian kimberlites. Geologiya Geofiz. Novosibirsk, No. 2, pp. 47-56.

SARSADSKIKH, N. N., 1970: Exploration for diamond deposits on the basis of mineral satellites. Inform. Sb. VSEGEI, No. 5, pp. 122-132.

SARSADSKIKH, N. N., 1970: The heterogeneity of the material in the upper mantle. Dokl. Akad. Nauk SSSR, 193, pp. 1392-1395.

SARSADSKIKH, N. N., BLAGUL'KINA, V. A. and SILIN, Yu. I, 1966: The absolute age of the Yakutian kimberlites. Dokl. Akad. Nauk SSSR, 168, pp. 420-423.

SARSADSKIKH, N. N., ROVSHA, V. S. and BLAGUL'KINA, V. A., 1960: Minerals of inclusions of pyrope peridotites in the kimberlites of the Daldyn-Alakit diamond-bearing region. Mater. VSEGEI, n. s., 40, pp. 37-55.

SASTRY, G. G., 1962: Determination of the end-member composition of garnets from their physical properties. Rec. geol. Surv. India, 87, pp. 757-880.

SATO, M. and HELT, R. T., 1971: Oxygen fugacity values of Apollo-12 basaltic rocks. Abstr. 2nd Lunar Sci. Confer., p. 144.

SATO, M. and WRIGHT, T. L., 1966: Oxygen fugacities directly measured in magmatic gases. Science, N. Y., 153, pp. 1103-1105.

SCLAR, C. B., 1970: High pressure studies in the system MgO-SiO_2-H_2O. Phys. Earth planet. Inter., 3, p. 333.

SEELIGER, E. A. and MÜCKE, A., 1969: Donathit ein tetragonaler Zn-reicher Mischkristal von Magnetit und Chromit. Neues Jb. Miner. Monatsh., 2, pp. 49-57.

SEIFERT, F. and LANGER, K., 1970: Stability relations of Cr-kyanite at high P-T. Contr. Mineral. Petrol., 28, pp. 9-18.

SHAFRANOVSKY, I. I., 1960: Lektsii po kristallomorfologii mineralov (Lectures on the Crystallo-morphology of Minerals), L'vov. Univ. Press.

SHAFRANOVSKY, I. I., 1961: Kristally mineralov (Mineral Crystals). Nauka Press, Moscow.

SHARP, W. E., 1966: Pyrrhotite: a common inclusion in South African diamonds. Nature, Lond., 211, pp. 204-403.

SHCHELCHKOVA, S. G. and BROVKIN, A. A., 1969: 'Titanolivine' from Siberian kimberlites. Zap. vses. miner. Obshch., 98, pp. 246-247.

SHEINMANN, Yu. M., 1968: Ocherki glubinnoi geologii (Reviews on Deep-Seated Geology). Nedra Press, Moscow.

274

SHIMIZU, N., 1971: Potassium contents of synthetic clinopyroxenes at high pressures and temperatures. Earth planet. Sci. Lett., 11, pp. 374-380.

SIIVOLA, J., 1969: On the evaporation of some alkali metals during the electron microprobe analysis. Bull. geol. Soc. Finland, 41, pp. 85-91.

SIMKIN, T. and SMITH, J. V., 1970: Minor-element distribution in olivine. J. Geol., 78, pp. 304-325.

SKINNER, B. J., 1956: Physical properties of end-members of the garnet group. Am. Miner., 41, pp. 428-436.

SMIRNOV, G. I., 1970: Protomagmatic phase of mineral formation in kimberlites. Geologiya Geofiz. Novosibirsk, No. 12, pp. 14-21.

SMITH, J. V., 1965: X-ray emission microanalysis of rock-forming minerals. 1. Experimental techniques. J. Geol., 73, pp. 830-864.

SMULIKOWSKI, K., 1964: An attempt at eclogite classification. Bull. Acad. Pol. Sci., ser. geol. geogr., 12, 1, pp. 27-33.

SMULIKOWSKI, K., 1968: Theoretical and geological arguments for eclogite occurrence in the upper mantle. Rept. Int. geol. Congr., 23, Abstr. Vol., Upper Mantle, pp. 34-35. Prague.

SNETSINGER, K. G. et al., 1967: Chromite from "equilibrated" chondrites. Am. Miner., 52, pp. 1322-1331.

SOBOLEV, E. V., LENSKAYA, S. V., LISOIVAN, V. I., et al., 1966: Some physical properties of diamonds from Yakutian eclogite. Dokl. Akad. Nauk SSSR, 168, pp. 1151-1153.

SOBOLEV, N. V., 1964a: Parageneticheskie tipy granatov (Paragenetic Types of Garnets). Nauka, Press, Moscow.

SOBOLEV, N. V., 1964b: An eclogite xenolith from ruby. Dokl. Akad. Nauk SSSR, 157, pp. 1382-1384.

SOBOLEV, N. V., 1964c: Orthopyroxenes from garnet peridotites and eclogites. Dokl. Akad. Nauk SSSR, 154, pp. 1096-1098.

SOBOLEV, N. V., 1968a: The xenoliths of eclogites from the kimberlite pipes of Yakutia as fragments of the upper mantle substance. Rept. Int. geol. Congr., 23, 1, pp. 155-163. Prague.

SOBOLEV, N. V., 1968b: Eclogite clinopyroxenes from the kimberlite pipes of Yakutia. Lithos, 1, pp. 54-57.

SOBOLEV, N. V., 1969: Peridotites with chromium-rich garnets and diamond-bearing eclogites as the most deep-seated xenoliths in kimberlites. Abstr. Vol., Sympos. Volcanoes and Their Roots, pp. 45-47. Oxford.

SOBOLEV, N. V., 1970: Eclogites and pyrope peridotites from the kimberlites of Yakutia. Phys. Earth planet. Inter., 3, pp. 398-404.

SOBOLEV, N. V., 1971a: The mineralogical criteria for diamond-bearing kimberlites. Geologiya Geofiz. Novosibirsk, No. 3, pp. 70-80.

SOBOLEV, N. V., 1971b: Some specific features of distribution and transportation of xenoliths in the kimberlitic pipes of Yakutia. J. geophys. Res., 76, pp. 1309-1314.

SOBOLEV, N. V., BARTOSHINSKY, Z. V. et al., 1970: The olivine-garnet-chromediopside association in a Yakutian diamond. Dokl. Akad. Nauk SSSR, 192, pp. 1349-1352.

SOBOLEV, N. V., BOTKUNOV, A. I. and KUZNETSOVA, I. K., 1969: A diamond-bearing eclogite with calcium-enriched garnet from the 'Mir' pipe, Yakutia. Geologiya Geofiz. Novosibirsk, No. 4, pp. 125-128.

SOBOLEV, N. V., BOTKUNOV, A. I. and LAVRENT'EV, Yu. G., 1974: A diamond-bearing corundum eclogite from the 'Mir' kimberlite pipe, Yakutia. Geologiya Geofiz. Novosibirsk, No. 5.

SOBOLEV, N. V., BOTKUNOV, A. I. et al., 1966: A new discovery of diamond-bearing eclogite in the 'Mir' pipe, Yakutia. Geologiya Geofiz. Novosibirsk, No. 11, pp. 114-116.

SOBOLEV, N. V., BOTKUNOV, A. I., BAKUMENKO, I. T. and SOBOLEV, V. S., 1972: Crystalline inclusions with octahedral facets in diamonds. Dokl. Akad. Nauk SSSR, 204, pp. 192-195.

SOBOLEV, N. V., BOTKUNOV, A. I., LAVRENT'EV, Yu. G. and POSPELOVA, L. N., 1971: Compositional features of the minerals associated with the diamonds from the 'Mir' pipe, Yakutia. Zap. vses. miner. Obshch., 100, pp. 558-564.

SOBOLEV, N. V., GNEVUSHEV, M. A. et al., 1971: The composition of garnet and pyroxene inclusions in Uralian diamonds. Dokl. Akad. Nauk SSSR, 198, pp. 190-193.

SOBOLEV, N. V., KHAR'KIV, A. D., LAVRENT'EV, Yu. G. and POSPELOVA, L. N., 1973: Chromite-pyroxene intergrowths from the 'Mir' kimberlite pipe, Yakutia. Geologiya Geofiz. Novosibirsk, No. 12, pp. 15-20.

SOBOLEV, N. V. and KUZNETSOVA, I. K., 1965: New data on the mineralogy of eclogites from the Yakutian kimberlite pipes. Dokl. Akad. Nauk SSSR, 163, pp. 471-474.

SOBOLEV, N. V. and KUZNETSOVA, I. K., 1966: The mineralogy of the diamond-bearing eclogites. Dokl. Akad. Nauk SSSR, 167, pp. 1365-1368.

SOBOLEV, N. V. and KUZNETSOVA, I. K., 1972: Chrome pyropes from xenoliths of peridotites from the 'Udachnaya' pipe, Yakutia, in Preblemy petrologii ul'traosnovnykh i osnovnykh porod (Problems of the Petrology of the Ultramafic and Mafic Rocks). Moscow.

SOBOLEV, N. V., KUZNETSOVA, I. K. and ZYUZIN, N. I., 1966: Chromium-bearing minerals in grospydites and new data on chrome-kyanite. Geologiya Geofiz. Novosibirsk, No. 10, pp. 42-47.

SOBOLEV, N. V., KUZNETSOVA, I. K. and ZYUZIN, N. I., 1968: The petrology of grospydite xenoliths from the Zagadochnaya kimberlite pipe in Yakutia. J. Petrology, 9, pp. 253-280.

SOBOLEV, N. V. and LAVRENT'EV, Yu. G., 1971: Isomorphic sodium admixture in garnets formed at high pressures. Contr. Mineral. Petrol., 31, pp. 1-12.

SOBOLEV, N. V., LAVRENT'EV, Yu. G., and POSPELOVA, L. N., 1972: Features of the content of trace-elements in the minerals of xenoliths from kimberlite pipes as a depth criterion. Tez. Mezhdunar. geokhim. Kongr., 1, pp. 442-462. Moscow.

SOBOLEV, N. V., LAVRENT'EV, Yu. G. and USOVA, L. V., 1972: Trace-elements from rutiles in eclogites. Geologiya Geofiz. Novosibirsk, No. 11, pp. 108-112.

SOBOLEV, N. V., LAVRENT'EV, I. K. and ZYUZIN, N. I., 1974: Minerals of the Al_2O_3-Cr_2O_3 isomorphous series in the 'Zagadochnaya' kimberlite pipe. Geologiya Geofiz. Novosibirsk, No. 10.

SOBOLEV, N. V., LAVRENT'EV, Yu. G., POKHILENKO, N. P. and USOVA, L. V., 1973a: Chrome-rich garnets from the kimberlites of Yakutia and their parageneses. Contr. Mineral. Petrol., 40, pp. 39-52.

SOBOLEV, N. V., LAVRENT'EV, Yu. G., POSPELOVA, L. N. and POKHILENKO, N. P., 1973b: The isomorphous content of titanium in pyrope-almandine garnets. Zap. vses. miner. Obshch., 102, pp. 150-155.

SOBOLEV, N. V., LAVRENT'EV, Yu. G., POSPELOVA, L. N. and SOBOLEV,

E. V., 1969: Chrome pyropes from Yakutian diamonds. Dokl. Akad. Nauk SSSR, 189, pp. 162-165.

SOBOLEV, N. V. and LODOCHNIKOVA, N. V., 1962: The mineralogy of the garnet peridotites. Geologiya Geofiz. Novosibirsk, No. 6, pp. 52-59.

SOBOLEV, N. V., POKHILENKO, N. P., LAVRENT'EV, Yu. G. and USOVA, L. V., 1975: Compositional features of chrome-spinels from diamonds in Yakutian kimberlites. Geologiya Geofiz. Novosibirsk, No. 11.

SOBOLEV, N. V., PUSTYNTSEV, V. I., KUZNETSOVA, I. K. and KHAR'KIV, A. D., 1969: New data on the mineralogy of the diamond-bearing eclogites from the 'Mir' pipe, Yakutia. Geologiya Geofiz. Novosibirsk, No. 3, pp. 113-116.

SOBOLEV, N. V., ZYUZIN, N. I. and KUZNETSOVA, I. K., 1966: The continuous series of pyrope-grossular garnets in grospydites. Dokl. Akad. Nauk SSSR, 167, pp. 902-905.

SOBOLEV, V. S., 1937: Features of the magmatic phenomena and the metallogenesis of platforms as exemplified by the associations of Siberian traps. Tez. Dokl. Int. geol. Congr., 17. Moscow.

SOBOLEV, V. S., 1944: The crystal chemistry of double salts and their role in petrology and mineralogy. Izv. Akad. Nauk SSSR, ser. geol., No. 5, pp. 69-77.

SOBOLEV, V. S., 1947: The importance of the coordination number of aluminum in silicates. Miner. Sb. l'vovsk geol. Obshch., No. 1, pp. 16-30.

SOBOLEV, V. S., 1949: Vvedenie v mineralogiyu silikatov (Introduction to Silicate Mineralogy). L'vov Univ. Press.

SOBOLEV, V. S., 1951: Geologiya mestorozhdeniy almazov Afriki, Avstralii, ostrova Borneo i Severnoi Ameriki (The Geology of Diamond Deposits in Africa, Australia, Borneo, and North America). Gosgeolizdat, Moscow.

SOBOLEV, V. S., 1955: The role of pressure in mineral formation. Miner. Sb. l'vovsk. geol. Obshch., No. 9, pp. 50-63.

SOBOLEV, V. S., 1960a: Conditions of formation of deposits of diamonds. Geologiya Geofiz. Novosibirsk, No. 1, pp. 7-22.

SOBOLEV, V. S., 1960b: The role of high pressures during metamorphism. Int. geol. Congr., 21, Dokl. Sov. Geol., Problem 14, pp. 36-45. Kiev.

SOBOLEV, V. S., 1962: Features of volcanic phenomena on the Siberian Platform and some general problems of geology. Geologiya Geofiz. Novosibirsk, No. 7, pp. 8-15.

SOBOLEV, V. S., 1963a: The problem of synthesis of minerals at the session of the International Mineralogical Association. Problems of Geol. at 21st Session of Int. geol. Congr. Akad. Nauk SSSR Press, Moscow.

SOBOLEV, V. S., 1963b: Characteristis features of the volcanism of the Siberian Platform. Pacific Sci., 17, pp. 452-457.

SOBOLEV, V. S., 1964a: The physicochemical conditions of mineral formation in the crust and mantle. Geologiya Geofiz. Novosibirsk, No. 1, pp. 7-22.

SOBOLEV, V. S., 1964b: The incongruent melting of minerals formed during changes in pressure. Dokl. Akad. Nauk SSSR, 156, pp. 341-344.

SOBOLEV, V. S., 1965: The effect of pressure on the limits of siomorphous substitutions. Dokl. Akad. Nauk SSSR, 160, pp. 435-437.

SOBOLEV, V. S., 1972: The problem of the formation of leucite rocks, in Problemy mineralogii i petrologii (Problems of Mineralogy and Petrology). Nauka Press, Liningrad.

SOBOLEV, V. S., KHAR'KIV, A. D. et al., 1972: Zoned garned from a kimberlite in the 'Mir' pipe. Dokl. Akad. Nauk SSSR, 207, pp. 421-424.

SOBOLEV, V. S., NAI, B. S. et al., 1969: Xenoliths of diamond-bearing pyrope serpentinites from the 'Aikhal' pipe, Yakutia. Dokl. Akad. Nauk SSSR, 188, pp. 1141-1143.

SOBOLEV, V. S. and SOBOLEV, N. V., 1964: Xenoliths in the kimberlites of Northern Yakutia and problems of the structure of the Earth's mantle. Dokl. Akad. Nauk SSSR, 158, pp. 108-111.

SOBOLEV, V. S. and SOBOLEV, N. V., 1967: Chromium and chromium-bearing minerals in deep-seated xenoliths from kimberlite pipes. Geologiya rudn. Mestorozh., 9, No. 2, pp. 10-16.

SOBOLEV, V. S., SOBOLEV, N. V. and LAVRENT'EV, Yu. G., 1972a: Inclusions in diamond from a diamond-bearing eclogite. Dokl. Akad. Nauk SSSR, 207, pp. 164-167.

SOKOLOV, G. A., 1946: The chrome-spinels of ultramafic complexes (fundamental features of their chemical composition and classification). Reported dedicated to Academician D. S. Belyankin. Akad. Nauk SSSR Press, Moscow.

SRIRAMADAS, A., 1959: Diagrams for the correlation of unit cell edges and refractive indices with the chemical composition of garnets. Am. Miner., 42, p. 294.

STAVITSKAYA, G. P., SMOLIN, Yu. I., TOPOROV, N. A. and PORAI-KOSHITS, E. A., 1959: The problem of the crystallography of hillebrandite under hydrothermal conditions. Dokl. Akad. Nauk SSSR, 126, pp. 616-618.

STEINWEHR, H. E. V., 1967: Gitterkonstanten in System (Al, Fe, Cr_2O_3) und ihr Abweichen von der Vegard-Regel. Z. Kristallogr., 125, pp. 377-403.

STISHOV, S. M. and POPOVA, S. V., 1961: A new dense modification of silica. Geokhimiya, No. 10, pp. 837-839.

SUBBOTIN, S. I., NAUMCHIK, G. L. and RAKHIMOVA, I. Sh., 1968: Mantiya Zemli i tektogenez (The Earth's Mantle and Tectogenesis). Naukova Dumka Press, Kiev.

SUTTON, J. R., 1921: Inclusions in diamond from South Africa. Mineralog. Mag., 19, pp. 208-210.

SWITZER, G. and MELSON, W. G., 1969: Partially melted kyanite eclogite from the Roberts Victor mine, South Africa. Smithson, Contr. earth Sci., No. 1, pp. 1-9.

THOMPSON, J. B., 1947: Role of aluminum in rock-forming silicates. Bull. geol. Soc. Am., 58 /abstract7, p. 1232.

TOROPOV, N. A., BARZAKOVSKY, V. P., LAPIN, V. V. and KURTSEVA, N. N., 1965: Diagrammy sostoyaniya silikatnykh sistem Spravochnik (Diagrams of State of Silicate Systems Handbook). Nauka Press, Moscow.

TROFIMOV, V. S., 1966: Diamond diatremes. Sov. Geol., No. 5, pp. 3-12.

TSVETKOV, A. I., YERSHOVA, Z. P. and MATVEEVA, N. A., 1964: Synthesis of a chromium silicate, analogous to olivine. Izv. Akad. Nauk SSSR, ser. geol., No. 2, pp. 3-14.

UKHANOV, A. V., 1968: Nickel in ultramafic inclusions from kimberlites in the North Yakutian pipes. Geokhimiya, No. 12, pp. 1470-1478.

UKHANOV, A. V., 1970: The geochemistry of chromium in the upper mantle based on data from an investigation of ultramafic inclusions in kimberlite pipes. Geokhimiya, No. 9, pp. 1053-1065.

URBAKH, V. Yu., 1964: Biometricheskie metody (Biometric Methods). Nauka Press, Moscow.

VAKHRUSHEV, V. A. and SOBOLEV, N. V., 1971: Sulfides in deep-seated xenoliths from the Yakutian kimberlite pipes. Geologiya Geofiz. Novosibirsk, No. 11, pp. 3-11.

VASIL'EV, E. K., 1969: Primenenie korrelyatsionnogo analiza pri izuchenii izomorfizma v olivinakh i granatakh (The Use of Correlation Analysis During the Study of Isomorphism in Olivines and Garnets). Nauka Press, Moscow.

VASIL'EV, V. G., KOVAL'SKY, V. V. and CHERSKY, N. V., 1968: Proiskhozhdenie almazov (The Origin of Diamonds). Nauka Press, Moscow.

VINOGRADOV, A. P., 1959: Meteorites and the Earth's crust. Izv. Akad. Nauk SSSR, ser. geol., No. 10, pp. 5-27.

VINOGRADOV, A. P., 1961: The origin of the material of the Earth's crust. Report 1. Geokhimiya, No. 1, pp. 3-29.

VINOGRADOV, A. P. and YAROSHEVSKY, A. A., 1970: The physical conditions of zone melting in the Earth's shells. Geokhimiya, No. 7, pp. 779-790.

VINOGRADOV, A. P., KROPOTOVA, O. I. and USTINOV, V. I., 1965: Possible sources of carbon in natural diamonds on the basis of $12_C/13_C$ isotope data. Geokhimiya, No. 6, pp. 643-651.

VINOGRADOV, A. P., YAROSHEVSKY, A. A. and IL'IN, N. P., 1970: A physicochemical model of the separation of elements during the process of differentiation of mantle material. Geokhimiya, No. 4, pp. 389-402.

VLADIMIROV, B. M., TVERDOKHLEBOV, V. A. and KOLESNIKOVA, G. P., 1971: Geologiya i petrografiya izverzhennykh porod yugo-zapadnoi chasti Gvineusko-Liberiiskogo shchita (The Geology and Petrography of the Igneous Rocks of the Southwestern Part of the Guinea-Liberian Shield). Nauka Press, Moscow.

VOSKRESENSKAYA, V. B., KOVAL'SKY, V. V., NIKISHOV, K. N. and PARINOVA, Z. F., 1965: The discovery of a titanolivine in Siberian kimberlites. Zap. vses. miner. Obshch., 94, pp. 600-603.

WAGNER, P. A., 1909: Die diamantführenden Gesteine Südafrikas. Ihre Abbau und ihre Aufbereitung. Berlin.

WAGNER, P. A., 1928: The evidence of the kimberlite pipes on the constitution of the outer part of the Earth. S. Afr. J. Sci., 25, pp. 125-148.

WATSON, K. D. and MORTON, D. M., 1969: Eclogite inclusions in kimberlite pipes at Garnet Ridge, northeastern Arizona. Am. Miner., 54, pp. 267-285.

WHITE, A. J. R., 1964: Clinopyroxenes from eclogites and basic granulites. Am. Miner., 49, pp. 883-888.

WILLIAMS, A. F., 1932: The Genesis of the Diamond. 2 vols. Benn, London.

WINCHELL, H., 1958: The composition and physical properties of garnet. Am. Miner., 43, pp. 595-600.

WITTKE, J. P., 1967: Solubility of iron in TiO_2 (rutile). J. Am. ceram. Soc., 50, pp. 586-588.

WOOD, B. J. and BANNO, S., 1973: Garnet-orthopyroxene and orthopyroxene-clinopyroxene relationships in simple and complex systems. Contr. Mineral. Petrol., 42, pp. 109-124.

WYLLIE, P. J., 1963: The nature of the Mohorovicic discontinuity; a compromise. J. geophys. Res., 68, pp. 4611-4619.

WYLLIE, P. J., 1970: Ultramafic rocks and the upper mantle. Spec. Publs miner. Soc. Am., No. 3, pp. 3-32.

WYLLIE, P. J., 1971: Role of water in magma generation and initiation of diapiric uprise in the mantle. J. geophys. Res., 76, pp. 1328-1338.

YAMAGUCHI, M., 1964: Petrogenetic significance of ultrabasic inclusions in basaltic rocks from Southwest Japan. Mem. Fac. Sci. Kyuchu Univ., ser. D., Geol., 15, No. 1, pp. 163-219.

YAROSHEVSKY, A. A., 1968: Zone melting of the mantle and some problems of a primary basalt magma, in Kora i verkhnyaya mantiya Zemli (The Crust and Upper Mantle), pp. 82-89. Nauka Press, Moscow.

YEFIMOV, I. A., 1961: The discovery of pyrope serpentinites in the Precambrian rocks of the Kokchetav Massif (Central Kazakhstan). Trudy kaz. Inst. miner. Syr'ya, 5, pp. 3-14.

YEFIMOVA, E. S., 1961: Solid inclusions in Yakutian diamonds. Tez. Soveshch. po Geol. almazn. Mestorozh. Yakutii. Yakutsk.

YERMAKOV, N. P., 1950: Issledovanie mineraloobrazuyushchikh rastvorov (The Investigation of Mineral-Forming Solutions). Khar'kov Univ. Press.

YODER, JR., H. S. and TILLEY, C. E., 1962: Origin of basalt magmas. An experimental study of natural and synthetic rock systems. J. Petrology, 3, pp. 342-529.

ZAVARITSKY, A. N. and SOBOLEV, V. S., 1961: Fiziko-khimicheskie osnovy petrografii izverzhennykh porod (The Physicochemical Bases for the Petrography of the Igneous Rocks). Gosgeoltekhizdat, Moscow.

ZHELYAZKOVA-PANAIOTOVA, M. D. and IVCHINOVA, L. V., 1971: The mineral species of spinels from the ultramafic rocks of Bulgaria. Geologiya rudn. Mestorozh., 13, No. 3, pp. 71-90.

ZIMIN, S. S., 1963: The composition and parageneses of chrome-spinels in ultramafic rocks. Geologiya Geofiz. Novosibirsk, No. 10, pp. 46-57.

ZYUZIN, N. I., 1967: The nature of the orientation of garnet inclusions in Yakutian diamonds. Geologiya Geofiz. Novosibirsk, No. 6, pp. 126-128.